The Species Problem

The Species Problem

Biological Species, Ontology, and the Metaphysics of Biology

David N. Stamos

LEXINGTON BOOKS
Lanham • Boulder • New York • Oxford

LEXINGTON BOOKS

Published in the United States of America
by Lexington Books
A Member of the Rowman & Littlefield Publishing Group
4501 Forbes Boulevard, Suite 200, Lanham, Maryland 20706

PO Box 317
Oxford
OX2 9RU, UK

British Library Cataloguing in Publication Information Available

Library of Congress Cataloging-in-Publication Data

Stamos, David N., 1957–
    The species problem : biological species, ontology, and the metaphysics of biology /
David N. Stamos.
        p. cm.
    Includes bibliographical references and index.
    ISBN 0-7391-0503-5 (cloth : alk. paper)
    1. Species—Philosophy. I. Title.

    QH83.S75 2003
    576.8'6—dc21   2002155405

Printed in the United States of America

♾™ The paper used in this publication meets the minimum requirements of American
National Standard for Information Sciences—Permanence of Paper for Printed Library
Materials, ANSI/NISO Z39.48–1992.

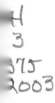

Dedicated to Professors David M. Johnson and the late Robert H. Haynes, my most esteemed mentors and friends in philosophy of biology, who taught me above all the true meaning of *Tentanda Via.*

Acknowledgments: In addition to David Johnson and Robert Haynes, I have greatly benefited from discussions with and wish to thank Joseph Agassi of Tel Aviv University philosophy, Jagdish Hattiangadi of York University philosophy, Peter Moens and Brock Fenton of York biology, Bernard Lightman of York humanities, Alex Levine of Lehigh University philosophy, Peter Stevens of the Missouri Botanical Garden, whose help was enormous, Marc Ereshefsky of University of Calgary philosophy, whose excellent anthology and many critical suggestions were of inestimable value, and last but not least Robert Tully of University of Toronto philosophy for his patient help on Russell.

Contents

Chapter 1

Introduction

The species problem has to do with biology, but it is fundamentally a philosophical problem—a matter for the "theory of universals."
—Michael Ghiselin (1974: 285)

1.1 The Species Problem and the Problem of Universals

In a sense, the species problem is really quite simple. Are biological species real, and, if real, what is the nature of their reality? Are species words merely operational conveniences made for the purpose of conveying various information and theories, or do species words refer to entities in the objective world with a real existence independent of science?

My purpose in writing the present work was basically fourfold and closely interconnected: (i) to fill a large void by weaving together the bulk of the more important (and much of the less important) of the vast literature on the modern species problem into one comprehensive, cohesive, and informative whole, useful for an interdisciplinary audience of professional scholars and students alike; (ii) to make an original contribution to both the criticism and interpretation of modern views, including that of Darwin; (iii) to show that the modern species problem is still today part of a much older problem in metaphysics, the problem of universals, and furthermore that the various solutions to the former largely fit within the framework of the traditional solutions to the latter; and (iv) to do to the modern species problem what Bertrand Russell did to the traditional problem of universals, namely to emphasize the reality of relations, and from there to develop an entirely new theory of species based on the ontology of relations. It is my contention that applying the relations category to the species category provides the best overall fit with the basic facts, theories, and practices in biology today. This is, of course, quite a strong and bold claim, and in addition to its defense it requires that each of the competing theories be closely and carefully scrutinized, for both their strengths and weaknesses.

But first, what is so important about the species problem? Certainly since "species" is a keyword in biology and part of its technical vocabulary, the species problem is a central problem for that science, and also for anyone trying to understand that science. But it is also important in a wider sense. The year 1859 marks the official beginning of a conceptual revolution that has proved to be of far greater importance than any other (cf. Mayr 1988a: 161). The renowned geneticist Theodosius Dobzhansky (1973) once stated that "nothing in biology makes sense except in the light of evolution" (125). But what makes the Darwinian revolution so exceptionally important is what there is outside of biology that also fails to make sense except in the light of evolution. Not only do we want to know what we are and where we came from (and are going), but also religion, sex, politics, greed, competition, skin color, lower back pain, and so much more, fail to make rational sense unless and until viewed from the Darwinian perspective of evolution. What needs to be underscored is that at the very core of that revolution is the concept of species. The fundamental importance of the species problem, then, resides in the fact that, as Alexander Rosenberg (1987) put it, when we talk about species "we have got to know what we are talking about" (193).

The species problem is also of great practical importance. In nineteenth-century America, for example, a matter of great debate among scientists and philosophers was over whether the different human races are subspecific or specific entities. The greatest practical implication of that debate concerned the institution of human slavery (cf. Gould 1981: 69–72). Underlying the debate was a debate over different species concepts. Of course today we no longer argue (at least among the learned) over whether the different human races are one or different species. (Indeed it is increasingly doubted that human races even objectively exist; cf. Gould 1977: 231–236.) Nevertheless serious practical implications remain for our treatment of other living things. For example, whether efforts should be made to protect and preserve near-extinct groups such as the Somali black rhino or the Florida panther ultimately depend on how species and subspecies are defined (cf. Ryder 1986, Wilson 1992: 67–68, 336, Eisner *et al.* 1995). Indeed the problem of conservation biology has become particularly pressing in recent decades, given the human-caused mass extinction that is currently under way. More and more it is being appreciated that the estimation of earth's biodiversity and the allocation of conservation resources rest largely on the concept of species that is used.

To this we might add a slightly older area of practical concern in biology, known as biological control, in particular pest control. Edward O. Wilson (1992) discusses the classic case of the vector (host) mosquito *Anopheles maculipennis*, originally thought to be a single species, now thought, instead, to be a group of sibling species:

> Following the discovery in 1895 that malaria is carried by *Anopheles* mosquitoes, governments around the world set out to eradicate those insect vectors by

draining wetlands and spraying infested areas with insecticides. In Europe the relation between the malarial agent, protozoan blood parasites of the genus *Plasmodium*, and the vector mosquito, *Anopheles maculipennis*, seemed at first inconsistent, and control efforts lacked pinpoint accuracy. In some localities the mosquito was abundant but malaria rare or absent, while in others the reverse was true. In 1934 the problem was solved. Entomologists discovered that *A. maculipennis* is not a single species but a group of at least seven. In outward appearance the adult mosquitoes seem almost identical, but in fact they are marked by a host of distinctive biological traits, some of which prevent them from hybridizing. . . . Once identified, the dangerous members of the *A. maculipennis* complex could be targeted and eradicated. Malaria virtually disappeared from Europe. [44–45]

But the species problem has an even more basic practical side. Quite simply, we may ask whether species words are really necessary for biologists to do biology. The answer would seem a clear *yes*, at least for most biologists. As the biologist Hugh Paterson (1985) put it, "When we trade in ideas in population biology, species constitute our currency" (137). One can only imagine the chaos that would prevail in the science of biology if the denotation of species words would vary from biologist to biologist. John Maynard Smith (1975) provides a telling case in point:

a few years ago I was studying the ways in which birds are adapted to different kinds of flight—soaring, gliding, flapping, and so on. For this purpose it was desirable to know the weight, wing span, and wing area of as many different kinds of birds as possible. Unhappily bird taxonomists usually do not measure any of these things; for example, they prefer to measure the length of the wing from the wrist to the tips of the primary feathers, rather than from wing-tip to wing-tip, because it can be done more accurately. In fact, the only place where I could find a large collection of the kind of measurements I wanted was in a charmingly entitled paper, 'The first report of the bird construction committee of the Aeronautical Society of Great Britain,' published in 1910. But the authors of this paper were unaware of the desirability of giving the scientific names of the specimens measured. Many birds whose measurements were given were identified by the single word 'hawk,' which might have meant the hovering Kestrel, the soaring Buzzard, or the fast-flapping Sparrow-hawk. Had the scientific name of each species been given, the list would have been of far greater value. The point of this anecdote is obvious. Despite the unavoidable imperfections of any system of classification, an internationally accepted system of naming does enable biologists to convey a fairly precise idea of the kind of animal or plant they have observed by the use of specific names, such as *Falco tinnunculus* for the Kestrel or *Accipiter nisus* for the Sparrow-hawk. In many cases such a name is insufficiently precise, and it is desirable to add the time and place where the individuals were collected, or the particular laboratory strain to which they belonged, but nevertheless a classification into species is a prerequisite for any accurate communication between biologists. [217–218]

In economics, of course, monetary currency is completely arbitrary, as proved by the many actual currencies in use. Perhaps biologists, then, need only decide upon a convention for determining and delimiting species. The mere need for such determination, of course, can hardly serve as evidence—let alone proof— that species themselves are extra-mentally real. But if species are not real, then much in biology would seem of a piece with storytelling, dealing largely with fictions. This in itself would not entail instrumentalism in biology, the view that scientific theories are neither more nor less true but instead only more or less useful, but it would make one of the oldest and still quite central concepts in biology a mere tool. Again, the difference would seem to be a real and important difference, which resolves into the issue of whether species are entities that are actually discovered or merely invented.

As one might expect, although the species problem has a long and varied history, the literature on it has mushroomed in recent decades. According to Marc Ereshefsky (1992b), in the past twenty-five years "well over a hundred books and articles have appeared on the nature of species. Those publications offer more than twenty definitions of the species category" (xiv). (If we add to this all of the related literature, the number of publications is in the thousands.) Surprisingly, however, although an enormous amount of literature has been written and continues to be written on the modern species problem, it is all divided among various articles, chapters, and anthologies. To date, not one author has attempted to provide a comprehensive study. Perhaps the task is looked upon as too large. Whatever the reason, the most that one can find is a number of anthologies. Aside from some old and out-of-date anthologies, the past decade, especially the second half, has seen a flurry of contributions, each of them important in their own way. Starting things off, Ereshefsky's (1992b) anthology is a very useful collection of some of the now classic papers on the species problem, but it is far from providing a comprehensive treatment and guide to the subject. Given the classic papers format, many competing species concepts and related issues cannot help but get either treated superficially or ignored altogether. In a very different approach, Claridge *et al.*'s (1997) anthology is a collection of papers from a recent symposium in which a wide diversity of specialists discussed the practical nature of the species problem, in particular from the viewpoint of biodiversity and conservation biology. Although of great value, it provides a seriously unbalanced treatment of the species problem. This is because it is quite strong on the biological aspects of the problem but very weak on the philosophical aspects (only one of the contributors is a philosopher, namely David Hull, who as we shall see is a partisan of the species-as-individuals view). The anthologies of Lambert and Spencer (1995) and Wheeler and Meier (2000) are even more parochial, the former devoted to the recognition species concept, the latter to which of five species concepts best conforms to the needs of phylogenetic systematics (all but one of the contributors are cladists). Parochial in a different way is the anthology of Howard and Berlocher (1998). This anthology provides a rich and detailed focus on recent empirical investigations into speciation. As such it is a sequel (which the editors admit) to the important anthology of Otte and Endler (1989). The main problem is that with both anthologies all of the authors are biologists, and only a few of the papers are devoted to the species problem *per se*. Since it is generally admitted that any speciation analysis presupposes a species concept, such analyses do not solve the species problem but merely serve to underscore it. Moreover many biologists (not to mention philosophers), such as Michael Ghiselin (quoted at the head of this chapter), have readily recognized that the species problem is not strictly biological but philosophical as well. Hence the need for an interdisciplinary approach. Richard Harrison (1998) notes that because of the largely philosophical nature of the species problem "the response of many evolutionary geneticists has been to ignore (if not disparage) these discussions" (19), while Daniel Howard (1998) adds that for many biologists studying speciation "the lack of interest [in the literature on the species problem] reflects discomfort with a debate that appears to have no end and seems to have become increasingly muddled" (439). Unfortunately, anthologies, by their very nature, cannot help alleviate this problem but can only add to it. A final case in point is the anthology of Robert Wilson (1999). While half of the twelve specially commissioned essays are by philosophers, there is little unanimity on the five main themes of the anthology, especially on "unanimity, integration, and pluralism" and "species realism" (the remaining three themes being "practical import," "historical dimensions," and "cognitive underpinnings"). Moreover, most of the competing species concepts are neither defined nor discussed, let alone analyzed, while much of the relevant biology and metaphysics are ignored altogether. As such, it can no more serve as a detailed introduction and analysis of the species problem than any of the other anthologies.

Given the importance of the species problem, then, and the enormity of the interest in it, surely the time has come for a detailed and comprehensive treatment of the problem, ideally an authoritative text as it were. The object of the present book is to fill this role. It is directed at principally four classes of readers: (i) Philosophers of biology, of course, should find this book of great interest, for the simple reason that the species problem is one of the core problems in philosophy of biology, as books on philosophy of biology readily attest (cf. Rosenberg 1985; Ruse 1988, 1989a; Sober 1993, 1994). They will perhaps, however, be surprised at the extent to which I take biology seriously. Virtually unexplored in the philosophical literature on the species problem are Lynn Margulis' theory of endosymbiosis, Sorin Sonea's bacteriology, and the use of state spaces in virology, to name a few examples. (ii) Biologists of many different specialties should also find this book of great interest and value not so much for the biology but for the conceptual or philosophical implications involved with different species concepts. Biologists almost always tend to treat these too lightly. The result of this indifference is usually conceptual confusion. For example, in a study of over a hundred species monographs Melissa Luckow (1995) found not only that most of the biologists were not explicit about the species concepts they used but that they often used conflicting criteria. Her ultimate recommendation was that "Scientists naming species should be familiar with and evaluate the assumptions inherent in each species concept. . . . The species concepts I have discussed rest on conflicting assumptions, and with real data these assumptions will lead to different species circumscriptions" (600). (iii) Historians of biology should also find this book of interest, for in my basic division of competing species concepts I explore historical antecedents and precursors, and from a critical point of view. Mostly they should be interested in my long discussion of Darwin, in which I attempt to take previous interpretation to a new level, and in my discussion of Aristotle, in which I attempt to defend the traditional essentialist interpretation against the new orthodoxy. Although they will find shorter discussions of important figures such as Buffon and Linnaeus, and although the species problem is rich in history, I should take this opportunity to stress that the present work is *not* a historical treatment of the species problem, nor was it ever intended to be. Instead, its design is that of "one long argument," to borrow Darwin's phrase. Consequently, much in the history of the species problem is overlooked. Where I have indulged in history, it has been for the purpose of correcting some of that history and adding new material and perspectives to the modern debate on the species problem. (iv) Advanced undergraduate and graduate students should also find this book very useful, biology and philosophy students alike. Although I wrote it for a primarily professional audience, that audience is an interdisciplinary one, so that students should find it unusually accessible. Indeed the present work could profitably be used as a text in a course on conceptual issues in science. Students of philosophy, however, in particular those with an interest in metaphysics, should find this book of extra interest given my unusual treatment of some of the basic problems in metaphysics, namely the ontology of classes (chapter 3.5) and the ontology of relations (chapter 5.3).

All in all, then, the present work is a very useful and greatly needed guide through a vast and imposing forest, a forest in which it is all to easy to get lost. But more than just a guide, it is a critical and provocative exposition with a specific destination. Confronting such a task, however, is not only the fact that the species problem is both a huge and genuinely interdisciplinary problem, nor that it is highly contentious, but more importantly the fact that it cannot hope to be solved unless biology and philosophy are taken equally seriously. Once again, although the species problem pertains mostly to biology, it is not, strictly speaking, a *biological* problem. True enough, empirical finds and categories—e.g., evolution, variation, selection, competition, niche, extinction, reproduction, genotype, phenotype, endosymbiosis—have a direct and important bearing on the subject, but by themselves they cannot decide the issue. Conversely, the species problem cannot be adequately argued, let alone solved, by purely philosophical analysis alone. Socratic dialectic will not serve us here. Nor in this matter can we extend the approach of the later Wittgenstein (1958) and attempt to determine the meaning of species words simply by how they are used. Even if we should restrict ourselves exclusively to modern scientific usage, the question of the ontology of the referents of that usage remains an open question. And that question is certainly a legitimate one which deserves to be answered.

More than just empirical or conceptual, then, the species problem is a hybrid of both. But what is the nature of the conceptual part? To say it is philosophical would be true, since much of the work that has been done and still needs to be done involves the uncovering of conceptual confusions. But the conceptual part is something more. What will become clear is that in trying to solve the species problem we invariably find ourselves dealing with other ontologies, such as that of individual, class, and relation, including similarity. And they, along with criteria of reality and minimum vocabulary, are *metaphysical* issues. The species problem, then, proves to be in large part not only empirical but metaphysical as well.[1]

That the species problem is in part metaphysical is also, it seems to me, indicated by the method of inference most appropriate to it. It is certainly not inductive, not only because of the theory-dependency of observation but because purely inductive procedures will invariably lead one to a variety of species concepts, depending upon which organisms one studies. Neither is it deductive, or rather hypothetical-deductive, since it is a curious feature of the species problem that contrary examples do not necessarily refute a species concept. Instead it is closest to what is commonly called *inference to the best explanation*, reasoning about "why this theory rather than that" (cf. Lipton 1991), reasoning that attempts to best account not only for as many facts, but for as many classes of facts, as possible, in spite of anomalies.

Most biologists, of course, are consumed with the practical side of biology, and give little thought to matters of ontology. Those who do so, however, and their number is increasing, find themselves forced into the world of philosophy as well. Likewise, interested philosophers find themselves forced into the world of biology. Not inaccurately, I think, has Ereshefsky (1992b) characterized the modern species problem as an instance of what he calls "practical metaphysics" (xvi). By itself a good indication of the truth of this characterization is the literature recently published on the species problem, for it has become increasingly rare to find a biologist write on that problem without any reference to the relevant literature by philosophers, and it has become virtually impossible to find a philosopher write on that same problem without any reference to the biological literature.

My own approach to the species problem, although quite sensitive to the empirical side, will prove to be more inspired by metaphysics than perhaps all the rest. It seems to me a valuable insight that the modern species problem maps onto, so to speak, the traditional metaphysical problem of universals—i.e., that the modern ontological characterizations of species basically fall within the different ontological categories traditionally given in the problem of universals—leaving, interestingly, one ontological category hitherto unexplored so far as the species problem is concerned (my own approach developed in chapter 5).

One of the main distinctions in traditional metaphysics is between universals and particulars. One way of looking at this distinction is through repeatability and instantiation. Universals are either repeatable or have instances, or at least are capable of such, whereas particulars are not repeatable and do not have instances. The problem of universals is the problem of determining whether there are in fact entities that are either repeatable or have instances. Or even more basically, it can

---

[1]　By "metaphysical" I mean good metaphysics, this-worldly metaphysics, neither empirically verifiable nor falsifiable but most definitely criticizable.

be stated as the problem of determining the nature of sameness, if any, between numerically different particulars.

Another way of looking at the distinction between universals and particulars is through language. At base, universals are predicates in language. In the stock example "Socrates is a man," "Socrates" refers to a particular, whereas "man" refers to a universal. The problem of universals is the problem of determining what it is, if anything, to which predicates refer. The problem is a central problem in traditional metaphysics, with a long and varied history, reaching back to Plato and Aristotle (who both, incidentally, approached the problem through predication).[2]

We may at this point distinguish three types of answers to this problem. Each constitutes one of the three basic positions in traditional metaphysics.

Realism is the view that predicate words refer to real, extra-mental entities in the objective world. These entities may include essences, either external to physical things (Plato) or internal to them (Aristotle), abstractions such as classes, or

---

[2]   Some philosophers might disagree with this way of stating the problem of universals. Most likely they will refer either to Gottlob Frege, whose writings on predication are notoriously unclear and subject to interpretation, or to F.P. Ramsey (1925), who, by using the statements "Socrates is wise" and "Wisdom is a characteristic of Socrates," attempted to prove that there is no logical distinction (underlying the obvious grammatical distinction) between subject and predicate—all of this, in spite of the fact that Frege is often described as a modern Platonist and that virtually all modern logic textbooks would represent the logical form of Ramsey's two sentences above as Ws and as entailing $(\exists x)Wx$. Of further interest is the difference in the way of stating the problem of universals between Bertrand Russell and W.V.O. Quine, both of whom were heavily influenced by Frege. According to Russell (1912), "nearly all the words to be found in the dictionary stand for universals" (53), and he sees the problem of universals as the problem of determining the denotation (he more than once uses the word "denotes") of words which express qualities and relations, "to consider what is the nature of this kind of being" (52). Quine, on the other hand, repudiating predicates as entailing any ontological commitment (for Quine, such entailment follows only from the use of bound variables), states the problem of universals (Quine 1961) as "the question whether there are such entities as attributes, relations, classes, numbers, functions" (9), and elsewhere (Quine 1987) as whether modern science is committed to there being "abstract objects, or *universals*—thus properties, numbers, functions, classes" (225). Notwithstanding, Quine, as we shall see in chapter 3.2, did not desist from claiming that predicate words can play a dual role, both as predicates and as names of classes! Indeed the relation between predicates and the problem of universals, it seems to me, in spite of various efforts to the contrary, has become somewhat entrenched. According to Anthony Quinton (1989), for example, "Two main sorts of universals can be distinguished: predicative universals, the properties and relations that are the meaning of general terms or predicates, and formal universals, the abstract entities of mathematics" (317). Of course, as David Armstrong (1989) points out, even for a realist "there is no automatic passage from predicates (linguistic entities) to universals" (84). Negative and disjunctive predicates are a case in point (cf. Armstrong 82–84). The road is not much better for what are often called two-place predicates, three-place predicates, etc., namely words for relations, which require at least two subjects. These will be discussed mainly in chapter 5. At any rate, the point is that predicates and predication remain a useful way of introducing the problem of universals.

repeatables such as properties and relations.

Conceptualism is the view that predicate words refer to mental entities only, namely concepts or abstract ideas.

Nominalism is the view that predicate words refer to nothing at all, that universals are real in name only. Hence the etymology of the word "nominalism," from the Latin word *nomen*, meaning "name."

It should be pointed out that these three views are not mutually exclusive. One need not be either a realist, a conceptualist, or a nominalist. One can be a mixture of any two or even all three. One might be a realist concerning one group of universals, a conceptualist concerning another group of universals, and a nominalist concerning yet another group of universals. Commitment to one position concerning one group of universals does not entail a similar commitment in position to all universals.

Having said this, it will be evident that the species problem is really part of the problem of universals. Species words are predicates (at least much of the time), and the species problem is the problem of determining what it is, if anything, to which species words refer.

Indeed there is something reminiscent in all of this with early Socratic dialogues. We are all reasonably competent at using species words (at least the common ones), but when pressed to explain or define what it is to which these words refer, the vast majority of us quickly become confounded and in the end resort to ostensive definitions.

I should say here that I do not expect the present study, even if favorably received, to sow a major seed of revolution in the theory and practice of biology, although I would be surprised if it did not make *some* difference. Steven Weinberg (1992), in an odiously matricidal chapter titled "Against Philosophy," has argued rather convincingly, it seems to me, that "we should not expect it [philosophy of physics] to provide today's scientists with any useful guidance about how to go about their work or about what they are likely to find" (167). I do not, however, think the same is true with philosophy of biology. As I hope will become evident from the present study, the fate of a number of ongoing research programs in biology are directly affected by the questions found and debated within the species problem.

But aside from that, like art for art's sake, the species problem seems to me an exceedingly worthwhile theoretical adventure just in itself, for it makes one think about and rethink a number of ontological and methodological questions that determine and shape the way we look at the world, questions such as criteria of reality, the nature of abstractions, the meaning of individuality, the metaphysics of language, and so much more.

Weinberg (1992) also complains that much of modern philosophy of science is "written in a jargon so impenetrable that I can only think that it aimed at impressing those who confound obscurity with profundity" (168). While I find myself largely in agreement when it comes to philosophy of physics (and many other areas in philosophy as well), I do not think his claim equally applies to philosophy of biology. Certainly philosophers of biology lack the level of knowledge and skill possessed by biologists *qua* biologists (indeed one is hard-pressed to find a single exception), so that the temptation is there to satisfy themselves (following most notably the analytic tradition) with trifling improvements in clarity, hair-splitting distinctions, esoteric formalizations, and a dizzying jargon. Nevertheless there is much in philosophy of biology that is clearly motivated by a sincere desire to understand core concepts in what is arguably the most philosophical of all the sciences, such as the nature of selection, of fitness, and of course of species. Moreover the legitimacy of philosophy of biology is strongly indicated by the fact that much of it, and much of the most interesting and best of it, is done by biologists themselves (e.g., Mayr, Ghiselin, Dawkins). In engaging in such "What is *x*?" questions, biologists are no less following in a philosophical tradition which, of course, goes back all the way to the pre-Socratics. The ideal, then, is to combine the best in empirical research with the best in conceptual tools. Accordingly I have attempted to take both biology and philosophy equally seriously, to make the present work respectable to both. Whether I have achieved the desired symbiosis, of course, remains for others to judge.

1.2   Ontology and Criteria of Reality

The world of the naive realist is a world populated solely by entities which philosophers have dubbed "concrete particulars." This world, of course, is easily shattered. What shall the naive realist make, for example, of Newton's Law of Gravitation? Surely it is not a concrete particular. And yet it is part of reality (cf. Weinberg 1992: 46–47). Or what, to take a simpler example, shall he make of, say, the average length of the necks of a group of giraffes? Surely it too is not a concrete particular. It is an abstraction. And yet it is part of reality, discoverable by simple empirical means, an abstraction that conceivably could contribute to the fitness of both the group and each of its member organisms (cf. Ereshefsky 1988b: 219). Indeed the world of the naive realist begins to fall apart when he begins to reflect on the science and math he learned in grade school. When once he comes to admit that reality includes more than just concrete particulars, he has begun (to borrow Plato's famous allegory) an ascent out of a dark cave into a world which includes not only concrete particulars but a host of abstractions, universals, and metaphysical entities which dazzle the imagination. Just what is to be included in this outside world is indeed the philosopher-cum-scientist's question.

The question is a matter of ontology. The etymology of the word "ontology" is from the Greek words *onta*, which means "the really existing things," "true reality," and *logos*, which means "the study of," "the science of." Hence, ontology is the study of what there is, more specifically the study of the categories and nature of what there is.

A matter usually taken for granted, and if not then rarely ever dealt with sufficiently, is the matter of criteria of reality. What is wanted is a criterion or set of criteria by which ostensibly real things can be judged. The making explicit of these criteria seems to me an absolutely necessary requirement for any discussion on ontology. Without prior agreement on such criteria, many discussions are destined for impasse, confusion, and outright error.

This point is well illustrated in a passage from Ernst Mayr (1964):

> What is true for the human species—that no two individuals are alike—is equally true for all other species of animals and plants . . . All organisms and organic phenomena are composed of unique features and can be described collectively only in statistical terms. Individuals, or any kind of organic entities, form populations of which we can determine the arithmetic mean and the statistics of variation. Averages are merely statistical abstractions, only the individuals of which the populations are composed have reality. [xx; cf. Mayr 1988a: 15]

This is a surprising statement for a scientist. Scientists are popularly known as dealing with abstractions, and Mayr's statement above would have us believe that all scientists (or at least all biologists) in dealing with abstractions are dealing with fictions. To this Elliott Sober (1980) seems to me to have provided the proper reply. Sober points out that not only individuals but also groups have properties. Why the former and their properties should be thought real while not the latter and their properties is by no means clear. As Sober put it, "Individual and group properties are equally 'out there' to be discovered" (352).

When turning to the literature on the species problem, the need for examining criteria of reality becomes especially important. Competing species concepts do not all employ the same criteria of reality. But even if they did, ontological presuppositions would still be part of the metaphysics of the species problem. Logically it is possible for a species concept to be fatally flawed right from the start simply because of its underlying criteria of reality. Accordingly, in a work devoted to examining the literature on the species problem, it is important that the issue at hand receive considerable focus, not only in general but also as it applies to various species concepts. The latter task, of course, the task of application, is best reserved for when each species concept is examined and evaluated. The former task, however, the general task, is best suited for the present Introduction.

To begin, it is important to examine a very basic way of looking at the species problem. This is to take a given species concept, specifically one with realist pretensions, and ask if it has empirical application. In other words, we can ask whether modern biology provides evidence and reasons to believe that this species concept has objective referents. This is why the creationist species concept, for example, is no longer a contender in the modern debate: there are no entities out there in the biological world for it to correspond to. What used to look like the required entities have been proven by modern biology, following the lead of Darwin, to not be the required entities at all. In effect the discovery of the fact of evolution has killed the creationist species concept. The concept itself, of course, is real, and many lay people still hold it, but its reality is conceptual only. As a purported realist species concept it needs something more, but that something more is just not out there in the world.

Some philosophers of mind, of course, such as Thomas Nagel and John Searle, have argued powerfully for the view that subjective experiences are irreducible and ought to be included in an objective account of reality. But for the species problem this debate is of no account. If biological species should turn out after all to be conceptual entities only, then they are little better than fictions, and to call them objectively real would be highly misleading. To be clear, we should agree that biological species, however defined, are objectively real only if they are extra-mental.

It is not enough, however, for a realist species concept to have extra-mental referents in the biological world. In addition, it must be largely applicable. For example, if in reality relatively few lineages correspond to the model of punctuated equilibria (according to which stasis is the norm for a species), then a species concept based on that model is not going to be a viable species concept, especially if it has competitors which have a much greater correspondence with reality.

But it is not enough for a viable species concept to merely have a high degree of correspondence with reality. A species concept based on the age of organisms, for example, might have a high degree of correspondence with the biological world, but it would fail as a viable species concept for other reasons. Principal among them is that the entities picked out by such a concept are not the kind of entities that evolve, let alone in the open-ended fashion required by modern biology and by the mechanisms so far discovered. Modern biology has established by empirical means that the entities deemed to be species do not remain the same but evolve over time principally by means of natural selection and genetic drift. Such facts rule out many older species concepts, including species concepts based on some form of orthogenesis (the idea that each species has an inner developmental program responsible for the direction of its evolution).

So there are empirical reasons for accepting some species concepts and rejecting others. But there are also logical reasons, reasons of a more philosophical nature. In the past, philosophers have put forward many different criteria of reality, some of which can be rejected outright for purposes of the modern species problem. Plato, for instance, thought that for something to be fully real it must be eternal and unchanging. For strict phenomenalists, on the other hand, only what is given is real, namely perceptions themselves. Neither of these criteria, for obvious reasons, will have applicability in the present study. In biology and its philosophy certain assumptions are essential, and these are abrogated by the two criteria above. Neither would it be appropriate to attempt an examination of the conventional geography of the word "real" (cf. Austin 1964: 62–77 for just such a study). Ordinary language can no more help us with the scientific meaning of this word than it can for the meaning of the word "species." Instead our discussion must be framed from the perspective of science alone, and this changes the nature of the discussion considerably. This is not to say, however, that for our present topic we should turn to sciences other than biology and import the concept of "real" that is to be found there, in particular physics, long touted as the paradigm science—what Panchen (1992), for example, appropriately refers to as "physics envy" (4). Notable in this regard is Nagel (1961), who as a result of his examination of the various senses of the word "real" in physics, and its honorific connotation therein, threw up his hands in despair and exclaimed that "it would be desirable to ban the use of the word altogether" (151). Even if true for physics, it by no means follows that such a ban would be equally desirable in biology. As Cracraft (1989a) pointed out, "one's particular ontological stance—that is, whether one takes species to be discrete, real entities or not—has major theoretical and empirical implications for systematic and evolutionary biology" (38). For example, on the theoretical side the issue of species reality pertains to the levels-of-selection controversy and by extension to theories of macroevolution. On the empirical side, it relates directly to the study of speciation mechanisms (if species are not real, there remains no reason to study their causes) as well as to other studies such as biological control, biodiversity assessment, conservation biology, and the topic of environmental ethics.

Turning now to more specific criteria of reality relevant to the species problem, one criterion of reality that is widely applicable is the one we have already seen above in Sober's reply to Mayr, namely discoverability. Fictions are something we create, not something we discover, although the distinction between these two activities is not always immediately clear. We think of the stars above as real, but we do not think of the constellations as real; instead we think of them as manmade groupings of stars, and arbitrary groupings at that. And yet there was a time when, for example, someone, in a sense, first "discovered" the Big Dipper. Moreover, the Big Dipper has objective spatial and luminosity relations, it uniformly appears to move across the sky, and accordingly, it has proved useful for activities such as navigation. But of course this could be said of *any* grouping of stars visible to the naked eye. When we ask the question "Why this grouping and not another?" it becomes evident that the above criteria do not make the Big Dipper extra-mentally real, which is the kind of reality required for a realist species concept. Even though the Big Dipper might readily suggest itself to tool-using humans on planet earth looking up in the northern sky at night, it is still nevertheless a conceptual entity only, for its existence, unlike its component stars and their relations, depends on minds. As such, the Big Dipper is a fiction, much like centaurs and unicorns, although the latter are even more obviously so. Nobody ever went into a forest and "discovered" centaurs and unicorns. Instead they were dreamed up in someone's mind (albeit using parts of organisms already discovered) and accordingly made into the stuff of myths and tall tales (which is not to deny that there are people who actually believe that centaurs and unicorns are real, but of course you won't find them among biologists). For species to be real in biology, on the other hand, they must not be fictions like centaurs and unicorns, or even like the Big Dipper; instead they must be "out there" in the biological world, either evolving, temporarily remaining in stasis, or going extinct, their reality independent of whether there are any minds making them out or not.

Discoverability, however, though indispensable, proves insufficient as the sole criterion of reality, for surely there exist many concrete and abstract entities which we not only have not discovered but presumably could never even in principle discover. At least we must allow for the possibility (which borders on certainty) that humans and human science as epistemological entities, and indeed all epistemological entities whatsoever in the universe, have inherent limitations and incapacities. Discoverability, then, will be one of our criteria of reality, but it cannot be our only one.

Another important criterion is discussed by Karl Popper (Popper and Eccles 1977). Popper accepts what most of us would, namely that the concrete particulars we can point to are the paradigm real things. By this he does not mean that they are the ultimate units of reality; only that for us, as human beings, they are the paradigm examples of reality. Popper then points out that we (or more properly we through science) extend this concept of reality to things that are too big for ostensive definitions, like planet earth and the Milky Way galaxy, to things that are too small for ostensive definitions, like atoms and electrons, and also to imperceptible things like fields and forces. What, he asks, is the ultimate criterion by which we accept these extensions? "We accept things as 'real,'" he says, "if they can causally act upon, or interact with, ordinary real material things" (10).

There are, to be sure, a number of problems with this criterion. Ever since Hume the concept of causality has come under intense fire. Causality in the sense of causal force is not something we perceive; instead it is something we infer. But is this inference valid? This is the question as Hume posed it, and his famous answer was in the negative.[3] Whatever our answer, I suggest that we must make a

---

[3]   Or at least this is how many have interpreted Hume. For example, according to Bertrand Russell (1946) "Hume's real argument is that . . . we never perceive causal relations, which must therefore, if admitted, be inferred from relations that can be perceived. The controversy is thus reduced to one of empirical fact: Do we, or do we not, sometimes perceive a relation which can be called causal? Hume says no, his adversaries say yes, and it is not easy to see how evidence can be produced by either side" (642). To this Russell adds that causation is something that Hume "condemns" (640), that for Hume "in causation there is no indefinable relation except conjunction or succession" (641). Others, more recently (cf. Denkel 1996: 238–240 and references therein), have argued that Hume was only giving a psychological analysis of our concept of cause, not that he was attempting to argue anything negative (or positive) about objective causation (causation actually in things). Part of the problem with interpreting Hume's argument is that it is often phrased by Hume himself in terms of "causal relations," and by this it is not always clear whether he is referring to extra-mental relations or to relations between mental impressions. At any rate it is evident that Hume makes the distinction between "power," "force," or "necessary connexion," which he says we do not perceive and moreover can have no idea of, and relations which we take to be evidence of them, such as the motion of one billiard ball following that of another. For example, he says (Hume 1748/51) "It is confessed, that the utmost effort of human reason is to reduce the principles, productive of natural phenomena, to a greater simplicity, and to resolve the many particular effects into a few general causes, by means of reasonings from analogy, experience, and observation. But as to the causes of these general causes, we should in vain attempt their discovery; nor shall we ever be able to satisfy ourselves, by any particular explication of them. These ultimate springs and principles are totally shut up from human curiosity and enquiry. Elasticity, gravity, cohesion of parts, communication of motion by impulse; these are probably the ultimate causes and principles which we shall ever discover in nature; and we may esteem ourselves sufficiently happy, if, by accurate enquiry and reasoning, we can trace up the particular phenomena to, or near to, these general principles" (30–31). It must be remembered, after all, that Hume was a subscriber to the Newtonian paradigm.

distinction between causal *forces* (whatever they are) and causal *relations* (cf. Hempel 1966: 53); these latter, it seems to me, are something that we in many cases do indeed directly perceive and otherwise may validly infer, just like many other relations, such as (relative) space relations and time relations. Indeed if this were not so we could never discern causality from mere correlation. Of course the fact is that we not only can but we often do. Accordingly causality conceived as causal relations would go far to preserve Popper's ultimate criterion of reality.

And indeed a concept of causality seems indispensable. The search for the cause or causes of mass extinctions, the search for the cause or causes of cancer, and on and on, are all legitimate searches. In comparison to these searches philosophical difficulties concerning causality seem, as indeed they are, little more than sophistic quibbling.

Nevertheless certain real difficulties remain for the causal criterion of reality. One difficulty is peculiar to Popper's version of that criterion. Once it is granted that we are epistemologically limited entities, it becomes evident that we should not want to preclude from reality entities which are causal but which only causally act upon and interact with entities which are themselves not paradigm real things. But then in admitting this we necessarily go beyond Popper's criterion.

There is a more important difficulty, however, which applies to all versions of the causal criterion of reality (i.e., the view that to be real is to have causal power; cf. Kim 1993: 348). Correlative with the concept of cause is the concept of effect. In other words it is true by definition that one cannot have a cause without an effect. What the causal criterion of reality forces us to infer is that an effect cannot be real unless in turn it is itself a cause. But this inference is not necessitated by the concept of cause itself. Instead all that that concept necessitates is that if a purported cause is real then its effects must also be real. Logically the concept of cause does not entail that an effect must also be causal in order to be real. That further claim must come from somewhere else, and if it does not come from the concept of cause it is indeed difficult to see where it could come from.

All of this has interesting consequences for the species problem. As we shall see in later chapters, many theorists hold that species on their view are causal entities, and they moreover either claim or imply that this gives their species concept a privileged position over other species concepts for which species are not causal entities. If the above analysis is correct, however, these claims are unjustified by themselves and therefore require further argument (which I may say has yet to be supplied).

There is a larger consequence for the modern species problem, however. A number of theorists (e.g., Levin 1979: 384; Cracraft 1983: 102–103, 1989a: 34–35; Kluge 1990: 418–424; Mallet 1995: 296–297; Luckow 1995: 589–590) have argued (and rightly, it seems to me) that all competing species concepts can be divided into two categories, one for species concepts according to which species are causal or process entities, and one for species concepts according to which species are exclusively effect or pattern entities (both categories are with respect to the processes of evolution). If the causal criterion of reality is employed to narrow down the field in the species debate, then the species concepts in the second category are immediately eliminated. But as we have seen above, the causal criterion of reality by itself is seriously flawed. Consequently, if species concepts according to which species are pattern entities are to be eliminated in the debate, some other reasons must be given for their elimination. Interestingly, as we shall see more fully in chapter 5.2, there is an advantage which these species concepts have over the causal concepts, namely that the latter, since they focus on only one or a few causal processes in evolution, cannot supply a species concept that is universal for all those research programs in biology that require a species concept. Consequently, causal species concepts either fail to satisfy the majority of biologists or, if they are to satisfy, they result in a sort of species pluralism. On the other hand, only pattern species concepts, since they are noncommittal on the causal mechanisms or processes involved in evolution, have the potential to deliver a truly universal species concept. Thus a common criterion of reality in philosophy finds itself turned upside down when applied to a real-life problem such as the species problem.

At any rate, we may conclude at this point that causal entities, both observed and inferred, must be thought real, but that a real entity need not be causal or inferred as causal in order to be real. Once again, then, the criterion of reality that we are examining must not be thought of as the only criterion but instead as one among others.

A further important criterion of reality is the one stressed by Bertrand Russell. Russell often claimed (e.g., 1918: 224) that something is irreducibly real if it would have to be included in a complete description of the universe.[4] This criterion would seem to be an advance on the other two, since it not only subsumes those criteria but by being less subjective avoids many of the difficulties resulting from their limitations. Indeed, this criterion has great intuitive appeal.

Nevertheless, Russell's criterion, though it seems to me an advance, suffers from at least one grave difficulty: it assumes that whatever is real can be described in words. The problem with this is that as limited epistemological entities we must make allowances for the ineffable. We must allow for the possibility that for all possible levels of epistemological entities throughout the universe there are entities which in principle are entirely ineffable.

In spite of this difficulty, however, Russell's criterion, though not sufficient by itself, is an important one, especially for the species problem. In biology the species category is generally thought to serve primarily two masters. On the one hand it is the basal category in taxonomy (increasingly the majority of biologists now consider subspecific categories as arbitrary and hence unreal, as the majority of them have considered superspecific or higher taxa for quite some time). Accordingly, the ideal of taxonomy is to produce a natural system, a system that carves nature at its joints (to use Plato's metaphor). An arbitrary system, of course, is not difficult to come by, and many systems have been offered in the past which are now considered arbitrary (e.g., the Quinarian system of William

---

[4]   It is this criterion, principally, that allows Russell to argue that relations are real and that classes are fictions. I will have more on this later in chapters 5.3 and 3.5, respectively.

MacLeay; cf. Panchen 1992: 23–25). No taxonomist, of course, would want to argue that an arbitrary taxonomic system of the biological world should be included in a complete description of the universe (or at least in a description of the biological world). But most taxonomists do believe that the biological world does divide into natural entities usually called species, that these entities and their historical and other relationships are out there, often waiting to be discovered, and that a description of the biological world would be incomplete without them. On the other hand, the species category is for most biologists also a basic category in evolutionary biology. Species are the entities that speciate, that have ranges, that evolve or remain in stasis, and that eventually go extinct. Accordingly, for most evolutionary biologists as well, a complete description of the biological world would be incomplete without a description of these evolutionary entities. (I of course exclude here those biologists who do not think that species are real. As we shall see, for some only individual organisms are real, while others go a little further and admit entities no larger than colonies or populations.) In all of this Russell's criterion of reality is presupposed by biologists, whether they are aware of it or not, and whether they are species realists or species nominalists. In either case the species problem can be recast in terms of Russell's criterion. The issue still remains, of course, which criteria are to be employed in satisfying that larger criterion. That is an issue, however, the specifics of which are best reserved for the chapter-by-chapter examination of competing species concepts (including species nominalism). So far the discussion has only been general, with the intention of serving no more than as a basic guide for the more specific discussions.

A further consideration, however, should be discussed before continuing. I mentioned above that the species category is usually thought to be a basic category for both biological taxonomy and evolutionary biology. Rarely, if at all, however, has a species concept proved to be equally compatible with both. Usually a species concept is designed to satisfy principally one or the other. The ideal, of course, is to provide a species concept that serves both maximally well. (Evolutionary biology, however, must be taken as fundamental, if only because, to quote Dobzhansky once more, "nothing in biology makes sense except in the light of evolution.") This ideal is, moreover, perfectly in harmony with a much larger scenario in the recent history of biology. In the 1930s–1950s the popularly-called Modern Synthesis was forged, namely the union of previously isolated subdisciplines in biology founded on the marriage of Darwinian natural selection with Mendelian genetics. Perhaps the greatest curiosity of the Modern Synthesis, however, is that although the term "species" has remained a fundamental term in biology, the Synthesis has yet to secure a unified species concept. Instead the situation has remained somewhat like it was in Darwin's day, with a number of different and conflicting species concepts enjoying more or less use. In fact, the situation has only worsened since Darwin's day, in spite of the fact that today's biologists all take evolution perfectly seriously. As the present book will attest, the plethora of species concepts in use in biology today is truly staggering. And yet the greatest paradox of it all is that in spite of this fact biology has proven to be the progressive science of the twentieth century *par excellence*. At any rate, the main desire of the present work, which will become especially evident in the fifth and final chapter, is to try to forge a unified species concept in the spirit of the Modern Synthesis. Whether this effort will have succeeded, of course, remains for others to judge.

1.3 Preliminary Assumptions and Concepts

Before proceeding further, certain preliminary assumptions and concepts should be clarified and made explicit.

Beginning with the former, since the present study is not only a work in metaphysics but in *practical* metaphysics, and in philosophy of biology at that, certain disputes and questions in metaphysics *per se* simply do not apply. For one, the issue of whether there is an external world will not be debated. Moreover, it will be taken for granted that mind, whatever it is, constitutes a very small and derivative part of the universe.

Another matter is whether biological evolution is a fact or merely a theory. I have no desire in the present study to engage in controversy with creationism, Christian or otherwise, nor is any such engagement necessary. As Carl Sagan proclaimed in the second episode of his award-winning video series *Cosmos*, "Evolution is a *fact*, not a theory—it *really* happened." And of course it continues to happen. The evidence for this conclusion is so massive and rationally overwhelming (for those who genuinely care to explore it) that it is on a par with the fact (what used to be thought a theory) that the earth is spherical and orbits around the sun. Virtually all modern biologists accept this as so (cf. Dobzhansky 1937: 8; Cain 1954: 12–13; Maynard Smith 1975: xvii; Gould 1983: 255; Eldredge 1985a: 13–14; Futuyma 1986: 15; Dawkins 1986: 287; Mayr 1988a: 192, 199, 262, 492; Ridley 1993: 57). Where biologists differ is on matters such as the mechanisms and rates of evolution, matters which are still a matter of theory and which I will examine mainly in chapter 4, and only then because it is relevant to the species concepts discussed therein.

Turning now to preliminary concepts, a short discussion on a number of these seems indispensable in the interest of avoiding confusion.

One crucial distinction is between the species category and species taxa. As Mayr (1970: 13–14, 1982: 253–254) has repeatedly stressed, the word "species" is ambiguous. It can refer either to species as a category or to species as a taxon. Failure to appreciate this distinction can easily lead (as indeed it has) to deep confusion.[5] Briefly, species taxa are the basal units of evolution and taxonomy. For example, *Homo sapiens* is a species taxon, as is *Tyrannosaurus rex*. The species category, on the other hand, is a class defined intensionally. It is the class of all species taxa, defined by what it takes to be a member of the class. Accordingly, the attempt to determine the basic ontological category for species taxa is a necessary preliminary in the attempt to define the species category. Consequently, if

---

[5]   Mayr (1988a: 321, 1987a: 149) cites three works which particularly confound these two concepts, respectively Sokal and Crovello (1970) and Kitcher (1984a, 1984b). Cf. Kitcher (1987: 187) for a reply to Mayr, and chapter 4.5n60 for my reply to Kitcher.

species taxa are in fact unreal, the species category is an empty class.

Another important distinction is between genotype, phenotype, genome, and phenome. These biological concepts, interestingly, in both biology and philosophy of biology, are not used as uniformly as one might wish.

The term *genotype* was coined by the Danish geneticist Wilhelm Johannsen (1857–1927). As Mayr (1982) explains Johannsen's concept, "'Genotype' refers to the genetic constitution of the zygote, formed by the union of two gametes" (782). Johannsen also, as Mayr points out, used the term in relation to populations and species. But this is not the way Mayr himself uses it. According to Mayr, the genotype is the "total genetic constitution of an organism" (1982: 958), its "genetic program, . . . unchanged in its components except for occasional mutations" (1988a: 16). The term has been used in a variety of ways. According to Edward Wilson (1992), the genotype is "The genetic constitution of an organism, either prescribing a single trait (such as eye color) or a set of traits (eye color, blood type, etc.)" (399). According to Alec Panchen (1992), it is "some part under study of the total genetic programme of an individual" (136). None of these definitions are idiosyncratic. Unfortunately, the situation is not improved by consulting science dictionaries. According to *The Penguin Dictionary of Science* (Uvarov *et al.* 1979), for example, which implicitly follows Johannsen, the genotype is "The genetic constitution of an individual organism or of a well-defined group of organisms. . . . A group of organisms that have the same genetic constitution. . . . A typical species of a genus" (182). I shall employ the term as does Mayr, meaning no more and no less than the total genetic program of an individual organism, since this seems its most common usage (cf. Eigen 1983: 106; Futuyma 1986: 43).

The term *phenotype* was also coined by Johannsen. Although he used the term to denote "the statistical mean value" of the observable physical characteristics of a sample of conspecific (species-specific) organisms (Mayr 1982: 782), it is rarely ever used in that way anymore. Today it is normally used to denote the total physical expression (morphological, physiological, behavioral) of an individual organism's genotype (cf. Weisz 1967: 846; Medawar and Medawar 1983: 116–117; Mayr 1982: 959). Although, of course, an individual organism's phenotype potentiality is completely determined by its genotype, its phenotype actuality is affected to some extent by environmental factors.

The term *genome*, perhaps more than any other, enjoys quite a variety of uses in the literature. According to *The New Penguin Dictionary of Biology* (Abercrombie *et al.* 1990), the genome is a purely quantitative concept: "The total genetic material within a cell or individual, depending upon context" (232), measurable in "picograms" (1 picogram = 10^{-12} g). According to Panchen (1992), the genome "refers to the total genetic complement of an individual, represented by all the chromosomes with their included genes in a somatic cell" (136). According to Wilson (1992), the genome is "All the genes of a particular organism or species" (399). Accordingly, the human genome (i.e., the genome of each individual in the species) consists, according to Wilson, of roughly 100,000 genes (76). On the usage of David Raup (1991), "genome" is synonymous with "gene pool": "a species is a group of organisms that share a common pool of genetic material (genome)" (14). Finally, according to Medawar and Medawar (1983), "A genome is the genetic apparatus of a species considered as a whole and as characteristic of it" (116). To a nonbiologist, it is not entirely clear whether some of these definitions are contradictory or merely complementary with different emphases. At any rate, I shall use the term *genome* to refer to what is common genetically to a group, population, or species, depending on the context of the discussion. I shall not use it in relation to particular organisms, reserving the term *genotype* for that. For example, the human genome has 46 chromosomes and roughly 100,000 genes (both in each somatic cell), but not all humans need have the same amount of chromosomes or genes. A human with Down's syndrome, for instance, has a genotype with an extra chromosome (and by implication extra genes). In other words, humans with Down's syndrome have a somewhat different genome (a subgenome) from humans without Down's syndrome.

The term *phenome* is rarely used in the literature. Panchen (1992) uses the term to refer to the "phenetic [outward] manifestation" (137) of the genome (as he uses this latter term above). I, on the other hand, shall use it as complementary to my use of the term *genome* above. Accordingly, individual organisms do not have a phenome, only groups, populations, and species. The phenome of a group, population, or species is what is common phenotypically of that group, population, or species. For example, both a polymorphic species and a sexually dimorphic species will have a phenome comprised of subphenomes, one for each subspecies in the case of the former and one for each sex in the case of the latter.

The final distinction I want to make is between the terms *set* and *class*, in philosophical contexts a distinction that is notoriously bad. One problem is that many philosophers use the terms synonymously. Bertrand Russell (1919), as a major example, uses them synonymously, as does, for the most part, W.V.O. Quine (1969c). Quine (3), however, points out that the term "set" has more currency in mathematical contexts. Moreover, Quine introduces a technical distinction. He reserves the term "set" for classes which themselves are capable of being members of classes, while classes not capable of being members of classes he calls "ultimate classes" (3). Quine's distinction, however, obscures the fact that he conceives of classes as abstract entities, and therefore by implication also sets. On the other hand, as Ruth Marcus (1974) amply illustrates, many philosophers use either or both terms to denote just the collections themselves, as in "the class or set of books in my study."

The problem is compounded by the fact that there are traditionally two ways of defining classes/sets. One way of defining them is by *extension*, by enumerating their members. The other is by *intension*, by enumerating the properties (including relations) which determine their membership. Russell (1919: 12) says the second type of definition is logically the more fundamental, and for basically two reasons: (i) a class defined by extension is reducible to a class defined by intension, while not *vice versa*; and (ii) extensional definitions cannot apply to classes which have indeterminate or infinite membership. While I can accept the second of these, Russell's first reason seems to me in error. What intensional definition, for example, can the set composed of my cat, the city of Toronto, and the number three be reduced to? This is an extreme example, but it illustrates the point that the intension/extension distinction is fundamental and that the words "class" and "set" need to be clearly defined accordingly. The problem is that philosophers rarely make the effort to define their meaning of these terms. Instead, as a whole they use them in quite a variety of ways, causing confusion for not only other philosophers but nonphilosophers as well (cf. Mayr 1987a: 148). The problem is especially acute in the literature on the species problem. As we shall see, many arguments, counterarguments, and outright confusions turn precisely on this issue.

In the interest of clarity, I shall use the term "class" for intensionally defined (therefore abstract) objects, whereas I shall use the term "set" for extensionally defined objects, namely collections. The members of a set, therefore, may, but need not, have common (nontrivial) properties, whereas the members of a class must have common (nontrivial) properties. (A collection, to be even more clear, is a set irrespective of whether its members do or do not have common (nontrivial) properties, while a class is a class whether or not it has set-membership.)[6]

1.4 Abstract of the Book

Aside from the rejection of the objective reality of species, it has become common to categorize modern species concepts as conceiving of species in either of two basic ways: (i) as concrete, physical entities; or (ii) as abstract entities. This division, as we shall see later in this book, is by no means clean, and I subscribe to it here only for the purpose of chapter division. It is also useful for setting apart my own view on species, evident in the quotation from Mayr at the head of chapter 5. In spite of the lack of a sharp division between the above two categories, the physicalist view is best divided, it seems to me, into basically four views (not altogether distinct), with species conceived of either as individuals, as sets (numerical universals), as clades, or as lineages. On the other hand it seems to me that the abstraction view is best divided into basically three views, with species conceived of either as strictly essentialistic elementary (element-like) classes, as loosely essentialistic cluster classes, or as even less essentialistic ecological niches. These seven realist views will be allocated to two chapters, one for the classes view and one for the physicalist view, the latter focusing mainly on the individuals view. My own view is a sort of hybrid between the abstraction view and the physicalist view and will be reserved for the final chapter.

Chapter 2 is devoted to the topic of species nominalism. "Species nominalism" will be the label used for the view that biological species are not objectively real—*real* in the sense of extra-mental—that they are fictions. Normally on this view only individual organisms are real (though sometimes also colonies or even

---

[6]   Interestingly, my particular distinction between sets and classes has a precedent. Cf. e.g., Gasking (1960: 1). Gasking remarks that one consequence of this distinction is that if the membership of a set changes, that set, properly speaking, is no longer the same, while, on the other hand, classes may remain the same even though their membership changes.

populations), so that groupings and classifications of organisms into species do not have an objective reality but instead are manmade, mind dependent, and ultimately arbitrary.

Species nominalism will be allotted the default position, and for a simple basic reason: For my part, I have never seen a species, nor can I imagine what one would look like. Instead, I have only seen examples of a this or a that. I shall take it, then, that the specter of nominalism haunts each and every chapter in which the species problem can be divided.

It is the nature of a default position, of course, that the burden of proof lies not on those who would hold the default position, but instead on those who would reject it. Nevertheless, species nominalists have historically provided evidence and argument in support of their position (as opposed to evidence and argument strictly against the positions of realists). These arguments cannot be overlooked.

Accordingly the arguments of some early naturalists and philosophers will be examined, not simply as an antiquarian interest or as an exercise in exegesis, but instead for their relation to the modern debate. Specifically, the views of William of Occam and John Locke will be examined, as well as the views of Buffon, Lamarck, and Darwin. Darwin's view is especially curious, and will be analyzed in light of the interpretations of Ernst Mayr (1982) and John Beatty (1985).

In addition, the species nominalisms of some modern biologists will be critically examined, including J.S.L. Gilmour (1940), Benjamin Burma (1949, 1954), J.B.S. Haldane (1956), Alan Shaw (1969), and Donald Levin (1979). In addition to these, each of which has a different approach, I will focus on the view that populations but not species are real, in particular the views of Donn Rosen (1978) and Mario Bunge and Martin Mahner (Bunge 1981; Mahner 1993; Mahner and Bunge 1997). The more popular and rising view known as species pluralism will also be examined, as well as the issue of cross-cultural tests.

In sum, a number of arguments will be provided for and against the position of species nominalism, while its status as the default position will be maintained throughout.

Chapter 3 is devoted to a discussion of the species-as-classes view. This view encompasses a number of related views, all of which share in common the idea that biological species are abstractions of some sort, abstractions existing objectively in the external world, with species membership involving some sort of essentialism. The primary position here is the view which conceives of biological species in the same way as natural kinds in chemistry, fundamentally as chemical elements in the periodic table of elements. On this view natural kinds are spatiotemporally unrestricted with kind membership determined by necessary conditions jointly sufficient for membership.

Although species essentialism is a view which has (to put it mildly) languished in recent decades, it has enjoyed a long and distinguished history, being traceable back, broadly speaking, to the views of Plato and Aristotle on the one hand and the Book of Genesis on the other. The combination of these two traditions found its culmination in Carolus Linnaeus, from whom modern biology received its binomial nomenclature for species. Although there are virtually no species essentialists alive today in biology, there is an impressive list of modern philosophers who maintain this view, including Irving Copi (1954), Carl Hempel (1966), W.V.O. Quine (1965, 1976), Saul Kripke (1972), Hilary Putnam (1970), and David B. Kitts and David J. Kitts (1979), each of whom will be briefly examined.

Due to the many difficulties involved with the element-like species-as-classes view, two related views have been proffered in recent decades. The first of these that shall be critically examined is the species-as-clusters view. On this view a species is a cluster class, with essential characteristics as above, but with kind-membership not determined by necessary conditions jointly sufficient. On this view there is no one essential characteristic that any member of a species must have in order to be a member of its species; instead, each member need only have a minimum number of the essential characteristics, a quorum. This view can be employed using either phenotypes (Morton Beckner 1959; Renford Bambrough 1960–61; Peter Sneath and Robert Sokal 1973) or genotypes (Arthur Caplan 1980; James Mallet 1995; M.H.V. Van Regenmortel 1997; Manfred Eigen *et al.* 1988; Eigen 1993). Each of these views will be critically examined.

A third variation of the species-as-classes view is the view which conceives of species either as inhabitants of ecological niches or as the niches themselves. On this view individual organisms are conspecific if and only if they occupy the same ecological niche. The views of Leigh Van Valen (1976) and David Johnson (1990) will be critically examined, in addition to partial or secondary defenders such as G.G. Simpson (1961), Ehrlich and Raven (1969), Ernst Mayr (1982), and Alan Templeton (1989). My reasons for including this view as a variation of the species-as-classes view will be discussed in chapter 3.4.

Chapter 4 is devoted to a critical examination of the species-as-individuals view. On this view a biological species is neither unreal nor an abstraction; instead it is a spatiotemporally localized, cohesive (concrete) individual. Accordingly individual organisms are not *members* of species (in the sense of class membership) but rather *parts* of species (in the sense of part-whole membership). Although this view can arguably be found in a nascent form in a number of earlier philosophers and biologists, such as G.W.F. Hegel, Julian Huxley, Karl Popper, and Ernst Mayr, and also contemporaneously in Hugh Paterson, it did not receive a fully explicit exposition and defense until the 1960s and 1970s, first in the work of the biologist Michael Ghiselin (1966, 1974), followed by the philosopher David Hull (1976, 1977, 1978). It is a view that has gained a large following among biologists and philosophers, notably among them Edward Wiley (1978, 1981), Edward Wilson (1992), Mary Williams (1985, 1987), and Elliott Sober (1984b, 1993).

Also critically examined will be the theory of punctuated equilibria proffered by Niles Eldredge and Stephen Jay Gould (1972). On their view the norm for species is not gradualistic evolution but instead long periods of stasis punctuated by relatively brief periods of rapid evolution. Since their theory seems to add substance to the characterization of species as spatiotemporally restricted entities, it was only natural that they should subscribe to the species-as-individuals view (Eldredge 1985a, 1985b; Gould 1982).

There are other physicalist views besides the species-as-individuals view. Prominent among them is the view of Philip Kitcher (1984a, 1984b, 1987, 1989), who conceives of species as sets of organisms extensionally defined (numerical universals). Much closer to the species-as-individuals view, however, either explicitly or implicitly, are cladistic species concepts, in which a species is conceived as a branch (clade) on the phylogenetic tree of life delimited by branching points. Examined in detail will be the views of Willi Hennig (1966), Mark Ridley (1989), and D.J. Kornet (1993). Closely related to cladistic species concepts is the recent plethora of so-called phylogenetic species concepts, which have been rising greatly in popularity. Although it would be pointless to examine them all in detail, I will briefly survey the field, providing a detailed analysis of two of its most important versions, namely the considerably different versions provided by Joel Cracraft (1983) and Brent Mishler and Robert Brandon (1987). Finally, I will examine the views of Bradley Wilson (1995) and Kevin de Queiroz (1999), who look at the base of the view of Ghiselin and Hull, as well as that of the cladistic and phylogenetic species concepts, and argue that although species are individuals they should more properly be conceived as lineages of organisms.

Each of these three basic paradigms,[7] then, namely species nominalism, species as abstract entities, and species as physical entities, enjoy various strengths and weaknesses. Each paradigm, accordingly, will be carefully examined in an attempt to assess as objectively as possible its various strengths and weaknesses. I should also mention here that there are many more biologists and philosophers than those named above whose views on the species problem will be examined, although their views will not be examined in as much detail.

In the same spirit of critical inquiry as that above, my own theory on the ontology of species will be developed and expounded in chapter 5, the view that species are relations. Although, of course, an attempt will be made to defend my own view as much as possible, this defense will not be undertaken at the expense of hiding its weaknesses, whether major or minor. Indeed as a theory it may turn out to be a dead end. At any rate, as it is a theoretical niche that no one as yet has ventured to explore, it seems to me excitingly novel, initially plausible, and fully worthy of serious consideration.

Whereas previous authors on the species problem have focused their solutions on one or another of traditional metaphysical categories, namely nominalism, universals, and particulars, virtually all have overlooked the fact that the universals category, traditionally conceived, is basically dichotomous. Universals come in basically two kinds: qualities, attributes, and properties on the one hand, and relations on the other.[8] Interestingly in both modern and traditional discussions on

---

[7]   I use the term "paradigm" loosely, in the sense of "a way of looking at things."

[8]   Cf., e.g., Russell (1913: 90–92). For my part, I make no distinction between qualities, properties, and attributes. Some philosophers, notably Quine (e.g., 1981), not only include classes in this list, but would have us think of each of the above entities *as* classes. But there are serious difficulties with this view. These and general difficulties with the ontology of the class category *per se* will be examined in chapter 3.5.

the problem of universals it is the former category that is normally focused on. Relations are either ignored altogether or given secondary treatment. As Russell often pointed out (e.g., 1914: 42, 48), this is part of the legacy of Aristotelian logic, in which propositions are properly of the subject-predicate form, corresponding to substances and attributes.

Though Aristotelian logic does not today enjoy the prestige and dominance that it long did in the past, on the matter of relations its negative influence, for the most part, still remains. Only recently has this imbalance begun to be redressed. As Armstrong (1989) pointed out, "It is not until the late nineteenth and the twentieth century with C.S. Peirce, William James, and Bertrand Russell that relations begin (no more than begin) to come into focus" (29).

It should come as no surprise, then, that the species problem, a problem that, though part of the traditional problem of universals, has only recently come of age, should reflect this imbalance. Indeed in this matter the modern (not to mention the traditional) discussion has pretty much ignored the relations category altogether.

Thus the stage is set for my own theory on the ontology of biological species, a theory that takes full advantage of the new logic developed largely by Russell, a logic that takes relations as nonreducible and ineliminable. As I shall argue in chapter 5, assuming that it is true that relations require relata for their existence, that there can be no *pure* relations—indeed I go so far as to say that the relata are part of what a relation is—it follows that in referring to real relations one is also referring to their relata. But it by no means follows that one is *only* referring to their relata. One is referring to their relata *plus something more*. It is that something more that is missed not only by the species nominalists but also by the realists who place their focus only on properties or on the organisms themselves. It is also missed by the remainder of the realists since they employ some relations but that employment is marginal and normally restricted in focus to no more than one kind of relation.

In saying that species are relations, I mean that when biologists correctly delimit species and when the rest of us correctly use species words (the words themselves, of course, are entirely arbitrary) we are all in effect referring neither to entities abstract or concrete nor to their members or parts; instead we are referring to the individual organisms and the relations between them that together constitute their reality as species. Their reality, then, is neither strictly abstract nor concrete, but is a sort of hybrid between the two. And it is because relations have not only a different ontology but also different existence conditions from both classes and physical entities that they make the most suitable candidates for the constituents of evolving species in the biological world. In short, I shall argue that thinking of species as relations leads to the novel view that a species is a complex of similarity relations (with organisms ultimately as the relata) objectively bounded or delimited by various causal relations (such as interbreeding relations, ecological relations, ontogenetic relations, caste relations, etc.). I call this the *biosimilarity species concept,* and I shall attempt to show that thinking of biological species from the viewpoint of this concept provides numerous advantages

over its many competitors, in sum that it avoids many of the difficulties and pitfalls from which they suffer and better fits the bare facts and basic theories of
modern biology.

In developing the above view, I have been guided by a number of fundamental distinctions which have become prominent in the recent literature on the species problem. One of these is the temporal distinction, which divides competing
species concepts into those which conceive of species as basically horizontal entities (e.g., Mayr, Paterson, Van Valen, Cronquist) and those which conceive of
species as basically vertical entities (e.g., Simpson, Hennig, Ghiselin, Eldredge
and Gould). To the pleasure of some theorists and to the utter dismay of others, I
argue that the horizontal dimension for species is logically and therefore ontologically prior to the vertical dimension so that horizontal species concepts have
priority over vertical ones. To drive home this point I use to full advantage Darwin's analogy of language evolution. Accordingly, in developing my own species
concept I conceive of species as primarily horizontal entities. The main consequence of this move is a much more powerfully realist species concept compared
with species concepts which depend on the vertical dimension for species reality
(typically the dimension focused on by modern species nominalists).

A second distinction that has become prominent in the recent literature on the
species problem is the division of species concepts into those which are primarily
process species concepts and those which are primarily pattern species concepts.
We have already encountered this distinction in §2 of this chapter and I will examine it more fully in chapter 5.2. Briefly, once more, process species concepts
(e.g., Mayr, Paterson, Van Valen) are based on only one or a few of the causal
processes or mechanisms involved in evolution, while pattern species concepts
(e.g., Sokal and Sneath, Cronquist, Cracraft) are based on the end-products or
effects of those causal processes. Process species concepts have the liability that
they tend to exclude certain kinds of organisms from species membership, so that
certain kinds of biologists are left out in the cold. The result is that many biologists either turn to other species concepts or embrace a sort of species pluralism.
Pattern species concepts, on the other hand, are not theory dependent as to the
different kinds of causal mechanisms and processes in evolution and are accordingly potential candidates for a truly universal species concept. The biosimilarity
species concept, it will be seen, is clearly a pattern species concept and is offered
in the spirit of the Modern Synthesis. It has advantages over other pattern species
concepts, however, that would seem to make it the prime candidate. Unlike traditional morphological species concepts and the more recent phenetic species concept which groups organisms according to overall similarity, the biosimilarity
species concept does not, for example, divide organisms into separate species
with radically different life cycles such as caterpillar-butterflies, nor does it unite
into a single species groups of extremely similar organisms that are reproductively isolated from each other (sibling species). This is because it takes seriously
causal relations which play a basic role in evolution, namely in the above cases
ontogenetic relations and interbreeding relations. Moreover, unlike phylogenetic
species concepts which claim to be pattern species concepts, the biosimilarity

concept is primarily horizontal. This not only gives its referents a stronger claim to reality, but it takes away any constraints imposed by monophyly (whether strict or loose)—which after all is a convention and not a causal process. The result is a species concept which allows for multiple origins (as in the case of repeated polyploidy speciation) and for extinction not being necessarily forever (which makes sense from the viewpoint of genetic engineering alone).

As hinted at already in the above, in developing my species concept I will also try to show how it helps to resolve many of the practical problems encountered by biologists in species delimitation (e.g., multiple origins by polyploidy, hybrid species such as the red wolf, ring species, anagenesis). This might be called its practical defense. But I will also attempt to defend it from a strictly philosophical or metaphysical point of view. My species theory clearly requires the objectivity of relations in general, and since their reality has long been controversial among philosophers (especially similarity), I will have to provide a lengthy discussion on the ontology of relations. Much of this will be based on the relations realism, including the resemblance universal realism, of Russell, who was the original inspiration for the development of my view on species. But I have more recently gained valuable insights from the resemblance thesis of Arda Denkel (1989, 1996), whose view on properties served as the inspiration for the final form of my view on species. Although neither Russell nor Denkel ever applied their respective resemblance realisms to anything like the species problem, and although their views cannot be seen as being perfectly congruent, my view on species may be seen as a biological application of their combined metaphysics.

In addition to the above two guiding distinctions, namely horizontal versus vertical species concepts and pattern versus process species concepts, I have been guided in the development of my own species concept by a further distinction which has become common in twentieth-century metaphysics but which has hitherto seen no light in the literature on the modern species problem. The distinction I have in mind is the distinction between internal and external relations. It is this distinction that has further guided me in my choice of which relations to include as actually constitutive of species (namely similarity) and which not (interbreeding relations, ecological relations, etc.). Combined with due consideration of the two distinctions above from the modern literature on the species problem, the natural outcome, I believe, is the biosimilarity species concept.

There is, of course, yet a further distinction, in addition to the first two above, that has become prominent in the modern literature on the species problem. As probably already guessed, the distinction I have in mind is the division of competing species concepts into those which conceive of species as classes and those which conceive of species as individuals. Although it is undoubtedly the most widely used in the literature, many will be surprised to find that it is a distinction which has not guided me in the development of my own species concept. For reasons that will be elucidated in chapter 4, I consider the class/individual distinction to be a false dichotomy, and a most damaging one at that. Deeply rooted in the ancient and now defunct tradition of Aristotelian logic, it has wreaked havoc in the modern literature on the species problem. Guided by misplaced analogies with individual organisms, philosophically-minded biologists and bio-logically-minded philosophers have taken the species problem into a myriad of avenues each of which would make Darwin roll over in his grave. Species the extinction of which is necessarily (logically) forever, species which cannot have multiple origins, species which remain numerically identical through unlimited evolution, species which can actually be seen with the eye, species which must have only sexual organisms as their members (parts), and most notably species which cannot possibly evolve, are only some of the mischievous results of a mis-conceived and misapplied distinction which cannot help but put modern evolu-tionary biology in a most unfortunate light.

In closing this introductory chapter I should perhaps briefly return to a point raised at the beginning of this book, namely that the species problem is not, so to speak, a problem unto itself. Whether one likes it or not, whatever species con-cept one decides upon will invariably entail consequences for other areas of con-troversy in biology and its philosophy. Moreover, difficulties, solutions, and de-siderata in those other areas may entail consequences for the species problem. Thus there is more to be taken into account when trying to solve the species problem than only what may be immediately obvious. For example, if it is agreed that biological entities have genuinely emergent properties (properties not merely additive), so that biological entities are not ontologically reducible to physical and chemical entities, and so that biology is neither theoretically nor predictively reducible to physics and chemistry but is an autonomous science, then that agreement exerts a sort of selection pressure on competing species concepts, a selection pressure to narrow the field in conformity with that ontological com-mitment. Another example is that whichever species concept seems independ-ently the most reasonable will invariably exert a selection pressure on competing taxonomies, which may in turn exert a reverse selection pressure on competing species concepts if one of the taxonomies is for reasons other than the species problem the dominant taxonomy. Similarly, different species concepts may entail a commitment one way or the other on the issue of the reality of higher taxa (gen-era, families, etc.), while a prior commitment about higher taxa may entail the rejection of certain species concepts.

As the debates concerning these three topics (reductionism, taxonomy, and higher taxa, among others) are in themselves quite large and require an extended discussion for each, I will not explore their relation to the species problem in the present work, principally for lack of space. However, the interrelations between them and the species problem should never be kept far from mind. For the record, my own solution to the species problem is motivated in large part by an antire-ductionism based on emergentism—specifically the emergence of relations, in-cluding especially the relation known as supervenience, which allows for a holis-tic organicism, and which in turn allows for real similarity between organism phenotypes. In chapter 5, I develop a unique view for how this might work for species. How it might work for higher taxa, however, I'm not entirely sure. Since I take similarity seriously, it may be thought that I take higher taxa to be real as well, but once again, I'm not sure how this would work (I suppose higher-order similarity would have to be involved, but this is an especially tricky concept), and at any rate it is not an issue to which I am committed in the present book. As for which of the three (or now four) main schools of taxonomy my view on species might support, I don't think it supports any of them, although one might naturally think that it would lend its support to a taxonomy that takes similarity seriously. Numerical or phenetic taxonomy involves overall similarity, but it takes similarity as reducible to properties, and hence as not real. Traditional evolutionary taxonomy involves both similarity and descent, in some ways giving priority to descent, but it has never provided clear criteria for how to fully incorporate these two. Cladistic or phylogenetic taxonomy involves only descent, namely branching points, and eschews similarity altogether, but we have seen and shall see in much more detail that the species concept I develop is anything but cladistic. Pattern cladism, an apparent hybrid between cladism and pheneticism, uses the methods of cladism to produce nested similarity classes, but in so doing it explicitly eschews history reconstruction. Since, as we shall see in chapter 5.2, the species concept that I develop takes history reconstruction quite seriously, it cannot fit in with pattern cladism either. It may be that one or more of these taxonomic schools can be altered at the level of the species category in order to incorporate the biosimilarity species concept, but I doubt it. These schools either take the metaphor of the Tree of Life seriously or they eschew it altogether, providing nothing in its place. Perhaps the ultimate lesson of this book is that real history reconstruction, .combined with a truly viable species concept, requires a completely new metaphor for best capturing the nature of the history of life (cf. chapter 5.3). As is so often the case, perhaps only time will tell.

Chapter 2

Species Nominalism

[T]his is a Man, that a Drill: And in this, I think, consists the whole business of Genus *and* Species.

—John Locke (1700: III.vi.36)

2.1 Preliminary Considerations

Species nominalism is the view that biological species are not real, that they have no objective reality outside of the mind, that they are ultimately arbitrary, man-made groupings of individual organisms conventionally bracketed together by general names for the purpose of linguistic convenience.

Although there are relatively few biologists and philosophers living today who subscribe to this view, it seems to me that species nominalism is a view that should be taken seriously and that deserves careful consideration. My reasons for this are basically fourfold:

First, as mentioned in the Introduction, even though I have been repeatedly told they exist, I have never actually *seen* a biological species, any species. Instead, I have only seen examples of a this or a that.

Second, species realists themselves cannot agree on the nature of species. Instead, they have provided a number of different and often incompatible theories on what they are.

Third, species nominalism, more than any other solution to the species problem, is the most parsimonious position. And parsimony, of course, is almost unanimously agreed upon as an important desideratum in theoretical work.

Finally, there are and have been species nominalists and, aside from my first three reasons above, they have provided evidence and argument for their position that has, at least, an initial strength and plausibility.

The first three of these reasons together seem sufficient to make species nominalism the default position. Interestingly, the concept of a default position in theory competition is akin to the presumption of innocence in Western law. In an entirely different context (*viz.*, atheism versus theism), Antony Flew (1984a) has pointed out that the presumption of innocence is not an *assumption*, but is instead a theory-free *presumption*: "The presumption of innocence indicates where the court should start and how it must proceed" (17). The importance of this distinction is that assumptions may in fact be defeated, but either way, whether the prosecution or the defense wins, the underlying presumption of innocence remains. Granting this, it would seem clear that the onus of providing evidence and sufficient proof rests upon those who would make a positive affirmation, not a denial. Of course, with a little word play either position can be expressed as a positive affirmation or as a denial. Flew surmounts this difficulty by putting the emphasis upon *knowledge*, in the sense of justified true belief. The prosecution claims positive knowledge of which the defense denies, not *vice versa*.

In the context of the species problem, the species realist claims that species words have objective application, of which the species nominalist denies. Insofar, then, as biology seeks knowledge, the underlying presumption must be that the onus of evidence and argument rests squarely on the shoulders of the species realist and that the nominalist's position constitutes the default position.

Nevertheless, rarely, if ever, have species nominalists relied merely on the position just outlined. Instead they have not only attacked the arguments of the realists but have provided positive arguments in support of their own position. To continue the analogy above, they have endeavored to establish an alibi. It is to an examination of these arguments that the remainder of the present chapter is devoted. The cases of the various species realists follow in subsequent chapters.

Briefly, the arguments of species nominalists can be broken down into three main and not entirely compatible divisions. First, in league with all universals nominalists, many species nominalists subscribe to an ontology of particulars, an ontology with strong intuitive appeal. (I myself used to subscribe to it, until I began work on the species problem.) Generally, it is the view that only particular objects and their properties exist (where properties are taken as particulars, not as universals). It involves the denial that there are objective abstractions existing in the world, that there are extra-mental abstract facts. For some, such as Occam and Locke, abstractions are a product of and exist only in the mind. For others, such as Berkeley and Hume, abstractions do not even exist in the mind.

Along with a rejection of extra-mental abstractions, many species nominalists (along with most universals nominalists) reject the reality of relations (both as particulars and as universals). For them only objects and their properties exist. Relations between objects and between their properties are not extra realities, extra facts. Needless to say, it is with relations that most nominalists run into their most serious difficulties.

The second of the three main divisions in the arguments of species nominalists has to do with the concept of a continuum in the natural world, specifically the living world. This can be further broken down into two concepts, the idea of an *ahistorical* continuum and the idea of a *historical* continuum.

The ahistorical continuum is what is known as *The Great Chain of Being*, the history of this idea receiving its supreme exposition in Arthur Lovejoy's classic work of the same name (Lovejoy 1936). It is the idea that the universe is

> composed of an immense, or—by the strict but seldom rigorously applied logic of the principle of continuity—of an infinite, number of links ranging in hierarchical order from the meagerest kind of existents, which barely escape nonexistence, through "every possible" grade up to the *ens perfectissimum*—or, in a somewhat more orthodox version, to the highest possible kind of creature, between which and the Absolute Being the disparity was assumed to be infinite—every one of them differing from that immediately above and that immediately below it by the "least possible" degree of difference. [59]

The analogy of links in a chain is quite appropriate, for it illustrates that on this view there is between every form of life an overlap between it and two other life forms, the one immediately below it and the one immediately above it (with the exception, of course, of the life form at the absolute bottom and at the absolute top), what Lovejoy has appropriately termed "twilight zones" (56). It can also easily lead to the idea of an infinite number of such links or forms, since there are no clear divisions between links.

The concept of the Great Chain of Being, or *scala naturae*, received its foundation, ironically, in both Plato and Aristotle. In Plato we find what Lovejoy (52) calls the principle of plenitude, the idea that in the physical world (or universe) all possible forms of life are and must be actualized.[1] In Aristotle, on the other hand, we find what Lovejoy (56) calls the principle of continuity and its introduction by Aristotle into natural history, the idea that there is a smeary continuum between kinds of living things.[2] Together these two ideas would eventually be-

---

[1] In his creation myth in the *Timaeus*, Plato says that all possible kinds of living things were made physically actual in order to make the universe complete, upon which its perfection depends (41b–c). What is overlooked in Lovejoy's account is that according to this myth man was made first, woman is a degeneration from man, and all other living kinds are successive stages of degeneration from woman (cf. 42b–c, 76e–77a, and 90e–92b).

[2] Although contraindicated in Aristotle's logical and metaphysical works, where the stated ideal is that of discrete essences, it is a curious feature of Aristotle's biological works that he seemed forced to conclude that the hierarchy of forms of living beings involves a smeary continuum, at least through some of its parts. The principal texts are: (1) "Nature proceeds little by little from things lifeless to animal life in such a way that it is impossible to determine the exact line of demarcation, nor on which side thereof an intermediate form should lie. Thus, next after lifeless things comes the plant, . . . there is observed in plants a continuous scale of ascent towards the animal. So, in the sea, there are certain objects concerning which one would be at a loss to determine whether they be animal or vegetable. . . . And so throughout the entire animal scale there is a graduated differentiation in amount of vitality and in capacity for motion" (*Hist. of An.* 588^{b}4–23). (2) "For nature passes from lifeless objects to animals in such unbroken sequence, interposing between them beings which live and yet are not animals, that scarcely any difference

come united into one idea and belief such that, as Lovejoy points out (though his point here seems to me exaggerated), "throughout the Middle Ages and down to the late eighteenth century, many philosophers, most men of science, and, indeed, most educated men, were to accept [it] without question" (59). Interestingly, since the Great Chain, of course, is not exactly a matter of observation, there eventually arose numerous "missing-link hunters" (235), whose alleged wares were exploited to the fullest on a gullible public by "that eminent practical psychologist, P.T. Barnum" (236).

It is indeed easy to see why species nominalists would be quick to make use of this popular idea. Infinite overlapping forms, like Archimedes' proof that a sphere can be constructed by continuously adding rectilinear sides to a square, leaves no distinct forms at all. Of course, how the idea of forms of life, even an infinite number of them, is compatible with an ontology of particulars (in this case concrete particulars) is by no means clear.

The historical version of the idea of a living continuum is founded, of course, upon evolutionary theory, and although it is part of the system of Lamarck it stems principally from the work of Charles Darwin. In his *On the Origin of Species* (1859) Darwin amassed not only overwhelming evidence in favor of biological evolution but he also went to great pains to argue that the main mechanism of evolution is natural selection. Acceptance of these two conclusions entailed for Darwin that evolution must be minutely gradational, that it must make no sudden leaps or saltations (contra T.H. Huxley and others), so that all throughout biological history, in spite of the fossil record, there must have been "an infinite number of those fine transitional forms, which on my theory assuredly have connected all the past and present species of the same group into one long and branching chain of life" (301).

Indeed Darwin, based mainly on different, more explicit passages, has often been interpreted by modern writers (wrongly, I shall argue) as a species nominalist. This is an issue of some importance and I will examine it closely in this chapter, since many of the problems that face modern species realists stem from Darwin.

The third and final of the three main divisions in the arguments of species nominalists has to do with a rejection of essentialism, either globally or restricted to biology. It is here that species nominalists make their weakest case. Throughout history many seem to have thought that a rejection of essentialism entails that species cannot be real. For example, one can find this view in an almost explicit form in Locke, whereas it is an underlying assumption in Darwin if one reads him literally. Even today the view is not without explicit adherents. I suggest it is a vestige of the dominance of the creationist point of view as revealed in Genesis and wedded with either Plato or Aristotle. At any rate, the fundamental error it involves is a surprisingly simple one. At bottom, it shows a lack of imagination,

---

seems to exist between two neighbouring groups owing to their close proximity" (*Parts of An.* 681^{a}11–15). (All Aristotle quotations are from Barnes 1984.) Cf. *Gen. of An.* 748^{a}13, and discussion in Lovejoy (1936: 55–59).

for it can be easily shown that the class category in the sense of strict essentialism is not the only possible category into which the reality of species may fall, as subsequent chapters in the present work will show.

It is from these three main divisions, it seems to me, that the arguments of species nominalists flow, even the modern ones in biology, whether partial or complete species nominalisms. Most, if not all of their arguments, it will be seen, are nothing really new under the sun, but stem in one way or another principally from the published views of Locke and Darwin.

2.2 Occam and Locke

If one studies the history of nominalism, two names come to the fore, namely William of Occam (*c.* 1285–1349) and John Locke (1632–1704). I shall examine their views not as a matter of exegesis but for what they have to contribute to the modern debate on species.

A study of Occam's writings is important, it seems to me, on principally three counts: (i) for the methodological principle that bears his name, known as Occam's Razor; (ii) for his ontology of particulars and his conceptualism concerning abstractions; and (iii) for his rejection of relations.

Beginning with Occam's Razor, it is *the* principle of parsimony in theory competition, usually stated as "Entities are not to be multiplied beyond necessity," although Occam never stated it quite in this way. Instead, he formulated it as "plurality should never be posited without necessity" (685),[3] which amounts to much the same.

The justification for this principle, however, is obscure. Occam seems to take it as an intuitive truth that "it is pointless to do with more what can be done with less" (687). But this is not intuitively true. It is by no means clear why luxury should not be valued more than parsimony. Aside from the fact that parsimony is fully consistent with monkish values (Occam was a Franciscan, which of all orders involved the strictest poverty), I conjecture that Occam's Razor can be traced to two sources, neither of which is sufficient for modern acceptance. First, it is a recurrent theme in Aristotle's writings that "Nature does nothing in vain" (*De An.* 434ª30), a mainly biological principle that can easily be applied to metaphysics. Indeed (second) it would be a small step for a Medieval Christian philosopher (especially one living in the wake of Aquinas and attempting an unpolluted interpretation of Aristotle) to apply this concept to God and metaphysics, so that God does nothing in vain, since doing more than is sufficient would seem to diminish His perfection. (Indeed this would later be precisely Bishop Berkeley's position in his argument against the existence of matter.)

At any rate, the only possible solid justification for Occam's Razor that I can see is that put forward by Bertrand Russell (e.g., 1918), *viz.*, that adherence to it "diminishes your risk of error" (280). Interestingly, many have thought Russell philosophically unsettled (and unsettling) for having changed his views so often

---

[3]   All references to Occam are to the translation and extensive anthology in Hyman and Walsh (1973).

throughout his career, but there is truth to his claim that it was a consequence of
an ever-increasing application of Occam's Razor, resulting in "a more clean-
shaven picture of reality" (Russell 1959: 49)—a desideratum that will guide the
present work. It should also be a desideratum for the heuristic of reductionism,
since the establishment of an absolutely minimum vocabulary in biology is a necessary first step if it is to be reduced to physics and chemistry.

Since species nominalism assumes less entities than any of the species realist
views, it has the initial advantage of running a lower risk of error. Each of the
species realist positions grant the reality of all that species nominalism involves,
namely the existence of individual organisms and a physical environment. But
each of the species realists assume something more. The ultimate question is
whether there is a *necessity* for anything more, whether the science of biology requires an expanded ontology to make sense of its empirical discoveries and theories as well as its day-to-day practice.

I believe we can say with certainty that modern biology does not require Platonic essences in a realm separate from and independent of the spatiotemporal
world. Indeed that is the type of universals realism that Occam was out to refute.
For Occam such entities are superfluous for explaining experience. According to
Occam cognition is either *intuitive* (that which is given in sense experience) or
abstract. Intuitive cognition is either of simple particulars or composites of particulars (670). Either way, "no one sees a species intuitively" (674), a point on
which Aristotle was in agreement (e.g., *De An.* 417^{b}22–23) and with which I will
side when I discuss problems with the species-as-individuals view (chapter 4.4).
Moreover, according to Occam "intuitive cognition can be accomplished through
the intellect and the thing seen, without any species" (673). Interestingly, Occam's argument for this is a causal one, and of especial interest, since it antici-
pates Popper's causal criterion of reality, as discussed in the previous chapter. For
intuitive cognition "no argument can prove that a species is required unless because it has efficient causality" (674). Plato's species, it should be noted, served
not an efficient but only a teleological causal role. His species are therefore superfluous in explaining intuitive cognition, since they play no efficient causal
role: "without any species, on the presence of the object with the intellect, there
follows the act of knowing, just as well as with the species" (674).

Plato's species are particulars, unchanging abstract essences existing outside
of space and time. But now what of Aristotelian species, wherein each species
essence is not separate from but is instantiated in each of its individual members?
Occam admits that we have abstract cognition. But must we posit extra-mental
abstract entities? Occam thinks not. According to Occam the more economical
yet sufficiently explanatory thesis is that abstractions exist only in the mind, as a
consequence of what he calls habit. First, as already noted, intuitive cognition
only provides cognition of particulars and composites of particulars, not abstractions. Second, abstract cognition requires intuitive cognition: "the intellect having
intuitive knowledge can perform abstractive cognition, and not having it, cannot"
(674). Clearly, then, the mind on Occam's view has a power enabling it to make
abstractions. Extra-mental abstractions, then, may exist, but they are superfluous

for explaining abstract cognition and are therefore to be excluded from our ontology. As he says, "Everything which can be preserved through a species can be preserved through a habit; therefore, a habit is required and a species is superfluous" (675).

There are, of course, a number of weaknesses in Occam's account. Clearly, if it could be proved that the mind is not capable of producing abstract cognitions, or if the concept of abstract cognition is itself incoherent, then Occam's argument fails. Then again, perhaps instead of producing abstractions we infer them, and as a matter of explanatory necessity. For example, surely it is an extra-mental abstract fact that a particular group of giraffes has an average length of neck, a length that may not be instantiated by any one member of the group, a length that is in principle discoverable, a length, moreover, that may partly determine the overall fitness of the group and therefore may be said to have a causal efficacy. There is also a difficulty in conceiving of causality only in terms of efficient causes. Try as we might, it seems we cannot escape from teleology in biology, as we shall examine in the next chapter concerning ecological niches. Moreover, it would seem a reasonable conclusion that if abstract cognitions are to have a truth value rather than a fictive arbitrariness, there must be something out there about particulars, something extra rather than just their particularity, something objective that allows for meaningful abstract cognition. Indeed individual particulars, particulars given in intuitive cognition, clearly do not seem to be on a par, but often seem to group themselves and to be groupable into classes, what modern metaphysicians such as Quine have argued are objective abstract entities in their own right. Moreover, individual particulars clearly possess properties and structures, which are either identical or similar to properties and structures possessed by other individual particulars. Properties and structures may therefore be said to express an individuated abstract identity through concrete particulars, an expression that may be perfect or imperfect (the latter being, for many theorists, similarity), all of which Quine would have us subsume under classes. As all of these are topics that will be taken up in chapters 3.5 and 5.3, I will reserve discussion of them for then. For the present it need only be kept in mind that on the criteria of reality discussed in the previous chapter an ontology of particulars and a conceptualism concerning abstractions face serious difficulties.

What is left is whether modern biology requires an ontology that includes relations.

An ontology of particulars does not normally include relations, for the simple reason that relations are neither substances nor attributes. (A relation is not a substance, on the traditional definition of substance, because as normally conceived it cannot exist by itself but must depend on relata. Nor can it be an attribute, since it requires at least two relata, whereas an attribute can exist with only one subject.)

Occam was acutely aware of the problems relations posed for an ontology of particulars and of the apparent difficulties encountered in any denial of relations: First, that it "destroys the unity of the universe, . . . since the unity of the universe is in the order of the parts to one another" (680). Second, that it denies part-whole relations, since "no whole will be really natural, since it necessarily requires a relation for its existence" (680). And third, "that it destroys all causality of second causes [causes other than God] . . . [since] second causes cannot cause unless they are proportionate and in proximity" (680). Against these weighty considerations Occam nevertheless thought that his arguments against the reality of relations outweighed those of his opponents, so that "a relation is either a name or a word or a concept or an intention, . . . and these certainly are not the same as external things" (682–683). Regarding concepts and intentions, Occam made it specifically clear that he did not think that relations, like abstractions such as genus and species, were produced by the mind and were "in the intellect" (684). Rather, he seems to think of relation words as linguistic conveniences, as a shorthand for referring to substances and common properties:

> And so, nothing exists in reality outside of absolutes [quantities, qualities, substances]. Since there are many absolutes in reality, the intellect can express them in divers ways: in one way expressing only that Socrates is white, and then it has only absolute concepts; in another way, that Plato is white; and in a third way, that Socrates as well as Plato is white. And this can be accomplished through a relational concept or intention, in saying that Socrates is similar to Plato with regard to whiteness. For it is altogether the same which is conveyed through these propositions: "Socrates as well as Plato is white," and "Socrates is similar to Plato with regard to whiteness." And hence it should be conceded without qualification that the intellect contributes nothing to the fact that the universe is one or that a whole is composite or that causes cause when in proximity or that a triangle has three, etc. And so it contributes no more concerning the others than to the fact that Socrates is white or that fire is hot or water, cold. [684–685]

Our error, according to Occam, is that we have been misled by words, imagining that "just as there are distinct names, so there are distinct things corresponding to them" (687).

Although I will have much more to say about the ontology of relations in chapter 5, I will briefly comment here on what seems to me a number of Occam's errors.

First, Occam clearly seems to think of properties as an all-or-nothing affair. Either Socrates is white of he is not, etc. But once it is admitted that properties are not a matter of all or nothing but of degree, then his arguments, at least against similarity, seem to fail. To accept Occam's view is to commit oneself to the view that, for example, one shade of blue is inherently no more related to a different shade of blue than any completely different color, which is absurd. The two shades of blue share a degree of similarity with each other that they do not share with other colors. In consequence of this, and since at least most properties admit of degrees, two individuals that share at least one property in common but to a different degree will have between them at least one similarity relation. Confining ourselves to what at least most of us would agree are the paradigm real things, the things we can point to, it would seem impossible to find any two individuals that do not have between them an overall similarity complex, composed of all the individual similarity relations between them.

Occam also seems to think that if relations are to be real, they must be absolutes of a sort. The question, as he understands it, is whether "a relation is a thing other than absolutes" (680). By absolutes he means "quantity, quality, or substance" (686).[4] Now certainly it seems absurd to say that a relation can exist without relata. But from this it by no means follows that relations must be unreal. Their ontology might be such that they are a mode of dependence, requiring relata for their existence. In arguing against the independent existence of relations, Occam probably had Plato in mind, for whom some of the Forms seem to be relations (e.g., absolute tallness, absolute equality, and other "relative magnitudes"; Phaedo 65c, 74a, and 75c, respectively).[5] But Plato's ontology, of course, is not the only possible one.

What still remains is the matter of relations existing in the spatiotemporal real world, either as particulars or as universals (repeatables). Can these be reduced to absolutes? Occam's reply seems to be that absolutes can be understood independent of one another so that relations are superfluous: "Every thing distinct in reality from another thing can be understood without that other thing being understood, and most of all, if neither is a part of the other" (680). Thus, talk of relations does not increase our understanding of absolutes at all, even when absolutes are in what we call spatiotemporal or causal or part-whole or similarity relations with one another. For example, "there is no inconvenience, nor does it include a contradiction, for that which is an effect to be understood without its cause having been understood" (681). Again: "whiteness is a thing for itself, however much it necessarily co-requires a subject in order to exist" (682).

Occam's claim, in other words, seems to be that relation statements do not state facts in addition to what can be stated in subject-predicate form; therefore, they can either be reduced or eliminated in favor of that form. That he is here continuing in a tradition begun by Plato and Aristotle and furthered later by Locke, Leibniz, Hegel, Bradley, and many others, seems beyond doubt. As will be discussed in chapter 5, no one did more than Bertrand Russell to expose the absurdities involved in this view. The new logic was developed in large part to make it compatible with modern science, as opposed to the either inconsistent or wild metaphysical systems resulting from the old logic. In the new logic relation

---

[4]   In reducing and eliminating relations in favor of absolutes, Occam seems to have thought that he was following Aristotle: "And hence according to the opinion of the Philosopher, there is nothing outside of those absolute parts, since the opinion of the Philosopher was that every imaginable thing is absolute" (685). Interestingly, in the *Metaphysics* Aristotle tells us that "the relative is least of all things a real thing or substance, . . . A sign that the relative is least of all a substance and a real thing is the fact that it alone has no proper generation or destruction or movement" (1088^a23–30).

[5]   Aristotle confirms this interpretation (cf. *Met.* 990^b15 and 1089^b6) and rightly rejects the view that relations could be expressed using monadic predicates (cf. *Cat.* 5^b16), not only because relations require at least two relata, but possibly for the reason that what he calls primary substances (concrete individuals capable of having proper names) cannot be part of a predication (cf. *Cat.* 3^a36–37), as in, e.g., "Plato is taller-than-Socrates."

statements are found necessary for stating facts that cannot be stated in subject-predicate form, resulting in an ontology that requires relations. That science is possible without relations proves absurd, for there are many objects and properties, especially in biology, that cannot be properly understood unless in relation to other objects and properties. Indeed part of the legacy of the new logic, unknown to Russell and rarely recognized by modern authors, is the discovery of new relations of great philosophical consequence for science. I speak here mainly of the relation recently christened as *supervenience*, the importance of which will be discussed in various places later on. Interestingly, it is a relation anticipated in different forms in both Aristotle and Locke, as we shall soon see.

It seems clear, then, as will be argued in detail in chapter 5, that in referring to real relations, particular relations, one is not only referring to or only saying something about the relata, but something more. Occam denies that something more, but in doing so he is also denying much of modern science and daily life.

Turning now to Locke and his *An Essay Concerning Human Understanding*, first published in 1689, we find that Locke also subscribed to an ontology of particulars: "all things that exist are only particulars" (III.iii.6). And like Occam, Locke's justification for this view is rather obscure. One possible reason for his ontology of particulars might stem from the fact that he strongly subscribed to Newton's philosophy, for whom it is a prime rule of reasoning that "Nature does nothing in vain, and more is in vain when less will serve; for Nature is pleased with simplicity and affects not the pomp of superfluous causes" (Thayer 1953: 3).

Another reason may be found in Locke's theory of ideas. According to Locke we do not directly perceive external objects but only ideas in our minds (cf. IV.i.1), which themselves are ultimately caused by external objects (II.xxx.2–3), such that ideas of primary qualities really resemble qualities of those objects (II.viii.15) but ideas of secondary qualities do not (II.viii.16). At any rate, the important point is that for Locke all our ideas, whether passively received or actively created by the mind, are each of them particular. In a concise passage he says

> Every Man's Reasoning and Knowledge, is only about the *Ideas* existing in his own Mind, which are truly, every one of them, particular Existences: and our Knowledge and Reasoning about other Things, is only as they correspond with those our particular *Ideas*. So that the Perception of the Agreement, or Disagreement of our particular *Ideas*, is the whole and utmost of all our Knowledge. Universality is but accidental to it, and consists only in this, That the particular *Ideas*, about which it is, are such, as more than one particular Thing can correspond with, and be represented by. [IV.xvii.8]

This latter comment about universality brings us next to Locke's theory about abstractions. But first it should be commented that, although still popular today (and with good reason, I should think), a causal theory of perception is not necessarily inconsistent with a theory of direct perception of external objects. One need only argue that direct perception is nothing else but a causal chain of the right kind. Moreover, there is what seems clearly a destructive difficulty with a view of perception that involves a veil of ideas. For Locke, words are "the Signs of our *Ideas* only, and not for Things themselves" (III.x.15)—a logical consequence of his view. The problem is that ideas are private, while external objects are public, so that if words signify ideas only, we seem locked in a private world, and all communication between minds is undermined. Since for Locke and the rest of us communication and society are fundamental desiderata, it follows that if we are to have these desiderata, words must signify more than just ideas. We must, therefore, even on a causal theory of perception, in some sense be able to be directly acquainted with external objects.

Consonant with his ontology of particulars, Locke, like Occam, was a conceptualist. For a conceptualist, abstractions are real, but their reality is not something extra-mental; instead, they are produced by and exist only in the mind. The process involved in the production of an abstract idea is on the surface rather simple. The peculiarity of various perceivables is ignored and what is common is retained and combined into one idea. In a classic passage on the process of abstraction, Locke says

> For let any one reflect, and then tell me, wherein does his *Idea* of *Man* differ from that of *Peter*, and *Paul*; or his *Idea* of *Horse*, from that of *Bucephalus*, but in the leaving out something, that is peculiar to each Individual; and retaining so much of those particular complex *Ideas*, of several particular Existences, as they are found to agree in? . . . by the same way the Mind proceeds to *Body, Substance*, and at last to *Being, Thing*, and such universal terms, which stand for any of our Ideas whatsoever. To conclude, this whole *mystery* of *Genera* and *Species*, which make such a noise in the Schools, and are, with Justice, so little regarded out of them, is nothing else but abstract *Ideas*, more or less comprehensive, with names annexed to them. [III.iii.9]

Although according to Locke "a definition is best made by enumerating those simple Ideas that are combined in the signification of the term Defined" (III.iii. 10), general or universal words (including species words) do not therefore signify those particular existences which were abstracted from in forming abstract ideas. In other words "they do not signify a plurality" (III.iii.12). Instead, that "which general Words signify, is a sort of Things; and each of them does that, by being a sign of an abstract *Idea* in the mind" (III.iii.12). "Whereby it is evident," he continues, "that the *Essences of* the *sorts*, or (if the Latin word pleases better) *Species* of Things, are nothing else but these abstract *Ideas*" (III.iii.12).

So biological species are nothing else but abstract ideas existing in the mind. No minds, no species. It might then be thought that for Locke species are entirely arbitrary. But this is only partly true. Certainly, since no two individuals have the exact same experiences, their abstract ideas are bound to differ. And yet Locke does not want to say that they are completely arbitrary and subjective. In another classic passage he says

> that Nature in the Production of Things, makes several of them alike: there is nothing more obvious, especially in the Races of Animals, and all Things propa-

gated by Seed. But yet, I think, we may say, the *sorting* of them under Names, *is the Workmanship of the Understanding, taking occasion from the similitude* it observes amongst them, to make abstract general *Ideas*, and set them up in the mind, with Names annexed to them, as Patterns, or Forms, (for in that sence the word Form has a very proper signification,) to which, as particular Things existing are found to agree, so they come to be of that Species, have that Denomination, or are put into that *Classis*. [III.iii.13]

In speaking above of *similitude*, and especially, as we have seen earlier, of the resemblance between ideas of primary qualities and the qualities of the external objects that caused them, one might naturally think that for Locke at least some relations are extra-mental. But this does not appear to be the case. Not only similarity but all relations whatsoever have "no other *reality*, but what they have in the Minds of Men" (II.xxx.4). They are nothing "but my way of considering, or comparing two Things together, and so also an *Idea* of my own making" (III.x. 33; cf. II.xxviii.19). Not being independent particulars, they are "not contained in the real existence of Things, but something extraneous, and superinduced" (II. xxv.8).

Again, as with my comments on Occam's rejection of relations, I will defer to chapter 5 for an extended discussion on this matter, allowing my previous comments on Occam to apply here as well. Suffice it for the present that Locke's error here (aside from that of theoretical incoherence) seems a common one. It is the fallacy often found in both ontology and epistemology that dependent entities are less real or even completely reducible to and can be eliminated in favor of the independent entities on which they depend, a fallacy I hope to fully explode later on when, in particular, I discuss supervenience.

In the case that one should happen to accept Locke's nominalism except for the part about relations, Locke provides two related arguments, both of them ostensibly empirical, for the unreality of biological species in particular. The first appeals to the Great Chain of Being: "That in all the visible corporeal World, we see no Chasms, or Gaps. All quite down from us, the descent is by easy steps, and a continued series of Things, that in each remove, differ very little one from the other" (III.vi.12). Connected with this, Locke discusses the criterion of reproductive isolation, specifically the lack of it (III.vi.23). On this criterion, according to Locke, we could only distinguish between "the Tribes of Animals and Vegetables." History, says Locke, testifies that "Women have conceived by Drills [baboons]," and he personally has seen "the Issue of a Cat and a Rat." Combined with more usual hybrids such as mules as well as monstrosities, Locke's conclusion would seem to be that it is the lack of reproductive isolation that is responsible for the phenomenon of the Great Chain.

Locke's argument, of course, rests upon a threefold confusion over hybrids, namely between infertile and fertile hybrids and speciation by hybridization. Although speciation by hybridization is common in plants, it is quite rare in animals (cf. Maynard Smith 1975: 267), and at any rate is irrelevant to Locke's claim unless the new species can backcross with either parental species, which in allopolyploidy is not the case (cf. chapter 4.1/4). Infertile hybrids, although nowhere near the degree that Locke suggests, are fairly common with certain groups in nature, but do nothing to reduce reproductive isolation. Fertile hybrids, of course, are even less common, and the existence of hybrid zones is nowhere near that required by Locke (cf. chapters 2.4 and 3.3). The effect of all of this is that it radically reduces the number of twilight zones mentioned by Lovejoy, in effect destroying the Great Chain.

Today, of course, reproductive isolation, and with it the existence of closed gene pools, is a major criterion by which biologists (more so zoologists, less so botanists) categorize species, although, as we shall later see, there are many biologists today who would side not far from Locke, in that they also do not think (though for different reasons) that reproductive isolation is a good or workable criterion for species delimitation. At any rate, as for the ahistorical Great Chain of Being, it is thoroughly discredited, for not only do we know today that there are discrete chemical and mineral kinds, but we also know that there are many discrete living kinds as well. Indeed biologists have discovered well over a million species alive today, while estimating that this is less than one-tenth of the total (cf. Wilson 1992: 132–133).

One interesting consequence of Locke's rejection of the extra-mental reality of relations is that, in spite of his rejection of Aristotle in so many matters, he is forced to accept the subject-predicate form of proposition as fundamental. Although Locke nowhere explicitly endorses this form of proposition, it is highly implicit throughout his writings (cf. III.viii.2) and necessarily follows from his view that all relations "*terminate in*, and are concerned about those *simple Ideas*, either of Sensation or Reflection; which I think to be the whole Materials of our Knowledge" (II.xxv.9). The particular difficulty Locke faces is in combining this view with, as we shall quickly see, his great reservations about substance or substratum, what he calls *something we know not what*. As Ayers (1991) put it, "Locke's position was a kind of mixture of the traditional view [that knowledge is ultimately about particular concreta] together with his radical explanation of substance-attribute logic as a mark of our ignorance" (II, 63). In rejecting objective relations as well as maintaining ignorance about subjects, one would seem left with a hopeless form of skepticism.

I want finally in this section to discuss a distinction fundamental in Locke which I think is of prime contemporary importance, although it has yet to be recognized as such. Discussions on Locke normally focus on his distinction between real and nominal essences, all too often as if he had only a twofold distinction (cf. Ayers 1981: 256; Ruse 1987: 346; Ghiselin 1987a: 134; Dupré 1993: 21–22). But Locke really had a fourfold distinction, and it is in this fourfold distinction that Locke may be found as a precursor of a most important idea.

First, according to Locke our species designations are determined by us from nominal, not real essences. In a classic passage he says

> This, then, in short, is the case: *Nature makes many particular Things, which do agree* one with another, in many sensible Qualities, and probably too, in their

> internal frame and Constitution: but 'tis not this real Essence that distinguishes
> them into *Species*; 'tis *Men*, who, taking occasion from the Qualities they find
> united in them, and wherein, they observe often several individuals to agree,
> *range them into Sorts, in order to their naming*, for the convenience of compre-
> hensive signs; under which individuals, according to their conformity to this or
> that abstract *Idea*, come to be ranked as under Ensigns: so that this is of the Blue,
> that the Red Regiment; this is a Man, that a Drill: And in this, I think, consists
> the whole business of *Genus* and *Species*. [III.vi.36]

In short, according to Locke "*each abstract* Idea, *with a name to it, makes a distinct Species*" (III.vi.37). But although on Locke's view nominal essences are made by humans and constitute species, he distinguished two kinds of nominal essence. The one is entirely arbitrary and manmade, like unicorns and centaurs. No one ever sees objects that conform to the combinations of qualities designated by these ideas. The other kind of nominal essence, however, still produced by the mind, does follow from nature. In this case the mind, he says, "in making its complex *Ideas* of Substances, only follows Nature; and puts none together, which are not supposed to have an union in Nature" (III.vi.28). A simple example would be each person's nominal essence of horse. Produced by different minds, how-ever, it follows that no two nominal essences of horse need be exactly the same. And this indeed is Locke's conclusion. It is evident, he says, that nominal es-sences "*are made by the Mind*, and not by Nature: For were they Nature's Work-manship, they could not be so various and different in several Men, as experience tells us they are. For if we will examine it, we shall not find the nominal Essence of any one *Species* of Substances, in all Men the same" (III.vi.26).

The interesting thing about this view is that membership in a kind is thus mind-dependent and subjective. To be a member of a kind it is only necessary to conform to an abstract idea. Descent does not count. Indeed Locke finds empiri-cal confirmation for this in the case of monstrosities born of human parents, since it was often in his time debated whether the creature is human and thus should be nourished and baptized (III.iii.14, III.vi.26). What is interesting is that although today we would all agree that descent is irrelevant for membership in chemical species, it is still a matter of debate in modern ontology and taxonomy whether it is required for biological species, as we shall see in subsequent chapters.

As with his division of nominal essences into two kinds, Locke similarly dis-tinguishes two kinds of real essences, only one of which he accepts. Accepting the corpuscularian hypothesis of Newton and Boyle as the most probable hypoth-esis (IV.ii.11, IV.iii.16), Locke thought that, like "the famous Clock at *Stras-bourg*" (III.vi.3), everything with outward manifest properties must have an inter-nal constitution or microstructure from which those qualities flow, in his own words "that real constitution of any Thing, which is the foundation of all those Properties, that are combined in, and are constantly found to co-exist with the *nominal Essence*; that particular constitution which every Thing has within it self, without any relation to any thing without it" (III.vi.6). And this is the only sense of real essence that Locke accepts, particularized essences, and then only as a probable conclusion. These essences, moreover, seem to Locke forever beyond the reach of finite minds, and principally for the reason that we were not designed to perceive them, but instead were fitted only "for the Conveniences of living" (II.xxiii.12). They are thus that *something we know not what* (II.xxiii.3), that something which it is reasonable to suppose, but not that something from which we determine kinds of things.

The other kind of real essence, which Locke rejects, is what he refers to as Scholastic or Aristotelian essences (III.iii.15), essences shared by different particulars which we can infer and which determine kinds. Though Locke must necessarily remain agnostic about the existence of these essences, he rejects them on basically two grounds. First, they are occult and unknowable and are not in fact the essences upon which we determine kinds (III.iii.17). Second, they don't produce what they are supposed to produce, namely fixed and clear boundaries (III. vi.27).

Beginning with the second ground, it involves what is clearly today a false premise: If species are made by Nature, then there should be precise boundaries. The fact of evolution refutes this premise. Species may be made by Nature, but evolution by natural selection, because it gives us numerous intermediate forms, must always be expected at any one period to give us some messy situations, as will be discussed later. The point is that because there are not everywhere in the biological world clear and determinate boundaries between species, it does not necessarily follow that species are not made by Nature.

As for Locke's first ground, Locke has been clearly disproved by subsequent history. Within this century man has with ever-increasing ability plumbed the depths of microstructure. In chemistry, chemical kinds are objectively determined by the number of protons in atomic nuclei. And in biology, DNA sequencing has been made a reality.

Shall we, then, agree with Ayers (1981), who credits Locke's *Essay* as the main reason why philosophers since Locke fell away "from the truth" and reverted "to a state of naivety" (248)?

Though it seems we must completely agree when it comes to chemistry, I suggest we need only partially agree, and even less so when it comes to biology. First, although it is a matter of controversy in modern biology, specifically in taxonomy, it seems to me that one important contribution of Locke to the modern debate, to which I subscribe, is the importance and priority of phenotypes over genotypes in biological classification. (Of course, where we mainly differ is that Locke was a species nominalist, whereas I am a realist.) Second, and of more importance (though related), in affirming particularized essences but rejecting shared essences Locke opened the way for supervenience, at least when applied to macro/micro dependencies.[6] What is not impossible on Locke's view, and in-

---

[6] There are a small number of passages in the *Essay*—namely III.iii.13 and III.vi.36, but especially III.vi.6, where Locke seems to admit that similar outward manifestations imply shared essences—which have made some commentators think that Locke did admit the existence of objective natural kinds (cf. Ayers 1981: 257n11). It is not so clear to me,

deed what is implied by it, is that two things can have the same manifest property or properties and yet have different (dissimilar) particularized essences.[7] Although in modern chemistry two different kinds of stuff can share a common property or properties (say, colorless liquidity), so that the common property may be said to supervene on different microstructures, two lumps of stuff cannot share all manifest properties without also sharing a common essence. So Locke was partly right and partly wrong. In biology, however, two organisms can indeed share all manifest outward properties and yet have different (dissimilar) DNA. Thus partly in chemistry, though completely in biology, Locke's view makes him an important precursor—along with Aristotle, as we shall later see in chapter 3.1n8—of supervenience.

Since I will have more to say about supervenience in later sections, and because of its importance in modern philosophy of biology, I should say a little more on the basic meaning of supervenience. Basically, supervenience is an empirically discoverable *relation* of a certain kind, specifically a mode of dependence between individual properties and families of properties such that a property or family of properties depends on a disjunctive base of subvenient properties (cf. Kim 1993: 55). (Instead of properties we may also speak of events, states, structures, facts, kinds, predicates, and even relations.) For example, though with chemical kinds the mode of dependency is one-one (one kind, only one microstructure), in biology it is one-many (one kind, possibly many microstructures). In other words, biological macrostructures are not only dependent upon physical microstructures but may be realized by a number of different physical microstructures. More specifically, we know today that different triplets of DNA (nucleotide) bases, different triplets of the four-letter genetic code (each triplet called a "codon" as transcribed on messenger RNA), may code for the very same amino

---

however, that Locke involves himself in any such admission. My disagreement is basically threefold: (1) Locke is committed only to the existence of particulars, not to objects with identity through multiple extension. (2) As we have seen above, in reference to particularized essences as the foundation of outward properties, Locke explicitly says that each particular essence has no other "relation to any thing without it" (III.vi.6). (3) In an important passage (III.iii.17) Locke claims that there are only two opinions concerning real essences, and he explicitly labels the opinion which subscribes only to particularized essences as the "more rational Opinion," thus implying that its immediate competitor, the opinion which subscribes to shared essences, is less rational. Indeed Locke proceeds to argue that it is not only less rational but is inconsistent with biological facts!—namely infraspecific variation (though of course he does not use this term) as well as the extreme case of monstrosities. In sum he says, "it is impossible, that two Things, partaking exactly of the same real *Essence*, should have different Properties, as that two Figures partaking in the same real *Essence* of a Circle, should have different Properties."

[7]   To my knowledge, the closest Locke comes to actually saying something like this is when he says "any one who observes their different Qualities [the qualities of conspecific individuals] can hardly doubt, that many of the Individuals, called by the same name, are, in their internal Constitution, as different one from another, as several of those which are ranked under different specifick Names" (III.x.20).

acid—it is this synonymy that makes the relation one of supervenience—sequences of which (codons) not only code for proteins but ultimately for phenotypes. For example, the amino acid arginine is coded for either by CGU, CGC, CGA, CGG, AGA, or AGG. While only two other amino acids have a disjunctive base of six codons, others have a disjunctive base of four, three, or two codons, while only two amino acids are coded for by only one codon for each (cf. Futuyma 1986: 45–46). Thus, not only individual amino acids but the whole family of twenty amino acids supervene (at least within living cells) on a base family of sixty-one codons (three codons, bringing the total number of codons to sixty-four, do not code for amino acids but instead are punctuation or "stop" codons). The upshot of this empirical find is that the same phenotypic trait may but need not be determined by one and only one DNA microstructure (an upshot that I will fully exploit in chapter 5). To use the metaphor of language, what this means is that *there are many different ways of saying the very same thing.*[8]

Interestingly, a strong case can be made for the view that not only phenotypes but all biological properties supervene on the physical (cf. Sober 1993), that supervenience does not entail reducibility (cf. Sober 1984a; Kincaid 1987), and thus that biology is autonomous and cannot be reduced to physics and chemistry. But as this, as noted at the end of chapter 1, is a topic that takes us much beyond the scope of the present work, I will not take it any further. It need only be mentioned that in Locke's opening up of the possibility that the macro can supervene on the micro, I find I must disagree with Ayers' negative assessment of Locke's contribution to subsequent philosophy and science. I think that in at least this one implication Locke helped open up for us both a very important and true way.

2.3 Buffon, Lamarck, and Darwin

Turning to the topic of species nominalism among biologists, three names invariably come to the fore when looking at the period since Locke up to the present century, namely Georges Louis Buffon, Jean Baptiste de Lamarck, and Charles Darwin. I shall examine the first two only briefly, on a few matters of interest,

---

[8] It is interesting to continue the comparison with language. In any given human language, there is only a finite number of different ways of saying the very same thing. Each meaning, then, supervenes on those different ways. (Cf. chapter 3.5n52 for my reply to the problem of synonymy for natural languages raised especially by Quine.) Similarly in biology, given the supervenience of amino acids upon codons, and given that the four-letter DNA (genetic) code is virtually uniform (it has some minor variations) for all life on earth, it follows that, though every phenotypic trait supervenes on a number of DNA sequences, that disjunctive base must also be finite. Returning to language, as soon as we realize that there is an infinite possibility of languages (at least in terms of semantics), it follows that any given meaning supervenes on an infinitely disjunctive linguistic base. Now if, as virtually all biologists agree, the present genetic code is not the only one possible but is ultimately arbitrary (cf. Crick 1968; Dawkins 1986: 270; Ridley 1993: 46–48), that life on other planets in the universe might therefore be realized by indefinitely different genetic codes, the possibility then becomes open that any given phenotypic trait may supervene on an ontologically indefinite (if not infinite) disjunctive molecular base.

while my main focus will be on Darwin. Darwin is normally interpreted as a species nominalist, with paradoxical consequences. In this I will attempt to go against the flow and prove that Darwin did indeed (appearances to the contrary) subscribe to the objective reality of species (both taxa and category) and that his main criteria were somewhat closely in accord with this author's own view.

Beginning with Buffon (1707–1788), throughout the writing of the forty-four volumes of his *Histoire Naturelle* Buffon's species concept underwent radical change. Originally a subscriber to the Great Chain of Being, in the "Initial Discourse" of the first volume of the *Histoire* (published in 1749) Buffon wrote: "in general, the more one augments the number of divisions of the productions of nature, the more one approaches the truth, since in nature only individuals exist" (Lyon and Sloan 1981: 115). Accordingly, Buffon held that "in order to make a system, . . . It is necessary to divide the whole under consideration into different classes, apportion these classes into genera, subdivide these genera into species, and to do all this following a principle of arrangement in which there is of necessity an element of arbitrariness" (Lyon and Sloan 1981: 102).[9] Interestingly, recognizing the need for a taxonomic system, in spite of arbitrariness, Buffon prescribed the grouping of organisms according to overall similarity, such that "If the individual entities resemble each other exactly, or if the differences between them are so small that they can be perceived only with difficulty, such individuals will be of the same species" (Lyon and Sloan 1981: 106). Extended criteria are immediately supplied for genera and classes, but in none of this are such entities implied to be real. Quite the contrary.

By the time Buffon came to write the second volume of the *Histoire* (also published in 1749), however, he had arrived at what seemed to him an objective criterion for the demarcation of species, namely the infertility of hybrids:

> We should regard two animals as belonging to the same species if, by means of copulation, they perpetuate themselves and preserve the likeness of the species; and we should regard them as belonging to different species if they are incapable of producing progeny by the same means. Thus the fox will be known to be a different species from the dog, if it proves to be a fact that from the mating of a male and a female of these two kinds of animals no offspring is born; and even if there should result a hybrid offspring, a sort of mule, this would suffice to prove that fox and dog are not of the same species—inasmuch as this mule would be

---

[9]   This was a view, likewise occasioned by a belief in the Great Chain, championed by not a few naturalists in Buffon's time. Cf. Lovejoy (1936: 230–231) for quotations from Charles Bonnet and Oliver Goldsmith. For Bonnet, "If there are no cleavages in nature, it is evident that our classifications are not hers. Those which we form are purely nominal, and we should regard them as means relative to our needs and to the limitations of our knowledge." Similarly for Goldsmith, all divisions of organic beings "are perfectly arbitrary. The gradation from one order of being to another, is so imperceptible, that it is impossible to lay the line that shall distinctly mark the boundaries of each." The interesting difference to notice is that while Bonnet's nominalism appears to be ontological, Goldsmith's appears to be epistemological.

sterile. [Lovejoy 1959: 93]

He would later (1765, volume XIII) write of this criterion that "This point is the most fixed which we possess in natural history. All the other resemblances and differences we can observe in comparing beings with one another are neither so real nor so certain; these intervals, therefore, are the only lines of demarcation which will be found in our work" (Lovejoy 1936: 363n9). Paradoxically, Buffon (volume XIII) defined species as "a whole independent of number, independent of time; a whole always living, always the same; a whole which was counted as a single unit among the works of the creation, and which consequently makes only a single unit in nature" (Lovejoy 1959: 101). I will return to this definition in chapter 4.1.

Returning to Buffon's reproductive criterion, the reason why reproductive relations would be taken as so important for the ontology of biological species should be rather obvious. Unlike chemical species, since all biological organisms are inherently mortal it is only through reproductive relations that biological species can perpetuate themselves and be maintained. Should these relations cease, all species would become extinct within one generation.

The effect and advantage of employing specifically interbreeding relations as the objective criterion of species demarcation might be thought to be basically threefold: (i) it allows for an objective test for the reality of species and species boundaries, (ii) it reduces questionable relations such as similarity to a secondary or minor role, and (iii) among species realists it reduces the number of recognized species while adhering to Occam's Razor.

The importance of the second and third points can be illustrated by comparing Buffon's species concept with what was a major competitor, namely what is often called the *typological species concept*, popular among what Mayr (1982) calls "the collector types" (263). For the typologist, any reasonably distinguishable specimen may serve as a type specimen for a species. Thus membership in a species is determined by a somewhat arbitrary degree of overall similarity. In the case of dogs this literally meant that every different breed of dog was accorded species status. In botany the situation was especially ludicrous, since every geographical variety was accorded species status (cf. Mayr 1982: 263). The main advantage of Buffon's concept was that it provided an objective criterion for the reality of species while at the same time adhering to a higher degree of parsimony. Instead of many species of dog there was only one, the entire species united by interbreeding relations.

Buffon's reproductive criterion marked an important turning point in the evolution of species concepts. Instead of admitting only *monotypic* species it now made allowance for *polytypic* species, the modern concept of species which includes subspecies divisions (including varieties and races). In making allowance for polytypic species Buffon's famous criterion allowed taxonomists to be "lumpers" as opposed to "splitters," which in turn, interestingly, helped pave the way for Darwin's theory of evolution since that theory requires species with varieties which are themselves incipient species.

Buffon's advance, however, took a long time before achieving wide acceptance. In the field of ornithology Mayr and Short (1970: 105) remark that even in
the 1920s many ornithologists continued to rank geographical races as full species, resulting in, for example, 922 species of North American birds compared
with (*ceteris paribus*) 607 species recognized today.

In effect Buffon made an important contribution to what is today the dominant
species concept, what Mayr has dubbed the *biological species concept* (a relational concept defined by internal isolating mechanisms and reproductive isolation, examined fully in chapter 4.1). But Buffon was not an evolutionist (cf.
Lovejoy 1959), so that we must agree with Mayr (1982): "By introducing this
entirely new criterion, Buffon had gone a long way toward the biological species
concept. Yet, by considering species as constant and invariable, Buffon still adhered to the essentialistic species concept" (262).

Turning to Lamarck (1744–1829), we find in him, particularly in his master-
work *Philosophie Zoologique* (1809), the first combination of truly evolutionary
thinking with species nominalism. But Lamarck, of course, was not always an
evolutionist. Indeed it is widely agreed that prior to 1800, having been engaged
for the previous 25 years chiefly in botanical pursuits, Lamarck was not only a
species realist (albeit a higher taxa nominalist) but an essentialist and fixist, notably providing for the *Encyclopédie Méthodique*, published in 1786, the following
standard definition in his entry on "Espèce":

> In botany as in zoology, the species is necessarily constituted by the whole group
> of similar individuals that perpetuate their kind through reproduction. By similar
> I mean in the qualities essential to the *species*, because the individuals that be-
> long to it often display accidental differences that are the basis of varieties, and
> sometimes display sexual differences. [Burkhardt 1987: 164]

In 1794, however, Lamarck was made Professor of the "insects, worms, and mi-
croscopic animals" at the Museum of Natural History in Paris (Burkhardt 1987:
162). In 1798 he took over the mollusk collection at the Paris Museum (a collection of modern and fossil mollusks sufficiently complete to trace phyletic lines).
And the rest is history, with his view radically and rapidly changing over to evolutionism, which he first publicly announced in 1800 (Burkhardt 1987: 167). Indeed as Mayr (1982) put it, "Probably no other group of animals was as suitable
for bringing about such a conclusion as the marine mollusks" (346–347).

As for species, on the one hand Lamarck (1809) seems to have been a realist,
since he held that "it is useful to give the name of species to any collection of like
individuals perpetuated by reproduction without change, so long as their environment does not alter enough to cause variations in their habits, character, and
shape" (44). Moreover, he seriously entertained the view that a given species
does not cease to exist by evolving into another species but continues to exist in
that new form: "Now, if a quantity of these fossil shells exhibit differences which
prevent us, in accordance with prevailing opinion, from regarding them as the
representatives of similar species that we know, does it necessarily follow that

these shells belong to species actually lost? . . . May it not be possible, on the other hand, that the fossils in question belonged to species still existing, but which have changed since that time and have become converted into the similar species that we now actually find?" (45; cf. Mayr 1972: 247). On the other hand, however, he seems to have really been a species nominalist, motivated apparently by a principle of plenitude and the view that however one chops up an evolutionary continuum is entirely arbitrary, a view that we will find again in the present century. For a start, Lamarck believed in a sort of *scala naturae*: "in each kingdom of living bodies [i.e., plants and animals] the groups are arranged in a single graduated series, in conformity with the increasing complexity of organisation and the affinities of the objects" (59). Moreover, aside from these two kingdoms he held that there are no gaps in nature: "The further we extend our observations the more proofs do we acquire that the boundaries of the classes, even apparently most isolated, are not unlikely to be effaced by our new discoveries" (23).[10] He also held that evolutionary "changes only take place with an extreme slowness, which makes them always imperceptible" (30). Finally, he held that classes, orders, families, genera, and species are "artificial devices in natural science" (20), since "nature has not really formed [such taxa] . . . but only individuals" (21). Indeed this was a view—that in biology only individual organisms are real—that Lamarck repeatedly stressed from 1800 to 1815 (Burkhardt 1987: 168–170), after which time, interestingly, he began to waffle, eventually reaffirming toward the end of his career the reality of species (Burkhardt 1987: 172–174), although he made no attempt to redefine the species category.

When we turn to Darwin (1809–1882),[11] we enter the modern era, for it is undeniably true that, roughly speaking, Darwin's theory of evolution by natural selection is the dominant paradigm in evolutionary biology today. Accordingly, it is with Darwin that the modern species problem really begins. The interesting thing about Darwin, however, is that instead of attempting to solve the problem, or even just to point us the way, his writings have proved highly paradoxical, so that the question of his species concept is usually overlooked and ignored. Unlike most authors on the modern species problem, however, I believe that a more than superficial look is worth the effort, and that a serious attempt to understand Darwin's real versus apparent species concept may shine new light on the ontology of species.

---

[10] Unlike Darwin (cf. Darwin 1859: 484), Lamarck did not hold a belief in common descent, the view that all life can be traced back to one or a few ancestors in the very remote past. Darwin, like biologists today, thought that the favorable conditions for the origin of life were long distant in the past (cf. Mayr 1982: 582–583). Lamarck, instead, shared the then widespread belief in constant and rudimentary spontaneous generation, so that on Lamarck's evolutionism the lowest forms of life are constantly being replenished as the rest evolve upward through ever-increasing levels of complexity, in the animal kingdom upward all the way to man, who is still evolving. Consequently, unlike the modern Darwinian view, Lamarck's Tree of Life is perhaps best thought of as a mixture of two (plant and animal) homogeneous forests (cf. Panchen 1992: 60).

[11] The following section on Darwin is drawn largely from Stamos (1996a, 1999).

Darwin, of course, did not begin his intellectual life as an evolutionist. In spite of having read his grandfather Erasmus' *Zoonomia* during his university years and having listened "in silent astonishment" to a university friend (Robert Grant) enthusiastically discuss the views of Lamarck, it all had, he tells us (Darwin 1876), no immediate effect on his mind (26). Indeed even throughout his voyage on the *Beagle*, completed in October of 1836, he remained a theist and a creationist (49). Shortly after, however, upon receiving in March of 1837 important feedback from the ornithologist John Gould, concerning in particular his mockingbird specimens from the Galapagos islands (Sulloway 1982: 22), Darwin began in July of 1837 his transmutation notebooks (Barrett *et al.* 1987; cf. Darwin 1876: 48). In September of 1838 he read Malthus (Notebook D: 134–135; cf. Darwin 1876: 71). And in 1839 he could say that his theory "was clearly conceived" (Darwin 1876: 74). Nevertheless, what is evident throughout the period of his transmutation notebooks (from July 1837 to July 1839) is that he continued to believe in the objective reality of species. In fact the species concept he held was based on the zoologically-minded criterion of reproductive isolation (cf. Notebook B: 197; Notebook C: 152, 161; Notebook E: 24), specifically the instinct to keep separate, which Mayr (1982) describes as reproductive isolation "maintained by ethological isolating mechanisms" (266). Such a concept, however, applies only to species reality at any one slice of time. In itself it need not presuppose evolution, as we saw with Buffon. What, then, about species reality through time, in terms of evolution?

There is scarcely any clear evidence to help us determine what Darwin thought on this matter (cf. Notebook C: 152). However, I like to think I have discovered a piece of evidence which suggests that Darwin at this time thought in terms of species realism in spite of his evolutionism. But first we must return to a classic passage we found in Locke's *Essay*:

> this is a Man, that a Drill [baboon]: And in this, I think, consists the whole business of *Genus* and *Species*. [III.vi.36]

Now consider an interesting passage to be found in one of Darwin's notebooks (1838):

> Origin of man now proved.—Metaphysic must flourish.—He who understands baboon would do more towards metaphysics than Locke. [Notebook M: 84]

According to Antony Flew (1984b), "Although this *Essay* [Locke's], first published in 1690, was enormously influential there appears to be no evidence that Darwin ever read any of Locke's more philosophical writings" (46–47).[12] At any rate, it seems to me that the above correspondence is too great to be dismissed as

---

[12]   Interestingly, according to Desmond and Moore (1991: 87–88), among the works that Darwin was examined on in his final examination for his B.A. at Cambridge (January 1831) was Locke's *Essay*.

coincidental. Darwin scholars typically interpret Darwin's note above as a reference to the metaphysics of mind. According to Herbert and Barrett (1987), for example, "Darwin believed the similarity of expressions in other animals to man strengthened, even 'proved,' the transmutationist case" (518). Of course, Notebook M (along with Notebook N) was explicitly devoted to and contains much in the way of speculation on expressions and the evolution of mind, and Darwin was notorious for ascribing in an anthropomorphic style human thoughts and emotions to animals.[13] However, if I am right, then Darwin, in the passage above, is not (or is not only) referring to the metaphysics of mind, but to ontology. His claim, I suggest, is that given evolution not only must Locke's ontology be expanded to include biological species in an extra-mental sense but these species are themselves the very units of evolution. Indeed it is interesting to note that Darwin in his notebooks thought that the baboon is a living ancestor of mankind (cf. Notebook C: 243; Notebook M: 123).

That Darwin's mind on the day he wrote the note quoted above (August 16, 1838) was focused on the issue of the ontology of species (particularly over time, as evolutionary units) is evidenced by a note he wrote on the very same day but which he placed in Notebook D. Writing of changes in geography and of corresponding changes in adaptation in the organic world, Darwin wrote that the view of "the world peopled with Myriads of distinct forms from a period short of eternity to the present time, to the future . . . [is] far grander than idea from cramped imagination that God created. (warring against those very laws he established in all organic nature) the Rhinoceros of Java & Sumatra" (36–37).

At any rate, when we turn to Darwin's *Origin* the situation changes dramatically. Indeed as Mayr (1982) put it, "When one goes to the *Origin* of 1859 and reads what it says about species, one cannot help but feel that one is dealing with an altogether different author" (266). In the *Origin* many passages could be cited in support, but the following are the two most often quoted:

> From these remarks it will be seen that I look at the term species, as one arbitrarily given for the sake of convenience to a set of individuals closely resembling each other, and that it does not essentially differ from the term variety, which is given to less distinct and more fluctuating forms. The term variety, again, in comparison with mere individual differences, is also applied arbitrarily, and for mere convenience sake. [52]

> In short, we shall have to treat species in the same manner as those naturalists treat genera, who admit that genera are merely artificial combinations made for convenience. This may not be a cheering prospect; but we shall at least be freed from the vain search for the undiscovered and undiscoverable essence of the term species. [485]

---

[13]   Unlike Descartes and many others, Locke, in subscribing to the Great Chain of Being, did much the same as Darwin (cf. *Essay* IV.xvi.12). Nevertheless, he denied that animals below man have the power of abstract thought and he affirmed that this "puts a perfect distinction betwixt Man and Brutes" (II.xi.10).

It is interesting to note that Darwin in the second passage above might appear to share with Locke, as we have seen earlier, the inference that if species do not have essences (cf. 45 and 484) then they must not be real. Indeed it was a view that was quite common in Darwin's day. Not only did Darwin's mentor Charles Lyell, for example, in the second volume of his highly influential *Principles of Geology* (1832), which Darwin studied assiduously while on his *Beagle* voyage, proclaim that "the majority of naturalists agree with Linnaeus in supposing that all the individuals propagated from one stock have certain distinguishing characters in common which will never vary, and which have remained the same since the creation of each species" (3; cf. 57), but he also stated the issue of species reality as depending on their fixity, as "whether species have a real and permanent existence in nature; or whether . . . . " (1; cf. 65). I have already commented on the falsity of this dichotomy. Whether Darwin also subscribed to it will be discussed shortly. What is most surprising is that there are philosophers who still subscribe to it today (and even an occasional rogue biologist). According to Kitts and Kitts (1979), for example, "To suppose otherwise [i.e., to suppose that there are no conspecific essences] is not to give reason to change our view of species, but to give reason to abandon the concept of species altogether. Biologists search for the underlying trait which explains the necessary relationship between an organism and its species in the genetic structure of the organism" (618; cf. Rieppel 1986: 312–313). As this is a topic that takes us much further than Darwin, I will reserve an extended discussion of it for chapters 3 and 4.

For the present, it is interesting to note that because of the two passages quoted above from Darwin's *Origin*, and many more (cf. 47, 59, 248, 276, 297, and 469)—many of which are reminiscent of Locke simply in their use of the words "arbitrary" and "convenience"—most scholars have simply taken Darwin to mean literally that species are not objectively real (e.g., Haldane 1956: 95; Ellegård 1958: 200; Hull 1965: 203; Mayr 1970: 13; Levin 1979: 382; Gould 1980: 205–206; Wiley 1981: 41; Howard 1982: 17, 37; Eldredge 1985a: 109–110; Rieppel 1986: 304, 307; Thompson 1989: 8; Ereshefsky 1992b: 190; Luckow 1995: 590). Since it is normally thought that if anything in biology is to evolve it is species, and given the title of Darwin's most famous work, the result on this reading is not a little paradoxical. Indeed, this was immediately felt subsequent to the publication of the *Origin*. The famous Swiss-born American naturalist, Louis Agassiz (1860b), for example, in spite of his irrational holdout against Darwin, wrote that "If species do not exist at all, as the supporters of the transmutation theory maintain, how can they vary?" (143). In our own time Elliott Sober (1993) has suggested that Darwin's masterpiece would be better titled "*On the Unreality of Species as Shown by Natural Selection*" (143).

I think this reading of Darwin is not only paradoxical but wrong. But first I want to briefly examine two theories on why Darwin would write what he wrote above. Mayr (1982), like many others, does not doubt the sincerity of those passages and accepts them at face value. Drawing mainly on Darwin's correspondence during the 1840s but especially during the 1850s, Mayr (267–268) surmises that Darwin's species nominalism resulted mainly from his friendships with botanists during that period and from the botanical literature, much of which was negative in influence on the reality of species.

For my part, however, I find this theory difficult to accept. It seems to me that Mayr here is merely taking his own difficulties with botanists in regard to his own criterion of reproductive isolation and projecting that difficulty back in time by blaming botanists contemporary with Darwin for a negative influence on Darwin. I don't think this will wash for the simple reason that Darwin tells us himself (as we shall see) that virtually all of his professional contemporaries were species realists.

Mayr's additional theory (269) is that in denying the reality of species Darwin was employing a clever strategy. Since the creationists were maintaining that species are distinct and immutable, and since Darwin in the *Origin* wanted to prove that species may change indefinitely, he wisely denied the reality of species and employed for the sake of convenience only a morphological concept based solely on "degree of difference." In thus treating species as "purely arbitrary designations"—arbitrary because it provides no objective criteria for a distinction between species and varieties on the one hand and species and higher taxa on the other hand—he could deny the objective distinctness of species: "When species are thus conceived, the origin of new species is not an insurmountable problem."

The problem with this theory is that it would seem to entail that species on Darwin's view are still in some sense objectively real. Indeed even Mayr admits that for Darwin "species continue to evolve" (269). But that contradicts Mayr's reading of the *Origin* and what he says about the influence of botanists. Thus Mayr contradicts Mayr.

A more plausible version of a strategy theory is provided by John Beatty (1985). According to Beatty, "Darwin made clear that it was to 'naturalists' that the *Origin* was addressed," a professional group "of which Darwin considered himself a member" (266). Within that community the term "species" was heavily theory-laden, so much so that it precluded evolution. Thus, to speak of species as evolving was a contradiction in terms. According to Beatty,

> He tried to get around this difficulty by distinguishing between what his fellow naturalists *called* "species" and the non-evolutionary beliefs in terms of which they *defined* "species." Regardless of their definitions, he argued, what they *called* "species" evolved. His species concept was therefore interestingly minimal: species were, for Darwin, just what expert naturalists *called* "species." By trying to talk about the same things that his contemporaries were talking about, he hoped that his language would conform satisfactorily enough for him to communicate his position to them. [266]

In adopting a nominalistic definition of species, Darwin could bypass their theory-laden definitions while at the same time, in retaining their *use* of the term "species," he could provide evidence to effect a reinterpretation of the referents of that term. Thus, according to Beatty (270), Darwin only *seemed* to deny the reality of species.

A clever strategy indeed. What strengthens Beatty's theory is that (271–274) he finds the presuppositions which underlie it explicitly discussed in the work of the British botanist Hewett Cottrell Watson, with whom Darwin was well acquainted. Not only, according to Beatty, did Watson in his writings discuss the difficulties in changing the meaning of an entrenched theory-laden term, but he also made clear the distinction between the definitional use of the term "species" and its referential use, and even suggested that the way to effect theory change is to deny the theory-laden definitional use while maintaining the referential use. Indeed, according to Beatty, that is precisely what Watson did in communicating the result of his experiments on the common descent of primroses and cowslips, two commonly regarded species that Darwin, interestingly, made much of in the *Origin* (cf. 49–50, 247, 268, 424, and 485). On Beatty's theory, then, Darwin found in Watson not only a clear discussion of the difficulties he faced in effecting so radical a paradigm shift, but he also found a plausible strategy for effecting that shift.

Although Beatty's theory is well thought out and argued, and may indeed best correspond with actual history, I find it difficult to accept for at least three reasons. First, it seems to me upon reading the *Origin* that Darwin intended it for a much wider audience than the community of professional naturalists of which he himself was a part. Not only is the *Origin* written with exceptional clarity and relative simplicity, but Darwin provides definitions of terms which would be entirely superfluous if his audience was as narrow as Beatty maintains. For example, on page 390 Darwin defines what an endemic species is, something even a rank amateur would not need done for him. Add to this his playing down of the coldness and horror of natural selection in the wild: "that the war of nature is not incessant, that no fear is felt, that death is generally prompt, and that the vigorous, the healthy, and the happy survive and multiply" (79). Contrast this with what he wrote in one of his letters to Asa Gray (May 22, 1860): "There seems to me too much misery in the world. I cannot persuade myself that a beneficent & omnipotent God would have designedly created the Ichneumonidae with the express intention of their feeding within the living bodies of caterpillars, or that a cat should play with mice" (Burkhardt *et al.* 1993: 224). That Darwin intended the *Origin* for a wider audience than just professional naturalists is also directly apparent from a number of his letters. For example, in one of his letters to John Murray (April 2, 1859), although Darwin admitted that "My volume cannot be mere light reading, & some parts must be dry & some rather abstruse," he immediately added that "yet *as far I can judge perhaps very falsely*, it will be interesting to all (& they are many) who care for the curious problem of the origin of all animate forms" (Burkhardt and Smith 1991: 277; cf. 278, 440; Burkhardt *et al.* 1993: 240). And given the previous considerations, it cannot be replied that he was being disingenuous here simply because Murrary was about to publish his *Origin*.

Second, it is simply not true, contra Beatty, that Darwin retained in practice what his fellow naturalists *called* species. This is especially evident in the case of dogs. Although virtually all naturalists from Buffon onward (as is the case today) categorize all the breeds of domestic dogs as members of a single species (*Canis familiaris*) because of their interfertility, there is nowhere, either in Darwin's published writings, notes, or correspondence, where he follows this practice. The main reason, apparently, is because he believed that the group as a whole was domesticated not from a single wild species but instead from a number of wild species, domestication eventually eliminating the sterility barriers inherited from the parental species. The issue often came up in his correspondence (cf. Burkhardt and Smith 1991: 357, 362–364, 384, 386, 392; Burkhardt *et al.* 1993: 170, 258, 261, 262–263, 320, 335, 366, 368, 378, 383–384, 393, 397, 399), but there is a particular passage in the *Origin* that quite nicely serves to illustrate Darwin's view that domestic breeds derived from more than one wild species should be accorded multiple species status. This is made explicit in the case of European cattle and humped Indian cattle, which though "quite fertile together," Darwin nevertheless thinks that they (because of their polyphyletic origin) "must be considered as distinct species" (254; cf. 18). The same conclusion follows (though implicit) for dogs, which Darwin discusses immediately above the cattle example as an analogous case.

Indeed the topic is only further highlighted by Darwin's view on convergence. In a letter to Darwin (January 3?, 1860), Watson suggested that Darwin allow "the hypothesis or inference that individuals *converge* into orders, genera, species, *as well as diverge* into species, genera, orders, through nepotal descent" (Burkhardt *et al.* 1993: 11–12). In reply to Watson (January 5–11, 1860), Darwin stated that "With respect to 'convergence' I daresay, it has occurred, but I should think on a very limited scale, . . . and only in case of closely related forms. . . . the same cause acting on two closely related forms (i.e., those which closely resembled each other from inheritance from a common parent) might confound them together so closely that they would be *(falsely* in my opinion) classed in same group" (Burkhardt *et al.* 1993: 18; cf. Darwin 1859: 427).

Thus we can see that, contra Beatty, far from employing the referential use in what his fellow naturalists called species, Darwin's own referential use sometimes put him at odds with those very naturalists. Instead of interbreeding as a criterion, he stressed common descent. In other words, his *real* species concept had a definite monophyletic component to it (though not in the strict sense as with many today, since as we shall see in chapter 4.4 he did not believe that extinction is necessarily forever).

Interestingly, these considerations also serve to discredit the common interpretation of Darwin that subsequent to the early notebooks he came to subscribe to a basically morphological species concept, based merely on degree of difference (e.g., Sulloway 1979: 37–38; Mayr 1982: 210, 1991: 30; Bowler 1985: 648; Ruse 1987: 344). The illegitimacy of this interpretation is evident alone from Darwin's many years of work on barnacles (Cirripedia), which spanned the period from his early notebooks to the beginning of his big species book. When Darwin discovered what he came to call "complementary males," males so small in relation to the hermaphrodites to which they were attached (such that naturalists had thought they were parasites), he nevertheless always thought of them, in spite of their being "utterly different in appearance and structure" from the hermaphrodites, as "yet all belonging to the same species!" (Darwin 1851: 293; cf. Darwin 1859: 424). Of course the argument against the common interpretation of Darwin's mature species concept as basically morphological is not new, having been masterfully argued much earlier by Ghiselin (1969: 89–102)—although it should also be added that Ghiselin (97), failing to perceive the possibility of a deliberate strategy on Darwin's part, thought that Darwin did at times lapse into an exclusively morphological mode of thought. And yet in spite of it all, many Horatios still remain. At any rate what I have begun to show and shall show further is that there are more things in Darwin's mature species concept than are dreamt of in even the philosophies of the Hamlets of the Darwin industry.

A further feature of Beatty's argument, and the third reason for why I find it difficult to accept, is that he draws (following Ghiselin 1969: 93) the consequence that if his theory is correct it means that Darwin, in his denials that the term "species" can be defined, denied the reality of the species *category*, not the reality of species *taxa*. Though the latter too are for Darwin not definable, they are nevertheless real, and for Darwin what naturalists call species are really "chunks with[in] the genealogical nexus of life" (278). This, says Beatty, "The suggestion that natural history could really get by without definitions of the categories of classification—especially a definition of 'species'—is admittedly hard to swallow" (280).

Never mind that subsequent evolutionists, as Beatty (280) points out, have since replaced (though their replacements are many) Darwin's nominalistic definition of the species category with real definitions, so that Darwin should be seen today as a sort of *liberation biologist* (if I may allude in analogy to modern theology). And never mind that Darwin, contrary to Beatty's "chunks with[in] the genealogical nexus of life," refused to think of every interbreeding group as a distinct species, or that he thought (as we shall see in chapter 4.4) that extinction is not necessarily forever. What is particularly disturbing about Beatty's interpretation of Darwin is that if he is right, it means that Darwin was fundamentally confused, for though he thought that species taxa are real without being essentialistic he nevertheless thought that the species category is not real because it's not essentialistic. Moreover, it means that although Darwin postulated real laws of nature for species *per se* (as we shall see), he nevertheless thought that the subject of those laws (the species category) is not itself real. These are conflicting, even contradictory, beliefs to ascribe to someone whom many consider to have been a genius.

It seems to me that the three above difficulties with Beatty's theory, combined with the principle of charity for the third difficulty, require us here to back up a bit and pursue another trail.

It is important, of course, to read Darwin not as one would read a modern biologist today, but to read him (as one might read, say, David Hume on natural religion) in the context of his time. In order to properly do this, we must be clear on what was the predominant species concept shared especially by Darwin's professional and theological contemporaries. Interestingly, Darwin himself in the *Origin* provides us with just about all that we really need to know. Indeed, it is almost as if he foresaw the future development of posterity and that the paradigm he sought to overthrow would quickly be forgotten. At any rate, it seems to me that the dominant species concept of Darwin's day can be divided among six points (at least for our purposes), although it will prove significant to also consider the view on higher taxa, in particular genera.

1) *Genera are not objectively real but instead are manmade and arbitrary.* Although Darwin tells us that "the points in which all the species of a genus resemble each other, and in which they differ from the species of some other genus, are called generic characters" (155), he also tells us that "Several of the best botanists, such as Mr. Bentham and others, have strongly insisted on their [higher taxa from genera to orders] arbitrary value" (419), and that many naturalists "admit that genera are merely artificial combinations made for convenience" (456).[14]

2) *Species are the product of divine creation, and it is this that constitutes their reality.* "Several eminent naturalists have of late published their belief that a multitude of reputed species in each genus are not real species; but that other species are real, that is, have been independently created" (482).

3) *An inherent quality and special endowment of species is that they are intersterile.* "The view generally entertained by naturalists is that species, when intercrossed, have been specially endowed with the quality of sterility, in order to prevent the confusion of all organic forms" (245).

4) *Each species has a species-specific essence, a set of characters which remain fixed and permanent throughout the duration of its existence, and which allow it to be captured in a definition.* Although Darwin tells us that "the points in which species differ from other species of the same genus, are called specific characters" (156), he gives us very little on the essentialism of his contemporaries, resting content to presuppose it (as we shall see) when elucidating his own alternative view. The view that species can be essentialistically defined is, of course, part of a tradition that goes back at least to Aristotle, found predominantly in his logical and metaphysical works, and which reached its zenith (as we saw earlier in the quotation from Lyell 1832) in the highly influential work of Linnaeus, all of which we will examine in chapter 3.1.

5) *Although species are immutable, this only means that they cannot change into other species, not that natural (secondary) causes (laws) cannot cause them to vary and produce varieties (subspecies, races, etc.). Thus, varieties are the limits of species variability.* "Generally the term ['species'] includes the unknown element of a distinct act of creation. The term 'variety' is almost equally difficult

---

[14] In addition to George Bentham, Darwin most certainly had H.C. Watson and J.D. Hooker in mind, two of his most important correspondents and among the best botanists in the world. In a letter to Darwin (March 23, 1858), Watson confided that "I look upon the orders & genera of plants as purely conventional arrangements, not natural groups,—that is, not groups *in nature*, but only as groups in books & herbaria" (Burkhardt and Smith 1991: 54). Similarly Hooker, in a letter to Darwin (February 25, 1858), stated that "Genera in short are almost purely artificial as established in Botany" (Burkhardt and Smith 1991: 35).

to define; but here community of descent is almost universally implied, though it can rarely be proved" (44). Again: "species, commonly supposed to have been produced by special acts of creation, and varieties which are acknowledged to have been produced by secondary laws" (469). Darwin also notes that naturalists suppose that there are definite limits to the variability of species via varieties: "It has often been asserted, but the assertion is quite incapable of proof, that the amount of variation under nature is a strictly limited quantity" (468). Finally, Darwin also notes that if intergrades are found between two reputed species the whole complex is then reclassified as a single species: "they grant some little variability to each species, but when they meet with a somewhat greater amount of difference between any two forms, they rank both as species, unless they are enabled to connect them together by close intermediate gradations" (297).

6) *Unlike species, varieties are not intersterile but instead are interfertile.* "A supposed variety if infertile [intersterile] in any degree would generally be ranked as species" (271).

7) *Varieties are not objectively real.* In spite of the two points above, and in spite of the fact that it was not clearly brought out by Darwin, it is important to point out that for the majority of naturalists in Darwin's day both the category *variety* and the taxa commonly ranked as varieties were thought to be ultimately arbitrary and unreal, since varieties lacked essences and were not specially created, being merely convenient groupings based on accidental (nonheritable, or when heritable nonpermanent) variations of conspecific individuals (cf. Ellegård 1958: 198–199; Larson 1968: 291).

Before we examine Darwin's reply to each of these points, it is important to note that in Darwin's eyes evolution was not, prior to the publication of the *Origin*, in some sense "in the air" but that virtually all naturalists subscribed quite strongly to all but the first of the above seven points, since as Darwin (1876) states "I occasionally sounded not a few naturalists, and never happened to come across a single one who seemed to doubt about the permanence of species. Even Lyell and Hooker, . . . " (73–74; cf. Darwin 1859: 310).

Now for Darwin's replies:

1) Although many naturalists thought that higher taxa were arbitrary and unreal, many did not, but to Darwin the criteria of these realists made their categories appear "almost arbitrary" (419). On the other hand, Darwin not only believed that "all true classification is genealogical," that classification "must be strictly genealogical in order to be natural," he also believed that "community of descent is the hidden bond which naturalists have been unconsciously seeking" (420), that genealogy has been "unconsciously used in grouping species under genera, and genera under higher groups, . . . only thus can I understand the several rules and guides which have been followed by our best systematists" (425).

2) For Darwin, the theory that species were divinely created by miracle made very little sense of the evidence from geology and biology. The evidence from geology, as Lyell had concluded, bespoke an extremely old earth, and an earth that was not static: its terrestrial environments were ever changing and (for the most part) ever so slowly (282–283). As species must be adapted to their envi-

ronments if they are to survive, they must change with their environments. Moreover, the theory of divine creation made little or no sense of immediately observable phenomena, such as vestigial organs (199–200), adaptive radiations (397–398) and other anomalies of geographic distribution (394), embryological evidence (449), and the evidence of the fossil record (333 and 465).

3) In studying the observations and experiments of zoologists and botanists, Darwin came to the conclusion that while, on the one hand, it is a fact that "hybrids [when they do occur] are very generally sterile" (245), it is also a fact, on the other hand, that when there is fertility between species the fertility (when all these species are considered together) grades insensibly from slightly above 0% to 100% (chapter 8). Darwin's conclusion, then, is that sterility between species can neither be considered a "specially endowed quality" (261) nor a "universal law of nature" (250). Since "the sterility of hybrids could not possibly be of any advantage to them" (245), it could not be something that could be selected by nature. (Natural selection, according to Darwin, operates principally only for the good of the individual, though sometimes also for groups; cf. 200–202.) Sterility, therefore, must be "incidental on other acquired differences" (245). And if incidental, it is arguable, as we shall see, that on Darwin's view it is not part of what a species is.

4) If species change and gradually evolve into other species, and if selection operates on favorable variations of conspecific individuals, it follows that any characteristic thought essential to a species need not be found in all of its member organisms, and that even if one characteristic is so found, it is only accidentally so, not necessarily so. This conclusion is evident throughout Darwin's writings, as in the following passage: "if every form which has ever lived on this earth were suddenly to reappear, . . . it would be quite impossible to give definitions by which each group could be distinguished from other groups, as all would blend together by steps as fine as those between the finest existing varieties" (432). Again: "the points in which species differ from other species of the same genus, are called specific characters; and as these specific characters have varied and come to differ within the period of the branching off of the species from a common progenitor, it is probable that they should still often be in some degree variable" (156; cf. 45). And yet again: "community of descent is the natural bond which naturalists have been seeking, and not some unknown plan of creation, or the enunciation of general propositions" (420).

5) The idea that species may vary only to a certain degree (varieties) but not indefinitely had, for Darwin, no basis in fact or logic (469). For this conclusion Darwin employed the evidence that may be garnered from breeders, what he called "the best and safest clue," remarking further that it is "very commonly neglected by naturalists" (4). If breeders making selections on individual variations could effect such remarkable changes over only a few centuries, how much more could natural selection from a changing environment accomplish over vast periods of time? As Darwin himself put it, "What limit can be put to this power, acting during long ages and rigidly scrutinising the whole constitution, structure, and habits of each creature,—favouring the good and rejecting the bad?" (469).

6) If species are to evolve, and if evolution must be quite slow and gradual, it follows that varieties must not be the limits of mutability but must instead be what Darwin called them, namely "incipient species" (52). Thus, it was extremely important for Darwin to prove that there is no fundamental distinction between species and varieties, which was contrary to what almost all other naturalists supposed. And once again, to support his conclusion he appealed to the evidence and controversies supplied by professional biologists themselves. Admitting "the almost universal fertility of [conspecific] varieties when crossed" (460), it can nevertheless "be shown that neither sterility nor fertility affords any clear distinction between species and varieties; but that the evidence from this source graduates away, and is doubtful in the same degree as is the evidence derived from other constitutional and structural differences" (248). To support this conclusion he appealed to numerous crossing experiments, most notably those by Kölreuter and Gärtner (270–272). He also appealed to the well-known fact, made much of also by Lamarck (1809: 36; cf. Mayr 1972: 231), that even the experts could not agree "whether certain doubtful forms should be ranked as species or varieties" (248). But the *pièces de résistance* were the anomalous cases which proved that the traditional criteria were inconsistent. Foremost among these were primroses and cowslips, "united by many intermediate links" (50) and consequently "generally acknowledged to be merely varieties" (485; cf. 268), and yet "according to very numerous experiments made during several years by that most careful observer Gärtner, they can be crossed only with much difficulty" (49–50), so that Gärtner "ranks them as undoubted species" (268). Thus, according to one widely used criterion they must be ranked as conspecific varieties, while according to another widely used criterion they must be ranked as distinct species. Obviously they cannot be both. In consequence of all the above, then, Darwin concluded that "I do not think that the very general fertility of varieties can be proved to be of universal occurrence, or to form a fundamental distinction between varieties and species" (271–272).

7) Although, as with "species," Darwin defined his use of the term "variety" nominalistically, as a term "applied arbitrarily, and for mere convenience sake" (52), it is clear that in practice he did not consider varieties as unreal. For one thing, as we have seen above, he considered them to be incipient species, and they could hardly be incipient if they were not real. Moreover, he often repeated his view that "varieties are generally at first local" (298; cf. 301 and 464), which again implies that he thought of them as in some sense real.

What Darwin more specifically thought about the reality of varieties will be examined a little later. For the present it is important to notice that of the seven replies above, his claim that varieties are not the limits of species mutability but instead are incipient species is the most important or pivotal point in what Darwin himself described as "one long argument" (459). Indeed, it is the Archimedean fulcrum with which he would move the world.

What needs to be further kept in mind is that in those various passages where Darwin overtly denies the reality of species and varieties, the context of his discussion is invariably the issue of whether there is a fundamental and universal distinction between species and varieties (cf. 44, 47, 49, 51–52, 248, 268, and 484–485). This is a crucial point, missed completely by Mayr and insufficiently explored by Beatty.

What all of the above suggests, then, it seems to me, is that in overtly denying the reality of species and varieties, Darwin was employing in the *Origin* a strategy that is best described as a sort of *consistency* argument. Accordingly, it is important to notice his characterization on page 424 of the importance of common descent as a grouping criterion in the work of practicing naturalists: "With species in a state of nature, every naturalist has in fact brought descent into his classification; for he includes in his lowest grade, or that of a species, the two sexes; and how enormously these sometimes differ in the most important characters, is known to every naturalist." Next: "The naturalist includes as one species the several larval stages of the same individual, however much they may differ from each other and from the adult." And finally: "He includes monsters; he includes varieties, not solely because they closely resemble the parent-form, but because they are descended from it."

In denying, then, the reality of species, it seems to me that Darwin was not necessarily trying to better communicate his own theory of evolution. Rather, he was perfecting a consistency argument: (P1:) If you employ common descent in grouping males and females and larvae and adults and monsters and varieties as conspecific, in spite of often great dissimilarity, and (P2:) if you have neither good reasons nor good evidence to believe that what are called varieties are the limits of mutability, but instead you have both good reasons and good evidence to believe that they are indefinitely mutable, then (C1:) you must as a matter of consistency conclude that what are called varieties are incipient species and that what are called species are indefinitely mutable. Moreover (P3:) if you also believe that varieties are not real because they are created by natural mechanisms rather than by divine fiat, and furthermore because they are variable and not fixed, then (C2:) *you must also conclude not only that varieties are not real but that species are not real either!*

Now that is a clever *reductio* against the creationist species concept. And I do believe that it was Darwin's strategy. But before we go on to examine what Darwin really thought about the nature of species, we must first examine a serious difficulty for my theory. In the *Origin* Darwin tells us that "every one admits that there are at least individual differences in species under nature. But, besides such differences, all naturalists have admitted the existence of varieties, which they think sufficiently distinct to be worthy of record in systematic works" (468–469). But is this, we must ask, in particular the part about varieties, true? According to Mayr (1982), "Essentialists did not know how to deal with variation since, by definition, all members of a species have the same essence" (288). Instructive in this regard is the highly influential case of Linnaeus. In his *Fundamenta Botanica*, published in 1736, Linnaeus stated that "There are as many varieties as there are different plants produced from the seed of one and the same species" (Leikola 1987: 49). While in plants Linnaeus focused mainly on sexual parts as providing essential characters for species, he tended to regard as variety characters, as Leikola (1987) put it, "a plant's size, colour, smell, taste, growthplace, time of blossoming, length of life, and economic use" (49). Moreover in the *Fundamenta Botanica* (and elsewhere) Linnaeus held that while genera and species are products of nature (whereas classes and orders are products of human art and nature), varieties are usually the product of what he called "culture"—"Naturæ opus semper est species & genus, Culturæ sæpius varietas, Artis & Naturæ classis ac ordo" (Leikola 1987: 56n9)—by which he meant not only human cultivation but variations in environmental conditions such as moisture, temperature, and soil (the exceptions to *sæpius* being sports or monstrosities). As such, variety characters are generally not fixed but vary in degree, not kind. Similarly, in the case of Buffon, he stated in volume XIII (1765) of his *Histoire Naturelle* that "The type of each species is cast in a mold of which the principle features are ineffaceable and forever permanent, while all the accessory touches vary" (Lovejoy 1959: 101). Although most variety characters within a species may vary widely, it is of course by no means necessary that every possible degree of every such character must be expressed. Indeed, much or most of the potential variation may in most cases not be expressed at all, ever. It is the gaps in expression, of course, produced by the fact that few if any plant and animal species have been allowed to grow in every possible environmental condition, that allow for the facility of naming varieties. The upshot, then, is that given that almost every variety character within a species is part of a potential continuum, the naming of varieties on an essentialist point of view must logically in the main be arbitrary and the varieties accordingly picked out unreal (the particular variations, of course, are nevertheless real). To see that this is so one need only ask from an essentialist point of view how many varieties there are for any given species, the case of humans being particularly instructive.

Turning now to Darwin's contemporaries, it is certainly true that naturalists, in their practice of naming and describing organisms, generally wrote and spoke as if they regarded varieties as real, but rarely if ever did they stop to defend their practice and argue for the objective reality of varieties *per se*. Indeed, if they stopped it was usually to express their doubts. For example, in 1838 Darwin noted that "Mr Yarrell.—says my view of varieties is exactly what I state.—&c picking varieties. unnatural circumstance" (Notebook C: 120). A little later he noted that "Dr Beck doubt if local varieties should be remembered, therefore do not consider it as proved that they are varieties" (Notebook C: 137). Both Yarrell and Beck were zoologists. As for botanists, instructive is what J.D. Hooker wrote in one of his letters to Darwin (March 14, 1858), namely that "The long & short of the matter is, that Botanists do not attach that *definite* importance to varieties that you suppose" (Burkhardt and Smith 1991: 49). I suggest that if pressed, species essentialists would have to admit that varieties are not real but arbitrary. What needs to be remembered is that (i) for the species essentialist a variety lacks an essence, (ii) variety characters often vary widely in degree and rarely if ever are all potential degrees ever actualized, and (iii) varieties accordingly are not fixed. Interestingly, Wallace (1858) tells us that it was generally assumed by naturalists in his day that "*varieties* produced in a state of domesticity are more or less unstable, and often have a tendency, if left to themselves, to return to the normal form of the parent species." From this it was concluded (wrongly, he holds) that "this instability is . . . a distinctive peculiarity of all varieties, even of those occurring among wild animals in a state of nature, and to constitute a provision for preserving unchanged the originally created distinct species" (10). Of further interest is that Charles Lyell (1832), as we saw earlier, stated the issue of species reality as depending on their fixity. Beatty (1985) points out that "Basically the same choice was offered by other non-evolutionists of the time" (270). Indeed Darwin added to the frontispiece of the second edition of the *Origin* a quotation from Joseph Butler's *Analogy of Religion*, published in 1736, according to which "The only distinct meaning of the word 'natural' is *stated, fixed,* or *settled*" (cf. Burkhardt *et al.* 1993: 334n11). This prevalent view on kinds, combined alone with what Wallace says above, forces the conclusion (for those who accept the premises) that what are called varieties are ultimately unreal or arbitrary. At least Agassiz, for one, for whom species have "essential features" and "the specific characters are forever fixed" (1860b: 150), made it explicit that "varieties, properly so called, have no existence, at least in the animal kingdom" (1860a: 410).

When Darwin, then, says that "all naturalists have admitted the existence of varieties," either he was speaking loosely or he deliberately misrepresented the facts. I believe it was the latter. But before giving my reasons why, it should be pointed out that many or most biologists today likewise believe (though generally for different reasons) that subspecific designations are arbitrary. Gould (1977), for example, points out that "Geographic variability, not race, is self-evident" (232) and concludes that "The subspecies is a category of convenience" (233), his main reason being that "Almost all the proponents of multivariate analysis have declined to name subspecies. You cannot map a continuous distribution [of the morphological characters of a species] if all specimens must first be allocated to discrete subdivisions" (234–235). Likewise Wilson (1992) tells us that "The uncertainty of the limits of populations combined with the discordance of traits [i.e., 'They change at different places and in different directions' (66)] means that the subspecies is an arbitrary unit of classification" (67; cf. Mayr 1982: 268). What is of particular interest is that Wilson sees a profound dilemma in all of this for evolutionary reasoning: "The dilemma is this: if subspecies are usually amorphous and cannot be defined by any single objective criterion, how can such an arbitrary unit give rise to the species, which is sharply defined and objective?" (68). I suggest that Darwin faced much the same dilemma. "What counts," says Wilson in answer to the dilemma, "is that somehow a group of individuals occupying some part of the total range evolves a different sex attractant, nuptial dance, mating season, or any other hereditary trait ['innate isolating mechanism'] that prevents them from freely interbreeding with other populations. When that happens, a new species is born" (68).

Could this solution also fit with Darwin's writings? The answer must be in the negative, not only, as we saw above with dogs and cattle, because he allowed for interspecific fertility, but also because he allowed for intraspecific sterility (cf. Darwin 1859: 269–272; Barrett 1977:II: 61, 128, 220). Indeed, as we shall see,

Darwin saw it as crucial for his theory that he argue against the received view that interspecific sterility is a special endowment of species and that he argue for his own view that sterility is merely "incidental on other acquired differences." Nevertheless Darwin faced a dilemma somewhat analogous to the one expressed by Wilson. Since for Darwin varieties are what he calls "incipient species," they need to be in some way objectively real. What he needed, then, is to argue not only that what are thought to be specific characters are indefinitely variable, but also that what are thought to be accidental characters (or at least many of them) have the potential for becoming relatively permanent, in other words that the heritable characters with which naturalists sometimes designate varieties (cf. Darwin 1859: 44–45) are not only indefinitely mutable but also indefinitely heritable (cf. Darwin 1859: 12), so that they are just as capable of becoming fixed as what naturalists designate as specific characters. And this indeed is the emphasis that one finds in Darwin's mature writings. For example, in the *Origin* (1859) he says "Those forms which possess in some considerable degree the character of species, but which are so closely similar to some other forms, or are so closely linked to them by intermediate gradations, that naturalists do not like to rank them as distinct species, are in several respects the most important for us. We have every reason to believe that many of these doubtful and closely-allied forms have permanently retained their characters in their own country for a long time; for as long, as far as we know, as have good and true species" (47; cf. Wallace 1858: 10–11). This was, significantly, Darwin's view for primroses and cowslips, thought by most naturalists to be conspecific varieties because of their intermediate gradations, but thought by Darwin to be "worthy of specific names" (485) largely because of their different adaptive characters (49–50), and in spite of his belief that there is "an overwhelming amount of experimental evidence, showing that they descend from common parents" (50). Indeed, on Darwin's view "species of all kinds are only well-marked and permanent varieties" (133), or to vary the expression "only strongly marked and fixed varieties" (155).

This, of course, makes Darwin's concept of variety quite different from that of the traditional and prevalent view of his contemporaries. And yet in spite of his being perfectly aware of this, he takes in the *Origin* the varieties in practice of his opponents, which are real in practice though not in theory, ignores their theory, and strongly implies if not outright states that for all naturalists varieties are objectively real. In this way the varieties in practice of the naturalists become in his one long argument the objectively real varieties that his theory requires for the role of incipient species via natural selection. When Darwin, then, says that "all naturalists have admitted the existence of varieties," I suggest that his words should be viewed as being little more than part of a clever argument using equivocation.

Assuming, then, that Darwin employed some sort of strategy beneath his overtly nominalistic definitions of species and varieties, and that he thought of varieties in a way quite different from that of other naturalists, it remains to be seen what he really thought about species. And indeed to answer that question is to jump onto center stage of the modern species debate. For we find in Darwin a distinction that has often been employed in modern writings, though usually without credit to Darwin, a distinction without which the question of the reality or unreality of species must remain utterly ambiguous. The distinction I am referring to is the distinction between *horizontal* species and *vertical* species. To be a true species nominalist, one must deny the reality of both. On the other hand, if one accepts either one as a real category (with members), one must then be a species realist.

What is the difference between a horizontal and a vertical species? The distinction borrows from the now entrenched metaphor of the Tree of Life, as inaugurated by Darwin (1859: 129–130). Horizontal species result from cutting through the Tree horizontally at any one level. Horizontal species, then, are the species that exist at any one time. By "any one time" virtually no one who ever defends a horizontal species concept means a single instant or a microsecond. What is meant is a *period* of time. This might sound arbitrary, and to some extent it is, but what is almost always meant is a period of time *less* than what would be required for gradual speciation to be completed (cf. §4n20). For my own part, the minimum time required would be the period of time that is necessary for the relations to obtain that delimit species, and the maximum time allowable would be the period of time that still allows for the species-membership relation to be transitive (i.e., if organism *a* is conspecific with organism *b* and organism *b* is conspecific with organism *c*, then *a* and *c* must also be conspecific).

Although I will not go so far and agree with Mayr (1982) that multidimensional species[15] are "in most cases biologically uninteresting, if not altogether irrelevant" (286), it is clear that horizontal species play the dominant role in biology today, for example in population genetics, in biodiversity studies, including food chains and food webs, in biological control, in ethology, and in medicine.

Turning to vertical species, they are, of course, species looked at vertically,

---

[15] Mayr's biological species concept is according to Mayr a purely horizontal concept, what Mayr also describes as a "nondimensional" concept: "The biological species concept, expressing a relation among populations, is meaningful and truly applicable only in the nondimensional situation. It can be extended to multidimensional situations only by inference" (Mayr 1982: 273). By "multidimensional" Mayr means populations compared at different times and different places. Whether members from two different populations completely separated in either or both space and time can fertively interbreed can never be determined by observation but can only be inferred. Such inferences result in a multidimensional species concept. However, since horizontal species refer to conspecific populations that may be separated geographically but not temporally, Mayr's nondimensional/ multidimensional distinction, contrary to what he thinks (cf. Mayr 1988a: 314), does not perfectly correspond to the horizontal/vertical distinction. At any rate, that the horizontal/ vertical distinction is pervasive in the modern literature on the species problem cannot reasonably be denied. Simpson (1961: 164–166), for example, uses the distinction to distinguish between his and Mayr's species concepts, while Mayr (1988a: 314) does the same in return. Sometimes, as with Salthe (1985: 225–226), Endler (1989: 627–628), and Ridley (1993: 399), the horizontal/vertical distinction is used to categorize most if not all modern species concepts. (Cf. Stamos 2002 for further discussion.)

employing the full dimension of time. For a particular species to be real vertically, it must have a temporal beginning and end (if only roughly). To return to the metaphor of the Tree of Life, a vertical species would be either a branch (of some sort) or a differentiated segment of a branch. There is, of course, a degree of arbitrariness here as well. Shall any twig, for example, be a species, or may twigs be conspecific? Since species produce varieties, there must be twigs, but when do we count them as species? Or in other words, what kinds of branching points are to count as speciation events? Moreover, what about nonbranching speciation, such as the *Homo* line?

Although vertical species seem to be more problematic than horizontal species, it must be kept in mind that whatever the problems are, they do not affect the reality or unreality of horizontal species. To illustrate this, we need only borrow and expand upon what proves to be one of Darwin's most appropriate analogies, namely his language analogy (cf. Darwin 1859: 422–423, Notebook N: 65, and chapter 4.4). English, for example, is no less real if it should prove impossible to say when it began, and it would be no less real if in the future it should prove impossible to say when it came to an end (cf. chapter 4.4n47 and accompanying text for a full discussion).

Indeed many biologists have thought and continue to think that species are just as real vertically as they are horizontally. A prominent example is Simpson (1961: 152–153), himself a paleontologist, according to whom species must be vertically real in order to be evolutionary. And then we have the cladists and the punctuationists. But as these are topics that are best discussed elsewhere, I will reserve them for later.

Now what about Darwin? I do not think it can be doubted that Darwin was a species realist at least when it came to horizontal species. He says, for example, "To sum up, I believe that species come to be tolerably well-defined objects, and do not at any one period present an inextricable chaos of varying and intermediate links" (177). Now certainly it is true that in the *Origin* Darwin often employed (in part) an overtly morphological species concept, which may easily be thought arbitrary and subjective. He says, for example, that as a consequence of his view "Systematists will have only to decide (not that this will be easy) whether any form be sufficiently constant and distinct from other forms, to be capable of definition; and if definable, whether the differences be sufficiently important to deserve a specific name" (484). Elsewhere he says that "varieties, when rendered very distinct from each other, take the rank of species" (114). Again, he says:

> Hence, in determining whether a form should be ranked as a species or a variety, the opinion of naturalists having sound judgment and wide experience seems the only guide to follow. We must, however, in many cases, decide by a majority of naturalists, for few well-marked and well-known varieties can be named which have not been ranked as species by at least some competent judges. [47]

In the same manner Darwin often uses the phrase "good and distinct species" (cf. 61 and 259), which should indicate that when he speaks of "well-defined" species

(171, 174, as well as above), he is not thinking of *definition* in the Aristotelian sense but is rather referring to the fact that the descriptiveness and status of such species *qua* species is clear and uncontroversial, a usage that has remained to the present day (cf. Mayr and Short 1970: 1).

All of this, of course, can be explained using some sort of strategy theory. But even without a strategy theory, what we saw earlier shows that Darwin employed other relations to bound the similarity relations, most notably common descent. But I think it becomes most obvious that Darwin was not a nominalist with regard to horizontal species (thinking that only individual organisms are real) once it is observed that he provides evidential criteria for the objective existence of horizontal species. One of these is the existence of sterility barriers. In spite of his evidence and arguments against the view that species are specially endowed with sterility barriers and that these are lacking between conspecific varieties—characterized by his injunction against *a priori* assumptions in favor of experiments: "No one can tell, till he tries" (265)—Darwin admits (what he need only admit, and certainly what no nominalist would admit) "that this is almost invariably the case" (268). It is the *exceptions* that make his case, which themselves prove that both interspecific sterility and intraspecific fertility "cannot, under our present state of knowledge, be considered as absolutely universal" (254). It is the evidence that these two criteria are *almost universal* that allows Darwin to steer a course right down the middle between species essentialism on the one hand and species nominalism on the other; in other words they provide evidence for both the existence of objectively real species and nominal species (the latter being messy situations in the horizontal dimension), all the while allowing for the indefinite mutability of real species, all of which are required by Darwin's theory of evolution.

None of this, of course, is meant to imply that for Darwin interspecific sterility is constitutive of species; there is too much evidence against such an interpretation (cf. Stamos 1999 for a full discussion). Nevertheless, I do not think that it can be doubted that Darwin uses it as preliminary evidence of species distinctness, in much the same way that, for example, Mayr and Ghiselin (as we shall see in chapter 4) use dissimilarity as evidence of species distinctness, although neither similarity nor dissimilarity are constitutive of species on their respective views.

Interestingly there is another evidential criterion that Darwin gives for the objective reality of horizontal species. Instead of reproductive relations, this one has to do with competitive relations. According to Darwin, "the struggle almost invariably will be most severe between the individuals of the same species, for they frequent the same districts, require the same food, and are exposed to the same dangers" (75). Of course, whether Darwin is right about intraspecific competition is beside the point.[16] The point we must notice is that no true species

---

[16] I myself am skeptical, but am of course incompetent to pass judgment. I think of, for example, members of *Pongo pygmaeus* (orangutans), which in the wild live an almost exclusively solitary existence. It seems evident that their competition for environmental

nominalist would assent to it, since it provides an objective criterion for inclusive entities higher than that of the individual organism.

To add to all of this, Darwin calls it "a great truth" in one of his published articles (May 9, 1863), irrespective of mechanisms, "that species have descended from other species and have not been created immutable" (Barrett 1977:II: 81). Although it is not clear from this passage whether Darwin meant horizontal or vertical species, the former would seem the more natural interpretation. The same is true for another important passage (of course outside of the *Origin*) where Darwin affirms the reality of species, namely his reply to the legitimate quip of Agassiz that we examined earlier. In one of his letters to Asa Gray (August 11, 1860), Darwin wrote "How absurd that logical quibble;—'if species do not exist how can they vary?' As if anyone doubted their *temporary* existence" (Burkhardt *et al*. 1993: 317; italics mine).

What then of the reality of vertical species? Certainly it was a common view by Darwin's time that species had a time of origin and that, following Cuvier's lead, the geological record showed that many have gone extinct. But it was also a common view in Darwin's day, among both philosophers and scientists, that if species evolve they cannot be real. For example, the Cambridge philosopher of science William Whewell (1837:III) gave voice to this view when he wrote "*species have a real existence in nature*, and a transmutation from one to another does not exist" (576). In this he seems to have been echoing Lyell (1832), who, as we have seen earlier, stated the issue of species reality as depending on their fixity, as "whether species have a real and permanent existence in nature; or whether . . . ." (1; cf. 65). On this matter Darwin's position in the *Origin* is not quite so clear. On the one hand, he not only clearly subscribed to but greatly furthered the historical version of the Great Chain of Being. He says, for instance, "looking not to any one time, but to all time, if my theory be true, numberless intermediate varieties, linking most closely all the species of the same group together, must assuredly have existed" (179). Moreover, as we have seen in the Introduction to the present chapter, he thought of the history of life on earth as "the great Tree of Life" (130), comprising "one long and branching chain of life" (301) with "an infinitude of connecting links" (463). Again, he often speaks of "the scale of nature" (cf. 88, 168, 208, 313, 345, 388, 406, and 468) and often employs the Linnean dictum "Natura non facit saltum" (cf. 194, 206, 210, 454, 460, and 471).

But why would Darwin maintain this view and repeatedly emphasize gradual-

---

resources comes mainly from man and not from each other. But this may indeed be an exception. Darwin might be right as a rule—his "almost invariably." Against my view, it seems almost a truism in ecology that species occupy niches and that if two species occupy the same niche, one species will eventually cause the extinction of the other. Known as the *competitive exclusion principle* (or Gause's axiom), it would seem to follow that "niche differentiation should reduce competition between species relative to that within species" (Law and Watkinson 1989: 273). However, as Law and Watkinson point out on the same page, there is almost no data to support this or any conclusion on the relative strengths of the two types of competition.

ness throughout the *Origin*? He was, of course, aware of monstrosities (cf. 44), and one could easily argue that in sexual species (which constitute the majority of species) evolution by sudden jumps would seem highly unlikely. If the monstrosity would be fertile (and most are not) but reproductively isolated from its parental species, it would require a suitable monstrosity of the opposite sex in order to breed. And even if, however improbable, one were found, the result would be subject to extreme inbreeding, and almost certainly would quickly die out (cf. 248–249). On the other hand, if the monstrosity would be interfertile with its parental species, whatever of its novel characteristics that confer a selective advantage (however improbable) would have to make their way through the species, resulting in gradual rather than saltational evolution. But interestingly, Darwin did not make this rather obvious argument.

Today, of course, virtually all biologists admit the occurrence of what is often called "instantaneous speciation" (cf. Maynard Smith 1975: 267), sympatric speciation either by hybridization or (unknown in Darwin's time) polyploidy. Polyploidy is a chromosomal accident resulting in at least three times the haploid (half) somatic chromosome number of the parent(s). (When only one parental species is involved it is called *autopolyploidy*, when the polyploidy is the result of hybridization between two parental species it is called *allopolyploidy*, although this distinction is by no means clean.) Although relatively quite rare in animals, instantaneous speciation is not uncommon in plants. Indeed at least one-third of all plant species have arisen by polyploidy speciation (Mayr 1970: 254). With sexual species, at any rate, the problem of inbreeding remains.

Darwin, of course, did not have the benefit of modern genetics. Instead, true to his Paleyan roots his fundamental reason for rejecting saltational evolution was because it failed to explain biological adaptations. Indeed, as Darwin wrote in a letter to Hooker (July 2, 1859), in criticism of a paper on evolution by Samuel Haldeman, "it did not attempt, I believe, to explain *adaptations* & this point has always seemed to me the turning point of the theory of Natural Selection" (Burkhardt and Smith 1991: 316). With such words by Darwin, of course, one naturally thinks of the beaks of the Galapagos finches, for example, or of Darwin's extended attempt in the *Origin* to explain the gradual evolution by natural selection of the eye (cf. 186–188). But Darwin was also strongly impressed by another kind of adaptation, namely the many delicate and intricate symbiotic relationships "which we see everywhere throughout nature" (132), what he called "coadaptations." Saltational evolution seemed to him to offer no natural explanation of these relationships:

> The author of the 'Vestiges of Creation' [Robert Chambers] would, I presume, say that, after a certain unknown number of generations, some bird had given birth to a woodpecker, and some plant to the misseltoe, and that these had been produced perfect as we now see them; but this assumption seems to me to be no explanation, for it leaves the case of the coadaptations of organic beings to each other and to their physical conditions of life, untouched and unexplained. [3–4; cf. 60 and Burkhardt *et al.* 1993: 574]

Moreover, Darwin was intent on providing a completely natural explanation of evolution. Because of the extreme improbability of adaptive evolution by saltation, a theory of such may have seemed to him as being no better than, or indeed as being a mere variant of, explanation by miraculous intervention, namely the view that "at innumerable periods in earth's history certain elemental atoms have been commanded suddenly to flash into living tissues" (483), a view that Darwin found not a little difficult to believe for reasons that we have seen earlier.

At any rate, does Darwin's version of a historical Great Chain of Being necessarily make him a species nominalist as regards vertical species? With some hesitation, I suggest not. To begin, there is Darwin's claim that "all true classification is genealogical; [and] that community of descent is the hidden bond which naturalists have been unconsciously seeking" (420). For a true species nominalist, on the other hand, there can be no such thing as a true classification of species, genealogical or otherwise, since there are no species in reality to classify.

In addition, Darwin, unlike Lamarck, did not think that all extinction is pseudo-extinction (the gradual evolution of one species into another), but that extinction is often real extinction. For instance, he tells us that as a consequence of natural selection (cf. 317) "those [species] which do not change will become extinct" (315) and that "Rarity, as geology tells us, is the precursor to extinction" (109). No true species nominalist, of course, would ever assent to these propositions. To echo Agassiz, if species do not exist at all, how can they go extinct?

Moreover, Darwin not only clearly subscribed to real extinction, but also to the view that extinction is probably forever. Anticipating what is today known as Dollo's Law, he concludes that "When a group has once wholly disappeared, it does not reappear; for the link of generation has been broken" (344). In addition to this he clearly thinks that even if a species should come to inhabit numerically the same niche of an extinct species, it would nevertheless be a different species, since it brings with it different ancestral characters (cf. 315).

Where the real difficulty arises is with speciation. According to the influential theory of Mayr (1982: 410–417), although Darwin was an early adherent of speciation by geographical isolation, by the time he wrote the *Origin* he no longer had a clear theory on speciation, so that his book was misnamed. A more likely scenario, it seems to me, is that Darwin evolved into a limited speciation pluralist (limited by its confinement to gradualism), accepting both nonbranching speciation or anagenesis (cf. 119–120) and branching speciation or cladogenesis, the latter including both sympatric speciation (in accordance with his Principle of Divergence; cf. 105–106, 112, 115–116) and allopatric speciation (cf. 399–404). Of course, that he considered sympatric speciation the most important (cf. 105) does not take away from this pluralism; it only increases his difference from modern speciation pluralists.

At any rate, given Darwin's insistence that all evolution must be extremely gradual (cf. 108) and that (almost) all species must evolve or face extinction (cf. 107 and 315), it might be argued that evolution and speciation on Darwin's view provides one with a continuum in which it is ultimately arbitrary where one chops it up and delimits vertical species.

What we have to remember, however, is that Darwin clearly thought of speciation, for the most part, as a branching pattern, as evidenced in his one and only diagram in the *Origin* (514–515), in his discussion of it in terms of species (116–125), and in his claim that "In a tree we can specify this or that branch, though at the actual fork the two unite and blend together" (432). And far from contending that all evolution must proceed at a constant rate, Darwin did accommodate the possibility of various rates of evolution, albeit very slow, including extremely slow rates and possibly even stasis—he did, after all, coin the term "living fossils" (107), possibly represented as species F in his diagram.[17]

In sum, it seems clear that Darwin, explicit claims to the contrary, thought that species taxa are objectively real, at least horizontally, and possibly in some cases also vertically. Moreover, in spite of his claim that "we shall at least be freed from the vain search for the undiscovered and undiscoverable essence of the term species" (485), there are compelling reasons for believing (contra Ghiselin and Beatty) that Darwin did not really think that evolution by natural selection renders the species category unreal. For one thing, like many modern species category realists today, Ghiselin included (cf. Ghiselin 1989, 1997: 127), Darwin held that there are laws of nature not for any particular species taxon but for species taxa *per se*. We have already seen that he held an early version of Dollo's Law, which stresses the extremely high statistical improbability of evolution ever precisely reversing or repeating itself. To this we could add his anticipation of what Van Valen (1973) has dubbed the Red Queen hypothesis, according to which a species must continually evolve just to keep its present position in the economy of nature and avoid extinction, because the species in its environment are continually evolving due to competition. As Darwin (1859) put it, "Hence we can see why all the species in the same region do at last, if we look to wide enough intervals of time, become modified; for those which do not change will become extinct" (315), which is clearly intended to be a statistical law given the anomalous forms he calls "living fossils" (107). But undoubtedly the most important law for species *per se*, according to Darwin, is that the main mechanism of their modification is natural selection. Although in the *Origin* (1859) Darwin only implies that natural selection is one of the "laws acting around us" (489–490; cf. 6 and 244), he repeatedly stressed in his correspondence that natural selection is a bona fide law of nature, a *vera causa*. For example, in a letter to Charles Bunbury (February 9, 1860) he wrote: "With respect to Nat. Selection not being a 'vera

---

[17] It is interesting to note that F^{14}—descended from species F 14,000 generations later, or better yet 114,000 generations later (cf. 117), "either unaltered or altered only in a slight degree" (124)—is twice called by Darwin a "new species" (124). Clearly, therefore, Darwin's species concept in this part of his discussion is horizontal and not vertical. Nevertheless, although this indicates that for Darwin the horizontal dimension has priority over the vertical dimension, it should not be taken to necessarily preclude the vertical dimension, as indicated by what he says about living fossils, *viz.*, that "These anomalous forms . . . have endured to the present day" (107). Moreover, it should be noticed that the "F" in F^{14} is capitalized, whereas none of the other forms on that line are capitalized, which suggests that F^{14} alone may be thought of as a vertical species with F (cf. also E and E^{10}).

causa'; . . . Natural selection seems to me in so far in itself not be quite hypothetical, in as much if there be variability & a struggle for life, I cannot see how it can
fail to come into play to some extent" (Burkhardt *et al.* 1993: 76). Indeed given
the prime importance that Darwin gave to natural selection for biological evolution, it would be extremely odd if his mature species concept, accepting now that
it is a realist concept, would not somehow rest primarily on his theory of natural
selection.

In addition to the above evidence that Darwin was not only a species taxa realist but also a species category realist, what needs to be kept in mind more specifically is that Darwin consistently and uniformly employed the same grouping
criteria (which inveighs against the rejoinder that he was a species category pluralist). We have seen this consistency and uniformity not only in the priority he
gives to the horizontal dimension but also in his use of common descent (as in the
case of domestic dogs and cattle) and his view on convergence, as well as his
view on intermediate forms. We have also seen, however, where the criteria of
common descent and intermediate forms get overweighed, as in the case of primroses and cowslips, and that the most fundamental criterion for ranking a group as
a species is distinct physiological characters of adaptive significance from the
viewpoint of natural selection (in contrast, interestingly, to the use of vestigial
characters for higher taxa; cf. Darwin 1859: 416, 425–426, 439–440, 449), which
accounts for why he did not include sterility as a species character and for why he
considered species to be nothing more than well-marked or relatively permanent
varieties. Incidentally, that varieties, on Darwin's view, eventually evolve into
species (providing they do not first go extinct) does not logically preclude the
above interpretation, any more than dusk and dawn invalidate the categories of
night and day. As we have seen earlier, Darwin's concept of definition allowed
him to hold that in most cases individual (horizontal) species taxa are real and are
definable, in spite of not having essences. And there is ultimately no good reason
to suppose, providing the premise that Darwin employed some sort of strategy
theory, that his nonessentialistic concept of definition for species taxa would not
also for him apply to the species category. Accordingly, I would say that on Darwin's mature view *a species is a primarily horizontal similarity complex of organisms distinguished by common descent (albeit not absolute), intermediate
gradations (when applicable), but most importantly relatively constant and distinguishable characters of adaptive significance from the viewpoint of natural
selection* (cf. Stamos 1999 for a much fuller defense of this reconstruction).

2.4 Modern Nominalists in Biology
In the final section of this chapter, I want to explore the various motivations, both
empirical and theoretical, of modern biologists who are skeptical of the objective
reality of species. One could also explore modern philosophical objections. In
this I will confine myself only to the philosophical objections that have actually
been employed by biologists, their employment being most important since biologists, after all, are those most intimately acquainted with the phenomena and
data upon which species concepts are based. The botanist J.S.L. Gilmour, for

example, is a good starting point, since he has attempted to import into biological theorizing the sense-data philosophy of the Logical Positivists. According to Gilmour (1940), only sense-data "are the real objective material of classification" (466). Everything else is a rational construction "depending on the *purpose* of the classifier" (465). Thus, "The classifier experiences a vast number of sense-data which he clips together into classes" (465), a natural group using many attributes in common, an artificial group using only a few, the former being "no more real than an artificial one; both are concepts based on experienced data" (467). Thus,

> A species is a group of individuals which, in the sum total of their attributes, re-semble each other to a degree usually accepted as specific, the exact degree be-ing ultimately determined by the more or less arbitrary judgment of taxonomists. [468–469]

The base of such philosophical objections, of course, has been the subject of volumes of philosophical debate, and it would take me too far afield to attempt a just treatment of the issue here. Suffice it to say that what seems most striking about Gilmour's approach is the undesirable consequences it entails. Not only, as he implies (464), are the "clips" based on Linnaean special creation on a par with the "clips" based on Darwinian evolution, but each individual organism, he ex-plicitly states, is likewise "a concept, a rational construction from sense-data" (466). From there it is but a small slide to the view (which Gilmour nowhere dis-cusses) that the external world, including other minds, is but a rational construc-tion and that all that is really real is my mind and my ideas (solipsism). Logically irrefutable, but if I may use the language of David Hume, such Pyrrhonian mus-ings can only be maintained at most for a few short hours while one is alone in one's study. Apart from that, one is dominated by what Hume scholars call *natu-ral beliefs*, beliefs so strong that in everyday life they force one into assuming a material, objective world with direct perceivables. It is in that world that biolo-gists *qua* biologists operate, and it is only with the assumption of that world that I am really interested in the species skepticisms of certain biologists.

The more common philosophical objections one finds among biologists skep-tical of species reality are to be found in the writings of Benjamin Burma. Ac-cording to Burma (1954), "that only is real which possesses extension in space-time" (196). On this view it follows that all abstractions, classes included, "have no real existence. They are . . . mental constructs, and as such lack actuality in the sense here defined" (205). He later tells us that "Whatever definition one may wish to use, species will be a defined class of some sort, and as such may be said, no more and no less, to exist in precisely the same sense that the class of unicorns may be said to exist" (205).

To Burma's concept of what is real, I would reply with what was said in chapter 1.2 concerning abstract facts such as the law of gravity and, more simply, the average length of the necks of a group of giraffes. All of this is taken up in more detail in chapter 3, especially in §5. I would further add that his concept of reality denies any objective reality to relations, creating for his view the same problems that beset the views of Occam and Locke as discussed in §2 (and further discussed in chapter 5.3). Suffice it to say that Burma's concept of reality is much too narrow, that it does not stand the test of close scrutiny, and that it is in conflict with much of actual science. But even if in spite of this we were to accept his narrow ontology, it does not necessarily follow that species must be classes, thereby making them unreal. As we shall see in chapter 4, species may be conceived as *physical* entities of a sort, which would circumvent his whole nominalistic approach.

So much for fundamentally philosophical objections. When we turn to fundamentally biological objections, we find, interestingly, that most biologists who doubt the objective reality of species come from a background either in botany or paleontology. But now one might find a fundamental disagreement over the proportion of modern "believers" versus "nonbelievers." According to Ernst Mayr (1988a), "Modern biologists are almost unanimously agreed that there are real discontinuities in organic nature, which delimit natural entities that are designated as species" (331). Similarly, according to Alec Panchen (1992), "Nominalism is a possible stance, but one that probably has no extant practitioners" (130). According to Niles Eldredge (1985a), on the other hand, "Indeed, at any moment, seemingly at least half the world's evolutionary biologists are perfectly prepared to deny that species—*any species*—even exist" (98).

How is it possible that these eminent authorities in their field could so fundamentally disagree on what is surely a relatively simple and straightforward matter of fact? The answer is not contextual, for I have not misleadingly quoted out of context. Instead the answer has to do with an often overlooked ambiguity underlying the word "species." The most famous ambiguity with this word, of course, is that between species as a category and species as a taxon, as we first saw in chapter 1.3. But what needs to be equally recognized is that the word "species" is also ambiguous in quite a different way, a way that is usually overlooked but which is just as important, namely (as we saw in the previous section) the distinction between horizontal and vertical species. What Mayr and Panchen say above is true if by "species" they mean horizontal species, whereas what Eldredge says above is true if by "species" he means vertical species. In no other way can the contradiction between them be resolved.

The obvious lesson is that any discussion on whether species are real, particularly from an evolutionary point of view, cannot afford to overlook the horizontal/vertical distinction, but must instead examine the issue from the perspective of both dimensions. To begin with the vertical dimension, there are basically two kinds of phenomena that have made many paleontologists (and other biologists) skeptical. The one type is perhaps best discussed by Alan Shaw (1969) in his presidential address to The Paleontological Society. Using a real example of "four critical characteristics of some upper Devonian platform conodonts" (1086), Shaw demonstrates in great detail that "the personal objectives of the individual paleontologist so completely control the manner in which he uses lower level taxonomy and binomial nomenclature that the system is wholly subjective and, therefore, incapable of serving as a means of communicating accurate paleontological information at the professional level" (1085). Shaw's example is too detailed to adequately describe here, but I have to say it is very convincing, at least for conodonts.[18] Suffice it to say that, of the four critical characters that Shaw examines, only six combinations of the variants of those characters are actually found in the strata, all of which is undisputed by paleontologists (1087). The difficulty with calling each of these six combinations a *species* is that, although each of these characters evolves, all are worldwide, most have multiple origins (polyphyly), many display convergence (comparing both Australia and Germany), and only some evolve together, so that how one designates a species lineage is ultimately subjective. What further complicates the matter is differing theories over modes of speciation (1095). Those taxonomists who believe that species replace one another will produce a different taxonomy from those who believe that ancestral forms may survive their descendant forms. Thus on Shaw's view, "Its non-objectivity makes the species concept useless for the purpose for which it was primarily intended—as a communications medium" (1096). Accordingly, Shaw's rallying cry is "HELP STAMP OUT SPECIES!" (1098). It is important to note, however, that Shaw is not concerned with the horizontal dimension (on that he is basically silent); he is concerned only with the vertical dimension (1085). Rejecting what he calls "synthetic" paleontology, paleontology which uses species concepts, Shaw favors what he calls "analytic" paleontology: "the detailed study of the exact form of each morphologic unit, its precise stratigraphic distribution, and the observed combinations with other morphologic units." He continues: "The minute we attempt to synthesize these groups of characters into species communication collapses, clarity vanishes, and all of the chaos we have just reviewed is upon us once again. *It is the synthesis that is the disastrous act.* The species concept is the *source* of the trouble" (1096).

I naively used to think that if biological evolution is a fact then there must be species to be the units of evolution. Shaw clearly refutes this. On Shaw's view, the units of evolution are not species but morphological characters. And indeed one often finds in biological literature a discussion of characters as units of evolution. Darwin himself (1859), for example, discusses the eye as a unit of evolution (186–189). Unlike most traditional taxa, of course, both species and higher, morphological characters often resist any criterion of monophyly (roughly, single origin). As Mayr (1988a) points out in the case of eyes, "eyes evolved in the animal kingdom at least 40 times independently" (409). On Shaw's view, then, evolution, with morphological characters and not species as the units, requires neither species nor monophyly.

Not every "collection of characters," of course, as Shaw (1096) prefers to call them, provides the same difficulties as Shaw's conodonts. Shaw tells us that

---

[18] "They are toothlike fossils found in black, silty shale deposited under anaerobic conditions from the late Cambrian to late Triassic, from about 580 to 200 million years ago. It is surmised that conodonts were used as 'teeth' to tear apart fungal hyphae and algal mats. . . . In the early 1980s conodont animals were assigned a (fossil) phylum of their own, Conodonta" (Margulis and Schwartz 1988: 192).

"What this little group of fossils shows in its most elementary form is mosaic evolution" (1090). Mosaic evolution simply refers to the phenomenon of different rates of evolution within a lineage or group, such that some characters evolve rapidly and others evolve little or do not change at all, all of which is a consequence of relative differences in the selection pressures. Hominid evolution is a classic example (cf. Mayr 1970: 381, 1982: 621–622). And as Futuyma (1986) states in general, "Mosaicism of evolution is the rule rather than the exception" (293). Where Shaw is highly misleading is in his suggestion that mosaicism in conodonts entails the same skepticism about species in other mosaicisms. It most certainly does not. As Simpson (1961) points out,

> conodonts, extremely useful horizon-markers, occurred in the whole animals as complex assemblages of markedly different small, hard plates, spines, combs, etc. Few assemblages have been found, and for the most part classification of conodonts has to be based on isolated parts of unknown relationship to each other. The inevitable result, as substantiated in a few cases by natural assemblages, is that different "species" and "genera" are routinely based on what were parts of one animal. [157]

So Shaw has not exactly based his generalizations on a representative example. In many other phyla, collections of characters, although they display a mosaic pattern, still nevertheless strongly suggest *species* evolution. Fossil horse species are perhaps *the* classic example, and it should be noted that they've been studied in more detail than any other group (cf. Maynard Smith 1975: 285–290). In spite of much mosaicism and a bush pattern of evolution stemming from one genus, *Hyracotherium*, no one ever seems to have doubted that there were species of horses which evolved and which left us with our present two species of ("true") horse, *Equus przewalski* and *E. caballus*. In other words, mosaicism does not of itself legitimize species (vertical) skepticism.

But now another problem arises. For a vertical species to be objectively real, it must have at least a vague beginning in space and time as well as (if it is not a living species) a vague ending in space and time. But it is here that many paleontologists (and other biologists) become skeptical. According to Burma (1949), for instance, himself a paleontologist, "a species, be it plant or animal, is a fiction, a mental construct without objective existence. Animal, and plant, lines of descent exist in a four-dimensional continuum. To set up species in this continuous line of descent, we must chop it into units, and in any such process the divisions are purely arbitrary" (369). Similar statements can also be found not only in Darwin, as we have seen, but in contemporary biologists following Burma such as A.J. Cain. According to Cain (1954), "forms that seem to be good species at any one time may become indefinable since they are successive stages in a single evolutionary line and intergrade smoothly with each other" (107). Similarly J.B.S. Haldane (1956): "Thus in a complete palaeontology all taxonomic distinctions would be as arbitrary as the division of a road by milestones. . . . the concept of a species is a concession to our linguistic habits and neurological mechanisms" (96). Similarly Arthur Cronquist (1978): "In such a successional line [the *Homo* line] there are no breaks, and the conceptually necessary organization of the phylad into separate species is purely arbitrary" (4). Similarly Olivier Rieppel (1986): "Under the aspect of continuity the species cannot be objectified except by the arbitrary delineation of some segment of the genealogical nexus. . . . Continuity dictates a nominalistic view of species—it emphasizes process, thus rendering pattern a matter of arbitrary lines of demarcation" (312–313). Even G.G. Simpson (1961), who argued vigorously for various rates of evolution, including what he termed *quantum evolution*, and who developed what he called the *evolutionary species concept*, had to concede that vertical species are usually arbitrary (an extended discussion on Simpson's species concept will be found in chapter 3.4). In all of this, of course, either the theory of punctuated equilibria proffered by Eldredge and Gould (1972), to the degree that it is true, or the cladistic view taken by Hennig (1966) and his followers would negate the charge of arbitrariness in delimiting vertical species, but both suffer themselves from being highly controversial (cf. chapter 4.3/5).

While not everyone who believes that vertical species are arbitrary accordingly argues that species *per se* are arbitrary (Burma, Haldane, Simpson, and Rieppel are guilty of this), I want to suggest that it does not really matter concerning the question of the objective reality of species. As I have argued in the previous section and elsewhere (Stamos 1996a: 139, 1996b: 185, and fully in 1999), it seems to me that horizontal species are logically and therefore ontologically prior to vertical species. My reasoning is simple. The reality of vertical species necessarily entails the reality of horizontal species. But the converse is not also the case. To see this one need only think of languages. I will have much more to say about the analogy between languages and species in chapter 4.4. For the present, consider the fact that although languages evolve, a language such as English is no less real even if it gradually evolved into existence and will just as gradually evolve out of existence. Indeed it will be just as real even if all languages evolve at a perfectly constant rate; in other words, it will be real even if there are no vertically real languages. The same is no less true for species. It is important to point out here, however, especially in view of chapter 4, that this is not to claim that species cannot in some sense have a vertical reality, only that, as in the case of languages, their reality does not depend on them being vertically real. It is also important to point out that I would not go quite so far as Mayr (1987b) when he says "I have always been somewhat troubled by attempts to place the species concept too strictly on evolutionary theory. . . . I am sure that it is possible to demonstrate the reality of species as aspects of nature, without invoking the theory of evolution" (220).[19]

Some might nevertheless question the coherence of a horizontal species concept. According to Burma (1949), for example, the notion of species existing "*at any one time*" necessarily involves the notion of "a discrete unit of time," which

---

[19] Cf. Mayr (1988a): "in 1942 I rejected this inclusion ['evolving'] as obviously not being a defining characteristic of species" (321).

he says is "arbitrary, and perhaps not a little ridiculous" (370), his reason being that it gives us "an infinity of species, time being infinitely divisible." This is, of course, to suppose a non-Einsteinian, in a word a Newtonian, concept of time. If instead we follow the modern view and suppose that time is in some sense dependent or supervenient on events, so that if there are no events there is no time, then time is no more infinitely divisible than are events themselves (i.e., events divided, infinitely as well as finitely, no longer retain their identity), so that real units of time can be based on whole events. But aside from arguments over the nature of time, it does not follow that a horizontal unit of time for species is arbitrary. One could define both the lower and upper limits in terms of events or processes of great biological significance from the viewpoint of evolution. For the lower limit, one might follow Cain (1954) in what he calls "the *time-quantum of the biospecies*," by which he means "a breeding season of either or both the forms under investigation" (102). For the upper limit, one could define it hazily but nonarbitrarily as being slightly less than whatever the minimum amount of time it takes for gradual speciation to be completed. If the theory of punctuated equilibria is basically correct (and I am not committed to saying it is), then we are provided with a relatively shorter period of time than normally thought. According to Eldredge (1985a), "the comfortable yardstick" for the punctuation in evolution, that is to say for speciation, is "5,000 to 50,000 years" (121)—a rubber yardstick to be sure, varying moreover from species to species, but a basically nonarbitrary one nonetheless. Such a yardstick is not only useful, but it provides a safe maximum range (5,000 years) by which to define horizontal species.[20]

The question now becomes whether horizontal species are real. To this question some have attempted to raise doubts by appealing to what are often called "messy situations." These come in basically three kinds, namely *rassenkreise*, ring species, and hybridizations. *Rassenkreise* and ring species may be thought of as vertical continua laid flat. A *rassenkreis* (the term was coined by Bernard Rensch in 1929 and in German literally means "race circle") is usually defined as Simpson (1961) defines it, as "a genetical species with a series of intergrading but distinguishable local populations" (160; cf. Cain 1954: 69; Beckner 1959: 61), to which Mayr (1970) adds that it is "not 'a circle of races'" (423). A ring species, on the other hand, is simply a *rassenkreis* that goes around some sort of large natural barrier, such as a mountain, a desert, or a lake. Like a river or stream that splits and then later rejoins, a ring species will have a geographical zone of origin from which over time it has spread around a natural barrier eventually meeting again on the other side. But unlike a river or stream, the two chains do not rejoin

---

[20] The earliest precedent that I've been able to find for my view here is in Simpson (1961). In the context of what he calls "contemporaneous" organisms, Simpson clearly thinks of "no appreciable time dimension," what I call the horizontal dimension, as "any span of time . . . too short to have involved determinable somatic change" (163). By the way, another possible way to define the horizontal dimension, as I define it above, is to use not years but generations. I am indebted to Brock Fenton for this suggestion, although I'm not sure what this number would be—a matter, at any rate, for biologists to decide.

to form a unity. Maynard Smith (1975: 228–230) discusses the circumpolar chain of herring gulls whose termini meet in Britain where they behave as two distinct species, named accordingly *Larus argentatus* and *L. fuscus*. With a ring species, then, as well as a noncircular *rassenkreis* as defined above, the interbreeding relation between the constituent populations, unlike other species, is intransitive. Mayr (1970), who normally has little or nothing to say about ring species (but cf. Mayr 1942: 180–185), dutifully confirms that "A more dramatic demonstration of geographic speciation than cases of circular overlap cannot be imagined" (293), all the while overlooking the difficulties such species pose for his *biological species concept* (cf. chapter 4.1). Cain (1954), on the other hand, points out that "one must either recognize the specific status of the end forms by making an arbitrary break in the chain and deciding that all forms in one direction must be given one specific name, and all the rest another, or one can avoid this by not making a break, but then good species will have to bear the same specific name" (142). In the latter case, Evan Fales (1982) notes how odd it is that "just by killing certain organisms [intermediates], I can instantly bring into being a (two?) new species" (84). But this only points to the fact that horizontal species are the product of, among other factors, *death*. It is the process of extinction, as shown by Darwin to have a central role in the process of evolution, that has given us our species today. It is the union of both processes, then, that refutes the ahistorical Great Chain of Being. As such it is interesting to think of death as having a *creative* aspect. Indeed from a biological point of view, were it not for death none of us would be here. At any rate, what is most significant is that Haldane (1956), in a paper devoted to arguing that species are not real, not only cites the phenomena which I call vertical continua laid flat, but quickly adds (95) that they are the exception rather than the rule, which indeed they are.

Much more significant for the question of whether horizontal species are real is the degree of hybridization that goes on in nature. Hybrid zones, for example, are zones of overlap in the reproductive relations between otherwise distinct species. (I will have more on hybrid zones in chapter 3.3). Hybridization can also result in the creation of new species. Speciation by hybridization, along with polyploidy, is normally cited as *instantaneous speciation* (cf. Maynard Smith 1975: 267), but as such can hardly be used as evidence against the reality of species. When hybridization is so cited, it is usually the type of hybridization that seriously blurs species boundaries, in which case the examples are normally taken from plants, in particular flowering plants. Cronquist (1978), for example, claims that "the biological species concept, so-called, works rather badly for plants" (4). He later sums up by saying "Too often it gives results that fly in the face of common sense, or it leads into a swamp in which it seems impossible ever to acquire enough data to come up with a definitive treatment" (14). Cronquist therefore provides what he believes is the unspoken species concept of plant taxonomists:

Species are the smallest groups that are consistently and persistently distinct and distinguishable by ordinary means. [15]

Although a vertical species nominalist, as we saw earlier, Cronquist does not believe that the above definition commits him or anyone else to horizontal species nominalism. "At any given time," he says, "there are gaps in the pattern of diversity among organisms, and these gaps are of critical importance in the delimitation of species" (4). Moreover, he tells us that his paper is "concerned with the recognition and definition of species as they now exist, within a time-frame of a few hundred or at most a few thousand years" (4). But the species definition above, itself basically a morphological definition, does seem to make species somewhat observer-dependent and accordingly subjective. Cronquist does, after all, tell us that "There is just enough truth in the latter definition ['an old in-joke, that a good species is what a good taxonomist says it is'] to make it funny" (15). Moreover, later in his paper he characterizes species as "potentially satisfying mental organizations" (19).

But more to the point, the species definition above would not only expand the number of species (contra Occam's Razor) by dividing polytypic species into separate species in accordance with the subjective whims of "lumpers" versus "splitters" (reminiscent of right- versus left-brain dominance), but it would preclude not only the discovery but the very existence of sibling species, species which are by definition reproductively isolated but nevertheless initially (at least) indistinguishable (cf. chapter 1.1). Cronquist, however, has the curious view (8) that the more sibling species zoologists claim that there are, the more they will find this an embarrassment to their system, since the less they will conform to common sense. "If continuing studies of animals," he says, "turn up more and more sibling species to confound the field naturalist and confront common sense, there will eventually come a time when some Young Turk will say, 'We have had enough of this nonsense! There is no use calling things different species if you can't tell them apart'" (19). Cronquist accordingly suggests that the time may come when zoologists will revert to the taxonomic concept of botanists.

This view fails, it seems to me, on a number of counts. First, no zoologist that I have ever read has expressed any sign of uneasiness by the phenomenon of sibling species. On the contrary, as with the *Anopheles* example cited in chapter 1.1, they all (including Mayr) seem to welcome with open arms any such discoveries. Second, Cronquist's suggestion as to the possible proportion of sibling to nonsibling species seems to me quite exaggerated. In ornithology, for example, Mayr and Short (1970) conclude that "sibling species are remarkably rare in birds" (89). Mayr and Ashlock (1991: 92–93) provide an excellent summary of the proportion of sibling species currently supposed for various higher taxa, warning that exact percentages are for the most part still premature, but also estimating that percentages may go as high as 50% in North American crickets and possibly even a little higher in *Paramecium*. Third, Cronquist clearly exaggerates the indistinguishability of sibling species. As Mayr and Ashlock (1991) point out, "once discovered and thoroughly studied, they [sibling species] are usually found to have at least a few previously overlooked morphological differences" (92; cf. Rosen 1978: 29). Even Robert Sokal (1973: 371), one of the dons of the phenetic species concept (cf. chapter 3.3), did not hesitate to suggest that sibling species are usually not as morphologically similar as readers are sometimes led to believe. Finally, to ignore sibling species is to ignore their immense importance in biological control. As David Rosen (1978) pointed out, "closely-related or sibling species of parasites may differ markedly in their host preferences, as well as in various other biological attributes that may determine their success as biological control agents. Recognition of such cryptic species may amount to the addition of new weapons to the arsenal of biological control" (28). Of course as he further points out, "reproductive isolation . . . [is] the ultimate criterion for the determination of their systematic status" (29). All in all, then, I highly doubt that zoologists will ever come to dismiss sibling species in favor of Cronquist's botanical species concept.

Returning to the problem of blurred species boundaries posed by hybridization, orchids provide a good topic for discussion. In an excellent article by Douglas Gill (1989) we find that not only do orchids comprise over 7% of all angiosperm (flowering plant) species, but that they are also highly interfertile, such that "over 300 new species of orchids are created by artificial intertaxonomic hybridization each month by countless orchid enthusiasts competing for big prizes" (465). Surprisingly, however, as Gill tells us, with orchids "hybridizations in nature are quite rare" (465). Either way, orchids are not only the most "speciose" (a term Gill deplores) but also, if I may add the term, the most "hybridizable" (cf. Cronquist 1978: 6) of all the families of flowering plants. Gill (1989) tells us that "The estimated number of known orchid species varies from the conservative 18,000 . . . to the seemingly liberal (but perhaps realistic) 30,000" (458). Nevertheless, in spite of this there is no questioning that orchids divide (more or less) into species based on reproductive isolation. Instead the debate is over mechanisms of reproductive isolation and other related issues. Cronquist's (1978) focus on a single angiosperm genus (*Gilia*) as a case to serve his point only misleads the unwary reader. Interspecific hybridizations often create problems, but not to the extent that Cronquist and some others would have us believe (cf. Richards 1986: 44, 46–47, 186–187).

A good case in point is Mayr's (1992) analysis of a well-studied local flora in Concord Township, Massachusetts. Mayr came to the conclusion that the biological species concept [based on reproductive isolation] faced serious difficulty only in 9.2% of the flora, which he suggested could be reduced to 6.6% given further research (236). Given that Mayr is the foremost partisan of the biological species concept (cf. chapter 4.1), his analysis might easily be passed off as being heavily biased. However, as we shall see in chapter 5.3, his results are close to the results of a recent study of botanical monographs covering almost 2,000 species. Mayr's (1982) earlier conclusion on messy situations in the botanical world, then, would seem vindicated: "To be sure there are 'messy' situations, . . . but I am far more impressed by the clear distinction of most 'kinds' of plants I encounter in nature than by the occasional messes. A myopic preoccupation with the 'messy' situation has prevented many botanists (but by no means all of them, perhaps not even the majority) from seeing that the concept species describes the situation of natural diversity in plants quite adequately in most cases" (285).

Returning to zoology, Templeton (1989) claims that the supposition common among zoologists that blurred reproductive isolation is much more common in plants than in animals is no longer tenable. His own DNA studies have revealed that in each of the four mammalian groups "there is evidence for naturally occurring interspecific hybridization" (11), wolves and coyotes being one of many examples. And yet even on such a view, his ultimate conclusion, with which I suspect most biologists will agree, is not that reproductive isolation should be scrapped as a criterion for species delimitation, but rather that "While the process [of speciation] is still occurring, the tendency is to have 'bad' species" (22). Given the fact of evolution, then, we should naturally expect bad species. (Indeed if there were no messy situations, I would seriously doubt that evolution is a fact.) But we should also expect good species. What is significant for the species problem is that the good species far outnumber the bad. (Indeed if messy situations were the norm, I would seriously doubt species realism.) As Edward Wilson (1992) put it, the biological species concept "works well enough in enough studies on most kinds of organisms, most of the time" (49), acceptance of which he adds is necessary "if we are to avoid chaos in general discussions of evolution" (65).

If one accepts the horizontal dimension as primary, however, and messy situations as besides the norm, it does not necessarily follow that species must therefore be real. Some biologists have argued that in addition to individual organisms and colonies, populations but not species are real. And indeed if one searches the literature, one can often find biologists refer to species and populations interchangeably. For example, Edward Wilson (1992) says "Evolution is absolutely a phenomenon of populations" (75), all the while admitting the reality of species (and as individuals no less, as we shall see in chapter 4.4) and telling us that "species are always evolving" (46).

So perhaps biologists and philosophers should just drop species talk altogether and talk only about populations. Paul Thompson, for example, expressed to me a few years ago his inclination toward this view. Among biologists it also has some adherents. One of the more notable is Donn Rosen (1978). Rosen concludes "that a species, in the diverse applications of this idea, is a unit of taxonomic convenience, and that the population, in the sense of a geographically constrained group of individuals with some unique apomorphous characters, is the unit of evolutionary significance" (176–177). As if to make his species nominalism even more clear, Rosen adds in a footnote that "If this view is accepted, it renders superfluous arguments about whether a 'biological species' is an individual or a class'" (177n6).

Rosen is often considered a precursor of the phylogenetic species concept. In chapter 4.5 we shall examine his definition of species (Rosen 1979) from this perspective. In the present publication, however, it seems quite evident that his view is that of species nominalism in favor of populations geographically and apomorphously conceived. But here is where the problems begin. The term "apomorphy" is from cladistic taxonomy and refers to a derived character, either unique to a taxon (autapomorphy) or shared with another taxon (synapomorphy). We shall see in chapter 4.5 that there are serious problems with defining species in terms of apomorphies. The problems do not go away if one takes populations as real instead of species. Perhaps the main problem is the degree of resolution. Just how fine-grained is one willing to go in identifying apomorphies? Is a family line with an extra finger to be considered as a population? Or a small group of organisms with a new allele? Of course populations, on Rosen's view, can be contained within larger populations. He states that some populations "are parts of larger populations that span two or more areas" (177). Rosen, however, wants populations apomorphously defined to be the significant units of evolution. But these two units need not always be congruent. All humans are part of a unit of evolution, as all gorillas are part of another. But within humans and gorillas there are numerous populations apomorphously defined, and none of these need be thought of as units of evolutionary significance. There is something about the larger, more inclusive populations, then, that needs to be recognized. And these are what biologists normally refer to as species.

A related problem with Rosen's concept of population is that it makes little sense of Mayr's founder principle, one of the best attested modes of speciation. On Mayr's view, as we shall see further in chapter 4.3, what he labeled peripatric speciation involves as a first stage the isolation of a *founder population*. Such a "population" is certainly of great evolutionary significance, but it would seem to be a meaningless nonentity from Rosen's point of view. This is because although it is a "geographically constrained group of individuals," it has no apomorphous characters. Nor need it ever evolve any. As Mayr (1988a) himself pointed out, "I did not claim that every founder population speciates. The reason is that the majority of them soon become extinct, but a second reason is that in the vast majority of them only minor genetic reorganizations occur" (444). Of course, Mayr does not mean by "speciation" the evolution of apomorphous characters. Nevertheless, the point remains that a founder population is an evolutionary significant unit that does not fit Rosen's definition of population but that is meaningfully "conspecific" (in something more than a populational sense) with its parental population while being itself a unique population. None of this biological reality is captured by taking species as not real and populations as real as defined in Rosen's sense.

Another approach to populations rather than species is that of Mario Bunge (1981). Although Bunge is a former physicist, I include him here because his views have been adopted by the German zoologist Martin Mahner (1993). They eventually collaborated on a comprehensive view of philosophy of biology (Mahner and Bunge 1997). Beginning with Bunge (1981), the title of his paper pretty much says it all: "Biopopulations, Not Biospecies, are Individuals and Evolve." According to Bunge, "there is no such thing as evolution of *species*, but only evolution of *populations* of organisms," so that "Darwin's major work should have been titled *The Evolution of Biopopulations*" (284). Acknowledging that "Nobody doubts that the notion of a species is genuinely problematic," Bunge claims that the species-as-individuals thesis of Ghiselin and Hull (cf. chapter 4.2) does not work and that his own alternative is the "correct solution" (284).

Bunge's "correct solution," interestingly, still retains a role for species. Acknowledging on the one hand that "if a species is a set, it cannot possibly evolve, for sets are concepts, not things," Bunge holds on the other hand that "species, genera, and other taxa are sets or collections, though not arbitrary groupings: they are classes formed on the basis of objective commonalities" (284). Thus a population, an objectively real thing, may possibly evolve (if it does not first meet with extinction) through one species and then another, each of the species conceptual only, and on up through different genera and other higher taxa, each of them conceptual only as well. This is an interesting variation (populations instead of species, taxa conceptualism rather than realism) on a theme that we shall encounter later with Caplan (chapter 3.3) and Rosenberg (chapter 4.2n21).

Although interesting, there are a number of serious problems with Bunge's view. To begin, it is not obvious that a biospecies, conceived in the sense of one or more biopopulations, cannot also be a unit of evolution. Mary Williams (1985) claimed that Bunge is wrong here, but failed to correctly say why. Saying that "each is an evolutionary unit . . . with respect to its own set of selection pressures" (585) only begs the issue concerning selection pressures, not to mention evolution, since it ignores genetic drift as well as the possibility of stasis, both of which are part of modern evolutionary theory. The correct reason why species may also be units of evolution is simply because species may be composed of more than one population. As Hull (1980) put it, "Although some species of sexual organisms are made up of a single population, most include several populations that are at least periodically disjunct" (324). As long as the component populations of a species evolve uniformly—i.e., in a parallel fashion—then the *species* also evolves.[21] If, on the other hand, the component populations begin to diverge in their evolution, then we have the beginning of the process of allopatric speciation, which may end in a number of ways. But in neither case, whether uniformity or partial divergence, is there any good reason to deny the extra-mental reality of the conspecificity of the component populations (this will become especially evident when I develop my own view in chapter 5.2) and to construe that

---

[21]   The cause of this uniformity is logically beside the point, although of course in reality it will be natural selection (since genetic drift is extremely unlikely to produce parallel evolution). What Futuyma (1986) points out in a different context is relevant here: "It seems likely that when a barrier separates two widely distributed forms, especially species that are ecologically generalized, the average forces of selection on either side of the barrier are similar, and so do not promote divergence" (244). What this means is that it is not only logically but biologically possible for two isolated though conspecific populations to evolve in a parallel fashion. On the other hand, the two populations, of course, might not even evolve at all, or only very little, so that they would remain conspecific for a very long time, which is also problematic for Bunge's view. Once again as Futuyma (1986) points out, "some populations that have been isolated for many millions of years have diverged hardly at all in morphology and remain reproductively compatible. For example, some plant populations that have been isolated for 20 million years, such as American and Eurasian sycamores . . . and American and Mediterranean plantains . . . form fertile hybrids. Likewise, European and American forms of certain birds, such as tits . . . creepers . . . and ravens . . . are so similar that taxonomists classify them as the same species" (244).

conspecificity, along with all speciation, as nothing but conceptually real.

Interestingly Mayr (1987a: 160) became very impressed with Bunge's focus on biopopulations and as we shall see in chapter 4.1 preferred to use the term "biopopulation" himself. But Mayr failed to see the import of this view. He dutifully pointed out that "One can recognize a hierarchy of biopopulations ranging from the local population (deme) [i.e., local interbreeding population] up to the species ('the largest Mendelian population' of Dobzhansky)." But then he cursorily dismissed this problem by saying "This is, of course, no more an argument against the adoption of the population terminology, than it would be in the case of class and individual, which can also be organized hierarchically" (163). The problem is not only that population terminology is hierarchical but that its meaning at whatever level in the biological world is far from agreed upon. Moreover, its application to real life (even when agreed upon) is notoriously vague, so that it is certainly not unproblematic as Bunge implies. Unfortunately, Bunge does not even clarify what he means by "biopopulation," which is simply inexcusable since the many problems with the concept of population are well known. One problem is that if Bunge means anything less than the Dobzhansky view, then we have a serious problem with delimitation. As Futuyma (1986) put it, "local populations are ephemeral" (247). The problem of delimitation is also connected with naming. Although Bunge does not require this, if biopopulations are the objective entities of evolution—the individuals—that he thinks they are, then they should be delimitable enough for naming. But as Mishler and Donoghue (1982) pointed out, "formally naming whatever the truly genetically integrated units turned out to be would be disastrous. There are certainly very many such units, they are at best very difficult to perceive even with the most sophisticated techniques and in the most studied organisms, and these units are continually changing in size and membership from one generation to the next. At any one time we can never know which units will diverge forever" (128).

Nor does Bunge clarify what kinds of organisms are to be included as members of biopopulations. He talks about the population concept of population genetics, but also of ecology. So does he include asexual organisms? Bunge does not say. The only kind of relationship for linking organisms that he specifically mentions (although he implies that there are others) is "that of reproduction" (284), which certainly does not help us decide the issue. The issue is important, for as we shall later see a number of biologists, such as Simpson (cf. chapter 4.4) and Templeton (cf. chapter 3.4), think that asexual organisms form populations.

Mahner (1993), on the other hand, is clearer on this. Agreeing with Bunge that "the unit of evolution is not the species, but the population" (111), and that species are sets (natural kind collections) and therefore conceptually real only (119), he adds that "A species concept in biology has to be applicable to all organisms irrespective of whether they live in biopopulations or not" (116). But what then is a population on this view? Mahner is not entirely clear, but he provides the following answer in summation of his view:

Granted, if species are natural kinds they cannot evolve but only succeed one another. Only real systems can evolve. Such real individuals do exist as systems called populations. It is for biologists to find out what populational system is the proper "unit of evolution" and to decide how that referent of evolutionary theory should be termed properly (i.e., deme, biopopulation, Darwinian subclan or else; . . .). The basic populational unit of evolution (if there is a single one at all) should, however, not be termed species . . . as biological fashion demands, because we need the species as a class/kind concept (i.e., as taxon, neither as biopopulation nor as category) if we strive for consistent concepts in biology and a sound philosophy of biology avoiding the prevailing metaphysical muddle due to the recent revival of nominalism [i.e., the species-as-individuals view of Ghiselin and Hull]. [121–122]

The view of Bunge and Mahner on species, of course, is developed in great detail in Mahner and Bunge (1997). Although there is too much there to discuss here, it will be profitable to examine a few points. One point of interest is their discussion on the meaning of "population" in biology. They point out that the term is "highly ambiguous: It can mean either (a) a *statistical population*, i.e., a mere collection of individuals, . . . [e.g.] 'the population of HIV-infected children in 1997'; or (b) an *aggregate* (or heap) of individuals, . . . [e.g.] 'the fish population in this pond'; or (c) a *system* of individuals, as is the case with reproductive communities or animal societies" (153). The first of these, they tell us, is a conceptual, not a concrete entity. The second is concrete, supposedly because "the organisms in question just happen to occupy the same locality," but they lack "bonding relations." For a population to be a "cohesive and integrated entity, i.e., a system," bonding relations are necessary, which "may, for instance, consist in mating relations, such as in reproductive communities, or in symbiotic or social relations" (153).

Whether the second kind of population should rightfully be conceived as "concrete," and whether the above list captures, for instance, Rosen's concept of population, are all beside the point. This is because on their view now the role of populations is greatly diminished in favor of organisms. Right at the outset they state that "we shall argue that species do not evolve, and that the most interesting, though not the sole, unit of evolution is the organism" (v). Later they tell us that "What matters are kinds of organisms, whether or not the latter compose a population" (318). They argue that "*evolutionary innovations are properties of organisms*. So every qualitative novelty must first occur in a single organism" (315).

This has a number of interesting consequences. To begin, "*evolution amounts to speciation*, that is, to the emergence of qualitatively novel organisms. Since natural kinds are collections, it suffices for the formation of a new species that there be only a single new individual" (238; cf. 320). From this it follows that "*organisms are the speciating entities*" (317). It also follows that organisms, and not populations or communities, are the main units of evolution, that the latter evolve only derivatively, and that there are as many modes of speciation as there are processes that produce qualitatively novel organisms: "there are as many evolutionary processes as there are speciating organisms. The same holds for biopopulations and communities as (derivative) units of evolution" (320). Thus they do not distinguish between anagenetic (phyletic) and cladogenetic (branching) speciation.

Mahner and Bunge state at the outset that their view is not part of mainstream biology (v). Why this is true is, I believe, because of the metaphysical assumptions with which they begin. It turns out that their view is a development of Occam's conceptualism. While one of their axioms is what they call *methodological dualism*, namely that every object is either a *concept* (a construct in a brain) or a *thing* (5), they make it clear that ontologically they are monists, in that "The world is composed exclusively of things (i.e., concrete or material objects)" (6), where a thing is "a substantial individual endowed with all its properties" (7). This brings for them many of the same difficulties that in §2 above we saw faced Occam, plus many more specific to modern biology. In particular, they cannot bring themselves to take relations as real. Thus, on their view, "lineages are not real entities and they neither change nor evolve. . . . the concepts of ancestry, progeny, and lineage are relational: there are only ancestries, progenies, and lineages *of* things. Real existents, by contrast, are absolute" (238). In like manner similarity is reduced, as we shall see with the phenetic taxonomists (chapters 3.3 and 5.2), to properties in common, so that "A qualitative measure of similarity is the intersection of the sets of properties of the similar objects" (216). They are also led to the apparent absurdity that the individual organism is the main unit of evolution, even though they acknowledge that "the developing organism does not change in kind or species" (316). Modern biologists, on the other hand, as we shall later see (chapter 4.4n48), are virtually unanimously agreed that the organism is not a unit of evolution, since it has ontogeny only, not phylogeny. Because of their metaphyics, however, Mahner and Bunge are forced into a contrary view. And, of course, on their Occamist metaphysics species are conceptual entities only, and hence not extra-mentally real, let alone capable of evolution, because they are not things (i.e., absolutes in Occam's terminology). They define a species as "a natural kind (rather than an arbitrary collection)" composed of "organisms (past, present, or future)" (154). To this they add that "like any other species [i.e., natural kinds], a biospecies is a collection, hence a conceptual object" (154). That a biopopulation should not be equated with a biospecies is because, on their view, the concept of the latter logically presupposes the concept of the former. This is because "aggregates and systems of organisms, i.e., concrete populations, may consist of organisms of either the same or different species. . . . This being so, the concept of a species is logically prior to, and independent of, any concept of aggregate and system of organisms—in particular that of population" (154; cf. 315). Thus, species do not exist out there in the biological world, discovered and waiting to be discovered, evolving and going extinct irrespective of human conceptualization.

All of this is so contrary to modern biology, so radically antithetical, that it seems, to me at least, much more desirable to reject the above conclusions and go back and begin with a different metaphysics, one that leads to conclusions much more in harmony with modern science. In chapter 5, I shall attempt to do this by taking relations, contrary to the Occamist tradition, as ontologically real and, moreover, as grounded in the world of concrete objects. The result, I shall argue, is both agreeably, in the main, descriptive of modern biology and prescriptive.

At any rate, in addition to all of the forms of species nominalism that we have examined thus far, one can sometimes find an argument for species nominalism based on the claim that the most famous species concept in modern biology lacks universality. The most famous species concept in modern biology, of course, in the sense that among practicing biologists (as opposed to professional taxonomists) it is the most widely subscribed to, at least in zoology, is Mayr's version (cf. chapter 4.1) of the biological species concept (cf. Cronquist 1978: 6–7; Wiley 1981: 23–24; Eldredge 1985b: 50; Wilson 1992: 38). Perhaps the most notable example of species nominalism based on the rejection of this species concept is the paper by the botanist Donald Levin (1979). Ostensibly confining his discussion to plant species, Levin argues that although the biological species concept "has met with some success when applied to animals, the characteristics and diversity of higher plant genetic systems and reproductive modes preclude the application of a universal plant species" (381). But Levin does not confine himself to botany. Upon reading the whole of his paper it becomes apparent that it is a surreptitious argument against the reality of species *per se*. This is apparent in his reply to what he discerns as being the "three features that allegedly make the species a unique evolutionary unit" (381). In reply to the first one, that species are apparently real as opposed to higher taxa, Levin countenances support from Locke and Darwin, both of whom he accepts were species nominalists ("Darwin concurs with Locke"), as well as structuralist philosophy, according to which "human behavior and perception are determined by our subconscious penchant to divide an assemblage of objects into clusters and form an abstract generalized concept of each resulting assemblage" (382)—never mind that later he wants to hold that "part of the order we perceive is the product of our choosing" (384). All of this is in explicit opposition to Mayr, according to whom species are apparently real.

That Mayr is the main springboard for Levin's species nominalism is even more evident in his (Levin's) reply to the second alleged feature of species reality, namely the apparent integration of conspecific populations. On the issue of integration by gene flow, Levin (382–383), following but exceeding the classic paper of Ehrlich and Raven (1969), summons empirical evidence based on an impressively long list of botanical papers, the sum of which show that the required gene flow is mostly absent in the botanical world. Moreover, as for reproductive isolating mechanisms, Levin argues that they "do not exist as properties of single species or single populations" and that they "are not the cause of divergent evolution any more than isolation by distance is the ['essential'] cause of divergent evolution" (383).

Finally, in reply to the third alleged feature of species reality, namely their apparent independence or unique evolutionary role, Levin argues not only that many plant species occupy multiple adaptive peaks and that many reputed genera and families also occupy multiple adaptive peaks, but that "With small species size and close inbreeding, a species might reside between peaks rather than on them" (383). (Of course for this latter clause to make sense Levin must not mean real species but only "book" species, i.e., reputed species.)

In replying negatively to the above three alleged features of species reality, Levin concludes that "As with all theoretical concepts, species concepts bear within themselves the character of instruments. In the final analysis they are only tools that are fashioned for characterizing organic diversity. Focusing on the tools draws our attention from the organisms" (384). From this it is clear that Levin by no means wishes to confine his species nominalism to botanical species but that, based mainly on his rejection of the front-running biological species concept (this is not to ignore his structuralist-type arguments, which I shall examine in general later in this section), he wishes to extend his nominalism to all species. But such a line of argument is clearly fallacious. It sets up a false dichotomy (i.e., either what are called species conform for the most part to the biological species concept, or else they are not real) and then argues that because one of the disjuncts is false (i.e., what are called species do not conform for the most part to the biological species concept) the other disjunct must be true (i.e., species are not real). The error is not in the reasoning from premises to conclusion but in the truth of the first premise, for there are many more alternatives for the reality of species than what is prescribed by the biological species concept.

What is most interesting about Levin's paper, however, is not this fundamental flaw in his reasoning, but that he shows us the way to a universal species concept without apparently even thinking to take that path himself. In the final paragraph of his paper he says "Similar products need not derive from the same processes. For this reason, we should avoid promulgating species interpretations founded upon a common underlying process or interaction. These species are formulated by edict. Species interpretations based on the products of evolution are not shackled with implicit or explicit assumptions of causation" (384). In chapter 1.2/4, I briefly touched on the interesting division of competing species concepts into causally-based species concepts (process concepts) and effect-based species concepts (pattern concepts) and intimated the view that the former lead either to species nominalism or species pluralism and that only the latter can lead to a species concept that is not only a realist but also a truly universal species concept. Levin has taken the species nominalism route. In chapter 5.2, I shall take a different route and attempt to use his insightful division of species concepts (in conjunction with other important distinctions) to fashion a universal species concept. As for species pluralism, we shall now turn to an examination of this increasingly growing view, an examination which shall occupy us for the remainder of this section.

Quite simply, *species pluralism* is the view that modern biology requires a number of different species concepts, so that the species category is heterogeneous (disjunctive). (In examining this view I shall expand my present scope and consider not only the views of biologists but also philosophers.) According to Mishler and Donoghue (1982), for example, "no single and universal level of fundamental evolutionary units exist; in most cases species taxa have no special reality in nature. We urge explicit recognition and acceptance of a more pluralistic conception of species, one that recognizes the evident variety and complexity of 'species situations'" (122). Similarly according to Philip Kitcher (1984a) there are "distinctions among organisms which can be used to generate alternative legitimate conceptions of species. . . . biology needs a number of different approaches to the division of organisms, a number of different sets of 'species'" (319).[22] In arguing that different kinds of biological investigation require different species concepts, Kitcher focuses on the distinction between structural and historical explanations, both of which he divides into further levels.[23]

Where these authors mainly differ is on the kinds of species concepts that pluralism needs, the former being more restrictive.[24] At any rate, I want to argue that their differences don't really matter since their views both lead to species nominalism and (what I shall argue at the end of this section) they both involve false premises.

Beginning with Kitcher, he evidently thinks that his species pluralism is compatible with species realism.[25] But if one is going to argue that different kinds of biological investigation entail different, sometimes incompatible species concepts, then, just like the division of stars into constellations, the division of organisms into species is going to be merely "for the convenience of classification, and having as little pretensions to reality" (Lyell 1832: 19; cf. Darwin 1859: 411). Even granting what Kitcher (1984b) tells us, that "each of the divisions corresponds to an objective pattern in nature" (626), so that species are not (Kitcher 1984a) "merely fictions of the systematist's imagination" (309), it does not follow that species on this view are *objectively* (extra-mentally) real; at best they are

---

[22]   Kitcher's own species concept will be properly examined in chapter 4.5. His overall view would seem to be that, although there are a number of equally legitimate species concepts, their legitimacy is strictly operational. Ontologically all species, on his view, are sets extensionally defined (i.e., collections). All that the different legitimate species concepts do, then, is pick out different sets. Given his species pluralism conjoined with this species-as-sets ontology, I can see no other way of interpreting Kitcher's view. His species pluralism, then, is operational only, while his species ontology is monistic.

[23]   For similar species pluralisms, cf. Cain (1954: 106, 124), Holsinger (1984: 303–305), Rosenberg (1985: 200–201), Endler (1989: 625), Johnson (1990: 67), Wilkinson (1990), Ereshefsky (1992a), O'Hara (1993), and Stanford (1995).

[24]   Mishler and Donoghue (1982: 130–133), and following them Mishler and Brandon (1987: 402–404), restrict their pluralism to phylogenetic (monophyletic) species concepts. There is a further, and what seems to me a less important difference. On Mishler and Donoghue's pluralism, for every different kind of biological investigation there is only one optimal classification of species, whereas on Kitcher's pluralism there may be more than one. Cf. the discussion in Mishler and Brandon (1987: 402–404). I have no desire to comment on this particular debate, since my disagreement with both sides goes much deeper.

[25]   Dupré (1981) precedes Kitcher in this and gives his own view the interestingly descriptive title "promiscuous realism" (82). Dupré (1993) provides a more detailed version, while Dupré (1999) provides a less radical version, modifed in the interest of reliable communication between biologists, such that taxonomic revisions should be avoided as much as possible, as well as overlapping taxonomies.

still only contextual and user-dependent, and hence only *conceptually* real.[26] To think otherwise is not only to muddy the waters concerning what is objectively real and not real, it is just plain ontologically false. For species to be thought objectively real all the while accepting different species concepts, those concepts *must not possibly* be incompatible, mutually exclusive, or only partially overlapping.[27] Instead they must be convergent, transtheoretic, consilient.[28]

---

[26]   In a different vein, from the viewpoint of the historical relativity of biological interest, Kyle Stanford (1995), using Cuvier's nonevolutionary species concept as his example, argues that there is a fundamental incoherence in Kitcher's combination of species pluralism with species realism. His argument, in short, is that "as the course of biological inquiry proceeds, we do not decide that we were previously *mistaken* about which groups of organisms were species; rather, as our explanatory and practical interests change, which divisions of organisms *actually are* species changes as well" (83). On Stanford's view "'Species' are thus the designations we use to pick out the significant and interesting distinctions between groups of organisms at a given time, that is, the differences that we wish to investigate. . . . Species are not independent of the states of particular minds, and are therefore not real or objective in the traditional senses of those terms" (86). Instead, "What counts as a legitimate species division among organisms depends upon what we find interesting about them" (89). While Stanford denies the reality of species taxa (and therefore by implication also the species category), other pluralists, however, such as Ereshefsky (1998), deny only the reality of the species category. But as we have seen in the previous section with Ghiselin and Beatty's attempt to attribute the same kind of realism/nominalism to Darwin, it involves a fundamental contradiction: no species category, no *species* taxa. Moreover, it bars the possibility that there could be laws of nature for species *per se*, since a law of nature requires a category as its subject, an abstract class that does not change even though its members come and go (cf. chapter 4.2n28 and 4.5n61).

[27]   According to Kitcher (1984a), "the patterning of nature generated in different areas of biology may cross-classify the constituents of nature" (330). Interestingly, one has to wonder (and one is never explicitly told) whether for Kitcher and other species pluralists the organisms commonly known as humans might legitimately "cross-classify" into different species. Either way the pluralists would seem faced with a sort of *reductio*.

[28]   In explicit reference to Whewell's (1840) criterion of the *consilience of inductions*, according to which a theory receives verificational support when an induction from one class of facts coincides with an induction from a different class of facts, Michael Ruse (1987) argues that consilience also serves to establish "why it is reasonable to think of species as natural kinds" (356). According to Ruse "There are different ways of breaking organisms into groups, and they *coincide!* The genetic species is the morphological species is the reproductively isolated species is the group with common ancestors" (356). (It will be noticed that Ruse leaves out the ecological species, not to mention others.) On the other hand, "doubts about the reality of species should arise when the various ways of defining species names come apart and fail to coincide" (357). Indeed one wonders how often favorable consiliences would actually obtain. At any rate, what Ruse fails to recognize is that consilience can only serve to establish that "good" species taxa are objectively real, such as *Homo sapiens* or *Gorilla gorilla;* it cannot also serve to establish the ontological status of those taxa, in other words the winning definition of the species category. Indeed Ruse seems to confuse taxa with category. He thinks consilience can establish that species taxa are natural kinds, but this is mistaken; at most, consilience can only establish that the species *category* is a natural kind (which is precisely the bone of contention with plural-

With Mishler and Donoghue, even though it appears that they only think their views entails species category nominalism and not species taxa nominalism (cf. Mishler 1999: 309, 314), it is arguable that their view entails species taxa nominalism as well. As we have seen in the previous section, in Ghiselin and Beatty's attempt to attribute a view of species taxa though not species category realism to Darwin, it involves a fundamental contradiction: no species category, no *species* taxa. In other words, if one believes that species taxa are real but not the species category, then the very phrase "species taxa" becomes a contradiction in terms.

Aside from the real/unreal discrepancy, it is interesting that both Mishler and Donoghue (1982) and Kitcher (1984a) appeal to *gene pluralism* to support their species pluralisms. According to Mishler and Donoghue (1982), "it is time for 'species' to suffer a fate similar to that of the classical concept of 'gene'" (132). More specifically, "The use of a disjunctive definition . . . allows a single term to designate a complex of concepts. However, this can become so confusing that it may be desirable to replace (at least in part) an old terminology with a new set of terms with more precise meanings" (134n5). Compare Kitcher (1984a): "One of the lessons of molecular biology is that there is no single natural way to segment DNA into functional units. . . . Even if we pretend that all genes function to produce proteins there is no privileged characterization of genes as functional units. Yet geneticists (and other biologists) manage their investigations quite well, and the use of a plurality of gene concepts does not generate illusions of agreement and disagreement" (326).

To this both Ghiselin (1987a: 136) and Hull (1987: 177–178) emphatically reply that the analogy to genes is false, that the modern concept of gene is not pluralistic or contextual but instead monistic in perspective. But when I turn to expert studies on the history of the concept of gene up to the present day, studies clearly not concerned with the species problem, I find that I must agree with Mishler, Donoghue, and Kitcher. In his detailed study on the concept of gene, in which he offers a classification of at least nine different molecular gene concepts used commonly today, Petter Portin (1993) concludes that "we now understand the concept of the gene to be a general, open, and abstract one, and its definition to depend on the context in which the term is used. Older terminologies no longer suffice, and no single definition is appropriate or adequate for every gene" (209).[29]

Of course, even if the gene analogy is a good one, it by no means proves the case of species pluralism, no more than it refutes the case of species monism.

---

ists). As stated at the beginning of this book, the species problem is twofold: whether species are objectively real, and if real the nature of their reality. Consilience can only serve to help solve the first part. Once satisfied that at least some putative species are real, the second part of the species problem can only be solved by submitting each competing species concept to rigorous empirical and philosophical analysis. This, of course, is the basis of the present book.

[29] Cf. Waters (1994): "Whether a sequence of nucleotides counts as a gene is context sensitive" (181).

More to the point is the issue of what are known as *cross-cultural tests*. Often these are used to support the view of species monism and the objectivity of (at least some) species taxa. Edward Wilson (1992) tells the now famous story of Ernst Mayr:

> In 1928 the great ornithologist Ernst Mayr traveled as a young man to the remote Arfak Mountains of New Guinea to make the first thorough collection of birds, . . . His species concept was that of a European scientist looking at dead birds, who then sorts the specimens in piles according to their anatomy, as a bankteller stacks nickels, dimes, and quarters. . . . In the end he found that the Arfak people recognized 136 bird species, no more, no less, and that their species matched almost perfectly those distinguished by the European museum biologists. The only exception was a pair of closely similar species that Mayr, the trained scientist, was able to separate but that Arfak mountain people, although practiced hunters, lumped together. [42–43]

Mayr (1988a) adds that "I have always thought that there is no more devastating refutation of the nominalist claims than the fact that primitive natives in New Guinea, with a Stone Age culture, recognize as species exactly the same entities of nature as western taxonomists" (317). Although Wilson (1992) relates that in his own case with ants in New Guinea the natives "could not tell one ant from another" (43), he nevertheless believes along with Mayr that cross-cultural tests "can distinguish cultural artifacts from natural units" (42). Similarly, Gould (1980): "In short, the same packages are recognized by independent cultures. I do not argue that folk taxonomies invariably include the entire Linnaean catalog. People usually do not classify exhaustively unless organisms are important or conspicuous" (207).[30] Indeed Gould (209) cites an interesting case of scholarly recantation on this issue, reminiscent of the attempt by Eddington to refute Einstein.

On the other side, Mishler and Donoghue (1982) point out that "New Guinea tribespeople are human too, with similar cognitive principles and limitations of language" (133n1). Similarly, Ridley (1993) claims that "the fact that independently observing humans see much the same species in nature does not show that species are real rather than nominal categories. The most it shows is that all human brains are wired up with a similar perceptual cluster statistic" (404). On this view we might have been "wired" differently and different species might now be wired differently from us, so that no one wiring can be said to be "true" or "veridical." Cross-cultural tests, then, on Ridley's view, "suffer from the general problem of phenetic classification" (405). Gould (1980), familiar with this line of reply, finds it "an interesting proposition, but one that I doubt" (212).

---

[30]   Cf. Holsinger (1984) and Hull (1987) who draw different conclusions from failed congruences. Holsinger (303–305) believes that failed congruences support species pluralism. Hull, on the other hand, is skeptical of the scientific value of folk taxonomies and believes that failed congruences only prove that cross-cultural tests are "mistaken in the extreme" (169).

I too find it difficult to believe. For a start, consider Scott Atran's (1999) cross-cultural study comparing the folk-biological concepts of students in Michigan with the forest-dwelling Itzaj Maya of Guatemala. Although this study corroborates Atran's thesis that humans have a "universal and possibly innate" (245) domain-specific mechanism for conceptualizing what he calls *generic species* (a hybrid of the Linnean species and genus), in a manner which he calls "teleo-essentialist" (and thereby not merely overall similarity), he nevertheless acknowledges that "Scientists have made fundamental ontological shifts away from folk understanding in the construal of species, taxonomy, and underlying causality. For example, biological science today rejects fixed taxonomic ranks, the primary and essential nature of species, teleological causes of species existence, and phenomenal evidence for the existence of taxa" (250). What this points to is the more general acknowledgment that even if our brains are wired in the manner of folk-biological theories, the wiring hypothesis fails to allow for the all-too-obvious element of *discovery*, especially when discovery goes against our intuitive wirings. For example, in spite of our initial intuitions, when we discover that a particular kind of butterfly is the adult stage of a particular kind of caterpillar, we do not continue to consider both as two distinct species. Instead, we change our view and conclude that what we have here is an extreme case of an ontogenetically dimorphic species of insect. Similarly, when we discover a new kind of bird such that we give it a new species name, and when in the same area we discover another new kind of bird though very different from the other such that we give it a new and different species name, even if all the samples from the first group are males and all the samples from the second group are females, we do not combine the two intuitive species into one unless and until we discover that the two interbreed. We then conclude that what we have discovered is an extreme case of a sexually dimorphic species of bird. Similar considerations apply in the case of sibling species, with the proof clearly in the pudding as with the *Anopheles* example. Theories and discoveries of phylogeny further extend the same point. The wiring hypothesis, then, seems to me to give the cognitive capacity of humans nowhere near the credit it deserves. Our observations, of course, may indeed be interest and theory dependent, and may indeed have to be, but this does not of itself preclude the possibility of objective discoveries. Instead, it need only imply that objective discoveries cannot be obtained without prior interests and theories (cf. Stamos 1996b: 189n6).

This, however, may still fail to satisfy. At any rate, what I find most conclusive against species pluralism, whether realist or nominalist, is that two of its essential premises can be shown to be quite false. First, the premise that no species concept can be privileged over another is falsified by the fact, as I argued earlier, that horizontal species concepts are logically prior to vertical species concepts. As such, the species concepts of paleontology and phylogenetic taxonomy must yield to the species concepts of neontology (the study of living organisms). This is not because of the fact that most of the work done by biologists is done at the horizontal level, as in population genetics, ecology, biological control, medicine, ethology, etc. My argument is not based on numbers or an appeal to popularity.

Neither is it based on the fact, in contrast to Shaw's (1969) claim for species concepts in paleontology, that species concepts in neontology convey a lot of information, as in the examples from Maynard Smith and Wilson examined in chapter 1.1. Indeed as David Rosen (1978) put it,

> Upon encountering an unknown pest, our first question usually is: "What is it?" Correct identification provides the key to any available information about the species, its distribution, biology, habits, and possible means of control, which would otherwise have to be independently investigated at a considerable expense of time and effort. [24]

No, my argument is not based on the greater amount of information content at the horizontal level either. Instead it is based solely on the logic of the ontology of the situation, as discussed earlier (cf. also Stamos 2002).

The second false premise of species pluralism is that every kind of biological investigation or interest requires its own species concept. As we saw earlier, Shaw (1969) demonstrated that a species concept need not always be necessary to do paleontology. The same can sometimes also be demonstrated at the horizontal level. For example, in his comprehensive survey of the literature on food webs, J.H. Lawton (1989) points out that "many of the components of published webs are not single species; they may be genera, groups of similar taxa (e.g., 'herbivorous gastropods' or 'algae'), or incredibly heterogeneous organisms that happen to feed roughly in the same way (one such group, by no means atypical, consists of dragonflies, spiders, and passerine birds! . . .). For simplicity, it is conventional to refer to all such groupings as 'species,' even though we know it is an oversimplification" (45–46).

In sum, although species nominalism is a minority viewpoint in biology today, it has been receiving new life in the form of a quickly growing view known as species pluralism, in spite of the fact that some species pluralists fail to recognize or acknowledge that their view all too easily leads to species nominalism. At any rate, what all species pluralists, whether realists or nominalists, fail to recognize is that some species concepts are logically privileged over others and that not every kind of biological inquiry requires its own species concept. These two points, along with the fact that species evolve and a viable species concept must be capable of capturing that fact, effectively destroy the very foundation of species pluralism, and with it the strongest support for species nominalism today. This, of course, does not guarantee that a single species concept will ever prove satisfactory in the future of biology as a whole, but it does remove what is arguably the most serious obstacle in the way of that goal.

Chapter 3

Species as Classes

But the disreputability of origins is of itself no argument against preserving and prizing the abstract ontology. . . . the abstract objects that it is useful to admit to the universe of discourse at all seem to be adequately explicable in terms of a universe comprising just physical objects and all classes of the objects in the universe (hence classes of physical objects, classes of such classes, etc.).

—W.V.O. Quine (1960: 123, 267)

3.1 Plato, Aristotle, and Linnaeus

I use the phrase "species as classes" to refer to all those views which conceive of biological species as objective abstractions of one sort or another. This is an approach, of course, that can be traced back at least to Plato. In fact the word "species" is the Latinized form of the Greek word *eidos*. Plato appropriated this common Greek word, which means "the look of a thing," and formalized it into a metaphysical category to refer to his abstract absolutes, his universals, each of them a *one over many* (*Phaedo* 76d, *Rep.* 476a), *universalia ante res* (to use the Latin tag). For Plato the true look of a thing is not something that can be perceived with the senses but rather only with the mind. It is only when the mind's eye is freed from the obscuring and blinding effects of the senses that it can see the unchanging abstract natures or essences of things, what Plato often referred to as *paradeigmata* (cf. Patterson 1985: 11–29). This is Plato's famous theory of Forms as found in many of his dialogues.

It has of course become a truism to say that Plato's metaphysical essentialism was based on mathematics and geometry. The essence of circles is eternal and unchanging and remains unaffected by the fact that circles in the physical world come and go and are never perfect. That this served as a metaphysical paradigm for Plato's reasoning on kinds other than those in mathematics and geometry is not only the traditional view received from Aristotle (cf. *Met.* 987ª29–ᵇ13, 992ª 29–34), but it seems clearly the case from close examination of Plato's texts (cf. *Rep.* 510d–e, *Epis. VII* 342a–343a).

It seems not without good reason, then, that Ernst Mayr (1982) called Plato "the great antihero of evolutionism" (304). More recently David Kitts (1987) has challenged and attempted to undermine this assessment, arguing extensively that Plato gives us no explicit doctrine on animal kinds, and concluding that "What biologists have objected to in Plato then cannot be some explicitly stated doctrine of species, but rather what they suppose to be the consequences of his metaphysics for a theory of animal kinds" (321).

Although Kitts is correct in stating that Plato provides us with no *explicit* doctrine on animal kinds, there is surely more to be said for Mayr's assessment than for Kitts' rejection of it.[1] For surely a particular doctrine on animal kinds *emerges* from a reading of Plato's dialogues (including *Epis. VII*), and one which warrants Mayr's assessment. Briefly, as can be distilled from Plato's dialogues, particularly his middle dialogues, for every kind of thing, expressed by a predicate, of which many things may be predicated, there exists a transcendental Form (*Rep.* 507b, 596a), existing outside of space and time, eternal and unchanging and fully real (*Phaedo* 79d), of which sensibles are only partly real, being merely reflections or images (*eidōla*), copies or imitations (*mimēmata*)—Plato used a variety of interesting metaphors (Patterson 1985: 30–31). Although Plato came later to express some doubts about certain kinds of things, such as "hair or mud or dirt" (*Parm.* 130c) or "barbarian" (*Statesm.* 262d), he never seems to have expressed any doubts that there are Forms of "the animals about us and all plants and the whole class of objects made by man" (*Rep.* 510a)[2] and that the dialectician in determining Forms according to the method of division (*diairesis*) should be like a good butcher and carve reality at its joints (*Phaedr.* 265e, *Statesm.* 262b, 287c; cf. Guthrie 1978: 129–133, 166–168).

At any rate, Mayr and Kitts (and others) aside, although Plato's Forms are abstract particulars, his theory is also arguably a class theory. What makes it a class theory, it seems to me, is that on Plato's view the *essence* is the *real* species, not the organisms which correspond in name (*Phaedo* 102b), although the latter may be said to constitute the extension of the class (*Rep.* 507b). Conspecific organisms may come and go, but for Plato the species remains utterly the same. This marks an important distinction between all the class views which I will discuss in this chapter and the views which I will discuss in the next chapter, the latter which conceive of species as physical entities of some sort.

A further distinction is that for Plato and many other class theorists the spe-

---

[1]   Cf. Mayr (1988b) for his reply to Kitts (1987) and Grene (1989a) for her reply to Mayr (1988b). In what follows I briefly defend Mayr against Kitts and Grene. In closely examining the trees, both Kitts and Grene have missed the forest and, to vary the metaphor, would have us look to only the literal rather than the foot of the letter when reading Plato.

[2]   Quotations are from the Hamilton and Cairns edition (1961) of Plato's works.

cific essence causally accounts for the nature of the conspecific organisms, all of which of course is denied by those who conceive of species as physical entities.[3] For Plato "participation" (*methexis*) in a Form is a causal (teleological) relation (*Phaedo* 100c–d).[4]

Yet another important distinction common to Plato's theory and many other class theories is that conspecific descent is neither necessary nor sufficient for class membership. To be, say, a dog, it is not necessary that one is born from a dog. It may in fact be the case that all dogs come from dogs, but it need not necessarily be so.

One respect in which Plato's theory is different from many other class theories, however, is that for Plato membership in a class, which for him is constituted by resemblance to a Form, is not an all-or-nothing affair. As different painted portraits of a particular person can be better or worse, different conspecific organisms can be more or less whatever their species resemblance. But as no picture can ever *be* that person, no organism can ever fully *be*, say, a dog. To say that an organism *is* a dog is to say no more than that organism resembles to some degree the transcendental Dog. This does not seem to me to preclude Plato's theory as a class theory, since there still remains both an intension and extension, although the extension may sometimes be fuzzy.

Although no biologists today adhere to anything like Plato's view, particularly with its otherworldliness, and although we may well reject it as idle metaphysical speculation and as contrary to the spirit and practice of modern science, as well as being ontologically utterly superfluous, it must be remembered that Platonic idealism in biology (Christianized in form) was a very strong force up to a little more than a hundred years ago (cf. Bowler 1989: 126–134). Its demise can be credited basically to Darwin.

Two names particularly stand out, both immediate contemporaries of Darwin, namely the Swiss-born American naturalist, Louis Agassiz, and the British comparative anatomist, Sir Richard Owen, the former forever an opponent of Darwin, the latter formerly a friend and later a bitter enemy. According to Agassiz (1860b), "while individuals alone have a material existence, species, genera, families, orders, classes, and branches of the animal kingdom exist only as categories of thought in the Supreme Intelligence, but as such have as truly an independent existence and are as unvarying as thought itself after it has once been expressed" (151). For Owen, with his theory of archetypes, which he specifically called

---

[3]   The cluster class views employ essentialism in a loose sense, while the ecological views employ it even less so and vaguely. As for causality, both the strictly essentialistic and the cluster class views (e.g., phenetic essentialism) need not necessarily impute causality to their essences, while the ecological views do so necessarily.

[4]   Armstrong (1989: 28–29) argues that when there is causal work to be done, it is done by things and their properties, not the classes to which they are thought to belong. Armstrong therefore thinks that classes, if they exist, are not causes. If Armstrong is right, then Plato's theory of Forms cannot be a class theory. Whether classes can be causes is an issue I will look at later in this chapter. It need only be mentioned here that Armstrong's discussion is implicitly only about efficient and not final causality.

"*ideai* of Plato" (Ruse 1979: 123), not species but only much higher categories of classification presuppose a transcendental archetype (Owen focused mainly on the vertebrate archetype), fixed in the mind of God, from which lower taxa such as species constitute divinely directed adaptive radiations to the various conditions of life. For Darwin, of course, evolution is contingent and Owen's archetypes are nothing more than ancestral forms (cf. Darwin 1859: 206, 435).

Turning to Aristotle, it seems to me that Kitts' thesis is better applied to Aristotle than to Plato. With Plato, the orthodox interpretation of his theory of animal kinds has still basically stayed the same, and not, I think, without good reason. With Aristotle the situation is quite different.

There is a tension in Aristotle's writings on animal species (cf. Sharples 1985: 119), a tension between, on the one hand, his logical and metaphysical works and, on the other hand, his biological works. The orthodox interpretation of Aristotelian species, which dates back at least to Porphyry, was based mainly on the former works, while what may be called the new orthodoxy, headed mainly by the works of David Balme and Pierre Pellegrin, is based mainly on the biological works, with the attempt to reinterpret the former in terms of the latter.

But first the orthodox interpretation (with references to Aristotle to support it). It is widely accepted that, whereas for Plato the essences (Forms) are outside (in a metaphysical sense) of their conspecifics, for Aristotle each essence is actually inside (in some sense) each of its conspecifics, each universal a *one in many* (*Pos. An.* 100^{a}8, *Met.* 1037^{a}29), *universalia in rebus* (to use the Latin tag). As Aristotle put it, "no universal exists apart from the individuals" (*Met.* 1040^{b}27),[5] and he clearly thought of this relation as one of interdependence (*Cat.* 2^{b}5–6, *Met.* 1071^a 18–21). This is true no less for bronze spheres and houses (*Met.* 1033^{b}20–23) than it is for animal kinds. His reasoning for so fundamental a reallocation is based primarily upon what is necessary for the possibility of generation, for "In some cases it is even obvious that the producer is of the same kind as the produced (not, however, the same nor one in number, but in form), e.g., in the case of natural products (for man produces man)" (*Met.* 1033^{b}29–32).

To this we must connect Aristotle's theory of change. To understand numerical identity in spite of change we must make, on his view, a distinction between essential and accidental attributes. An accident is "something which may either belong or not belong to any one and the self-same thing" (*Top.* 102^{b}6). A good example is whiteness, "for there is nothing to prevent the same thing being at one time white and at another not white" (102^{b}9–10).

On the other hand, "the essence is what something is" (*Met.* 1030^{a}3–4). It turns out in Aristotle's writings that essences are species and genera. In the *Categories* Aristotle calls concrete individuals "primary substances" (since they can only occur as subjects, not as predicates) and species and genera "secondary substances" (since they both can occur as subjects or predicates). Species and genera, as secondary substances, indicate of primary substances *what they are*, and thus their essential attributes: "only they [species and genera], of things predicated,

---

[5]   Quotations are from the Barnes edition (1984) of Aristotle's works.

reveal the primary substance. For if one is to say of the individual man what he is, it will be in place to give the species or the genus (though more informative to give man than animal)" (*Cat.* 2^{b}30–34). It follows, then, that if a particular primary substance were to undergo a change in either species or genus, it would undergo a *substantial* change and would lose not only its sameness (in species) but also its numerical identity—it would no longer be the same *this* but a numerically different *this* (*Met.* 1033^{a}30–^{b}25). As Aristotle put it in the *Topics*, "it is impossible for a thing still to remain the same if it is entirely transferred out of its species, just as the same animal could not at one time be, and at another time not be, a man" (125^{b}36–38). This follows from his view that the identity of each primary substance is a combination of matter and form (*Met.* 1069^{b}35–1070^{a}13), as Callias, for example, "is formula together with matter" (*Met.* 1058^{b}10). It follows that if Callias were to change in form, he would no longer be Callias.

That Aristotle thought that (what we today clearly recognize as) biological species each have their own essence or form is clearly indicated in a number of passages: he mentions, for example, man and horse (*Met.* 1058^{a}3), ox (*Cat.* 2^{b}27), dog (*Gen. An.* 747^{b}30–33), and three species of thrush (*Hist. An.* 617^{a}19), among many other examples. That each has its own essence or form is indicated in particular by what he says of man, for example "this is Callias or Socrates; and they are different in virtue of their matter (for that is different), but the same in form; for their form is indivisible" (*Met.* 1034^{a}6–7).[6]

Indeed there is a convergence in Aristotle's works in the terms translated as "species" (*eidos*), "essence" (*ousia*), "form" (*eidos*), and "universal" (*katholou*), the different translated words depending largely on the context of Aristotle's discussion: species distinguished from genera, essences from accidents, form from matter and as final cause,[7] and universals from particulars. Since Aristotle often refers to man as a species (e.g., *Cat.* 2^{a}17, *Hist. An.* 490^{b}17, *Met.* 1058^{a}4), it is evident from the above quotation that in spite of different contexts "form" is really synonymous with "species." But Aristotle also tells us that "form" is synonymous with "essence" (and therefore also "species" with "essence"), when he tells us that "By form I mean the essence of each thing and its primary substance" (*Met.* 1032^{b}1–2; also 1033^{b}7, 1035^{b}32). Since "a definition is a phrase signifying a thing's essence" (*Top.* 101^{b}35), that "the definition of man ought to be true of every man" (*Top.*139^{a}26), and since "definition is of the universal and of the form" (*Met.* 1036^{a}28) and that "the universal is common" (*Met.* 1038^{b}10–11) and

---

[6] Problematic is how their shared form can be indivisible when Socrates may die before Callias? Clearly one instance of the form may perish while the other persists (Barnes 1995b: 98). According to Sharples (1985), an increasing number of scholars interpret Aristotle as saying that "the form of man in one individual is *numerically* distinct from that in another, . . . but in no other way" (119; cf. 127n12). However, since Aristotle's forms are not constituted by matter, as we shall see, it should be noted that he adds that "all things which have *no* matter are *without qualification* essentially unities" (*Met.* 1045^b 23–24).

[7] Cf. Pellegrin (1985: 97), who distinguishes four different uses of the word *eidos* in Aristotle.

is "whatever is one and the same in all those things" (*Pos. An.* 100^{a}8), it follows that "universal" is synonymous (at least some of the time and in spite of different contexts) with "species," "essence," and "form."[8]

All of this is important to keep in mind when reading Aristotle. However, we do not yet have a class theory in the abstract sense until something else is added. A natural candidate is to think of Aristotle's forms as eternal, or rather, since they are in a sense in the world, everlasting (sempiternal).[9] Certainly Aristotle thought that the universe and the world (it must be remembered that Aristotle's universe is finite and geocentric; cf. *De Caelo* 286^{a}20–^{b}10) is everlasting. He says "the universe is permanent" (*Meteor.* 352^{b}18) and that "there will be no end to time and the world is eternal" (353^{a}15; cf. *De Caelo* 287^{b}25, *Met.* 1039^{a}29, 1071^{b}6–9), all the while allowing for large-scale cyclical changes in the earth's geography,

---

[8]   A further synonymous term should be added, namely "soul" (*psychē*). According to Aristotle "the form of the living being is the soul" (*Pa. An.* 641^{a}18), is what "constitutes the essential character of an animal" (641^{a}23), and "is what enables us to classify animals" (*De An.* 414^{a}1). Although interpretation is controversial, it may be gathered from the *De Anima* (esp. 412^{b}12–23) that Aristotle conceives of soul not as a vitalistic or substantial entity in its own right (substance dualism) but rather as a functional entity (property dualism) akin to modern functionalism in philosophy of mind. Indeed elsewhere he seems to define all essences (forms) in terms of functions: "What a thing is is always determined by its function: a thing really is itself when it can perform its function; an eye, for instance, when it can see. When a thing cannot do so it is that thing only in name, like a dead eye or one made of stone, just as a wooden saw is no more a saw than one in a picture" (*Meteor.* 390^{a}10–12). Similarly "a dead man is a man only in name" (389^{b}31–32; cf. *De An.* 412^b 18–22, *Pa. An.* 640^{b}35–641^{a}5). On this interpretation, then, a biological species would be a kind of organism the essential function of which supervenes on a disjunctive base of physical structures.

[9]   The Greek term that Aristotle normally uses for everlasting objects is *aïdios*, which commonly means *everlasting*, as opposed to *aiônios*, which is strictly applied to the timelessly eternal (Urmson 1990:12–13), although *aïdios* is often translated as "eternal" in modern translations of Aristotle. James Lennox (1985), however, argues that Aristotle's forms should not even be thought of as everlasting. His argument seems to be (I say this because it is somewhat obscure) that because Aristotle argues (contra Plato) that the forms do not have independent being, unlike his sun, moon, stars, and planets which are independent and everlasting (*Met.* 1040^{a}29–30, 1073^{a}30–35), they (forms) are therefore not entities in their own right and cannot be everlasting. (Interestingly we find much the same argument in many philosophers against the reality of relations.) Lennox's argument seems to me unconvincing. Clearly Aristotle's forms are for Aristotle something and not nothing, in spite of being grounded in and dependent on (supervenient on) physical being. And since for Aristotle they explicitly have a sort of unity (as per my n6 above), and since their physical members constitute an everlasting succession (*De An.* 415^{a}30–^{b}8, *Gen. An.* 731 ^{b}24–732^{a}1), this entails that they too must be everlasting, their physical dependence notwithstanding. Indeed there is a passage in the *Metaphysics*, completely overlooked by Lennox, wherein Aristotle, in the context of a criticism of Plato's Forms and with clear reference to his own conception of forms, says that "these substances must be without matter; for they must be eternal" (1071^{b}21–22). At any rate, what follows later in my text should hopefully add to this conclusion.

such as the extremely gradual replacement of a large land mass by a sea (*Meteor.* I.14). But to be an abstract class theory (and this will be argued for in §5), what needs to be added is that the forms (essences, species, universals) are incapable of change. And *that* Aristotle adds. That they are not capable of *accidental* change follows from his view that accidents are constituted by matter (*Met.* 1026^{b}32–1027^{a}15) and forms are not so constituted: "I call the essence substance without matter" (*Met.* 1032^{b}14–15). Indeed for Aristotle change is only possible in things that have matter as a constituent: "Nor has everything matter, but only those things which come to be and change into one another" (*Met.* 1044^{b}26–27; cf. 1069^{b}25–27).[10] That the forms are not capable of *essential* change, of course, follows tautologically from the fact that they themselves are the essences. But not only are the forms, according to Aristotle, incapable of change, they are also incapable of either coming to be or ceasing to be. His conclusion in fact is that "neither the matter nor the form comes to be" (*Met.* 1069^{b}35–36). Instead only individual things, concreta in the sense of primary substances, themselves composites of matter and form, are ever created or destroyed: "it is a 'this' that is made, i.e., the complex of form and matter that is generated" (1043^{b}18–19). His argument for this is based on the need to avoid an infinite regress: "E.g. we make a bronze sphere; and that in the sense that out of this, which is bronze, we make this other, which is a sphere. If, then, we make the sphere itself, clearly we must make it in the same way, and the processes of making will regress to infinity. Obviously then the form also, or whatever we ought to call the shape of the sensible thing, is not produced, nor does production relate to it,—i.e., the essence is not produced; . . . But that there is a *bronze sphere*, this we make. For we make it out of bronze and the sphere; we bring the form into this particular matter, and the result is a bronze sphere" (1033^{b}2–10; cf. "the being of house is not generated, but only the being of *this* house" 1039^{b}25–26). Indeed for Aristotle it is only in reproduction that an individual living being can partake of the eternal, namely the eternality of its species (*De An.* 415^{a}30–^{b}8, *Gen. An.* 731^{b}24–732^{a}1).

Moreover, it is this changelessness of essences that apparently allows for the possibility of science (knowledge). Indeed the connecting notion for Aristotle between knowledge and essence seems to be that of necessity. Highly implicit in what we have seen is that it is necessarily of the nature of essences (because of their immateriality) that they cannot change. Instead, contingency, the opposite of necessity, requires, or is a consequence of, matter, and this makes it unfit as a subject of knowledge. As Aristotle put it, "scientific demonstrations are about what belongs to things in themselves, and depend on such things. For what is accidental is not necessary" (*Pos. An.* 75^{a}30–32). Again, "it is opinion that deals with that which can be otherwise than as it is, clearly there can neither be definition nor demonstration of sensible individuals" (*Met.* 1040^{a}1–2).[11]

---

[10] It would appear for good reasons, then, that Aristotle scholars are generally agreed that Aristotle's view precludes species evolution; cf. Preus (1979: 343), Balme (1980: 300), Guthrie (1981: 222).

[11] The relation between definition, science, and essences (forms, universals, species),

What we have thus far in Aristotle, then, is clearly an abstract class theory of animal species, since each animal *eidos* (as a physically dependent abstract entity) remains thoroughly unchanged even though its physical members (individual organisms) come and go.

Problems begin to arise, however, when we begin to wonder about the nature of Aristotle's essences, specifically in living beings, whether (to use a modern distinction) their intensions are genotypic or phenotypic. Much of what Aristotle says suggests that he thought of them as phenotypic. First, essential definition is according to Aristotle a matter of naming the genus and the differentiae, such as "two-footed animal" (*Met.* 1037^{b}29–1038^{a}5). Elsewhere he mentions "feet, feathers, scales, and the like" as examples of "essential attributes" (*Pa. An.* 645^{b}2–5).[12] Moreover, when criticizing the Platonic method of bifurcate division, his main criticism is that it breaks up natural groups (*Pa. An.* 642^{b}10). Instead he asserts that, depending on the case, anywhere from one (642^{b}8) to a multiplicity (643^{b}24) of differentiae may be sufficient. And then there is that interesting passage in the *Politics* where, while drawing an analogy to the classification of political constitutions, he says:

> If we were going to speak of the different species of animals, we should first of all determine the organs which are indispensable to every animal, as for example some organs of sense and the instruments of receiving and digesting food, such as the mouth and the stomach, besides organs of locomotion. . . . the possible combinations of these differences will necessarily furnish many varieties of animals. (For animals cannot be the same which have different kinds of mouths or of ears.) And when all the combinations are exhausted, there will be as many sorts of animals as there are combinations of the necessary organs. [1290^{b}25–36][13]

---

according to Aristotle, would seem to be as follows. As he simply put it, "how will it be possible to know, if there is not to be some thing common to a whole set of individuals?" (*Met.* 999^{b}26–27). Thus, "there is knowledge of each thing only when we know its essence" (*Met.* 1031^{b}6). Moreover, since accidents are infinite in number (1007^{a}15), if all attributes are accidents, "predication, then, must go on *ad infinitum*" (1007^{a}30–1007^{b}1). Since definitions involve predications and science involves definitions (*Pos. An.* 99^{a}23), it follows that there can be no science of accidents (*Met.* 1064^{b}30–1065^{a}5), but only of essences. Thus, not only is definition "of the universal and of the form" (1036^{a}28) and of the "essence" (1030^{a}6), but "every science is of universals and not of particulars" (1059^{b}26–27) and "all science is either of that which is always or of that which is for the most part" (1027^{a}20–21). The "for the most part" is explained by the fact that for Aristotle forms (essences) play a teleological role in nature and nature is not perfect in its operations, the latter due to the fact that the forms must be realized in matter and "matter in some sense resists the imposition of form" (Hankinson 1995: 116; cf. *Phys.* 199^{a}11, *Gen. An.* 778^{a}8).

[12] It should be noted that for Aristotle whatever is different only in degree is not different in kind (cf. *His. An.* 486^{a}22–^{b}17, *Pa. An.* 644^{a}16–20, *Pol.* 1259^{b}36–38).

[13] Interestingly, in the *History of Animals* Aristotle indicates that classifications may be made not just according to the parts of living beings but rather "in their modes of subsistence, in their actions, in their habits, and in their parts" (487^{a}11–12). It is clear from Ar-

On the other hand, there are passages which strongly suggest that Aristotle thought of conspecific essences as genotypic (or more properly genomic, as per chapter 1.3). These passages come from his exposition of his theory of inheritance. According to Aristotle, in the generation of animals it is the male who provides the form (soul), while the female provides the matter (*Gen. An.* 729^{a}9–11, 738^{b}23–26). And to this we must remember that for Aristotle form is not potentiality. Instead matter is explicitly potentiality and form actuality (*De An.* 412^{a}9, 414^{a}12–30, *Met.* 1045^{a}23–24, 1050^{b}2). This then would entail a genotypic as opposed to a phenotypic conception of essence in living beings (cf. *Met.* 1014^{b}16–18). And this is in fact how many modern commentators interpret Aristotle.[14]

Greater difficulties arise when we turn in Aristotelian studies to the new orthodoxy (Hankinson's 1995: 123 term). I do not wish to enumerate the main features of what is properly characterized as a heterogeneous trend, but only to mention in brief some of the difficulties it poses for the traditional view outlined earlier, and where I think it fails. To begin, Balme (1962) points out that in the logical and metaphysical works the *genos/eidos* distinction is relative and not absolute, such that at any level (except the logically lowest, the ultimate or indivisible, the *infima* or *atomic* species) an *eidos* in its turn can function as a *genos* if it can be further subdivided, each subdivision being an *eidos*, and so on. Indeed as Pellegrin (1985) put it, "Aristotle never uses the word εἶδος, whether in the biological works or elsewhere, *in an absolute way* to designate a class or collection of objects: an εἶδος is always a sub-class or more exactly a subdivision of an expressed or understood γένος" (98). One would still nevertheless expect in the biological works that what are clearly biological species would be designated as *infimae* species. "But the surprising fact," says Balme (1962), "is that he makes least use of it [*infimae* species] in this field" (84). Moreover, "The traditional interpretation assumed that Aristotle did actually classify animals into genera and species, but this assumption is not supported by the evidence. He accepts as data

---

istotle's ensuing discussion, however, that he does not think that these different modes might result in different classifications, but rather that they will result in a convergence of classifications.

[14] E.g., Preus (1979: 344), Mayr (1988a: 56), Hankinson (1995: 133). Preus (344) attempts to obviate the above genotypic/phenotypic difficulty by arguing that, since the form in sperm is unobservable, phenotypes are used by Aristotle for classification as *evidence* of genotypic essences—if true, this would make him a precursor of modern population genetics—so that phenotypes are not therefore *constitutive* of those essences. This solution, however, does not seem to me entirely satisfactory, for the simple reason, as we have seen, that Aristotle in a number of passages characterizes forms and essences as *functional* entities based on gross phenotypic characteristics (e.g., the seeing of an eye). Preus' genotypic interpretation does not conform to this. It may be, then, that the form in semen should be thought of as a potentiality instead of as an actuality. Indeed in the *Generation of Animals* Aristotle tells us that semen has soul "potentially but not actually" (737^{a}13). But again, as pointed out above, Aristotle explicitly denies that form is a potentiality; only matter is a potentiality, while form is and must be on his view an actuality, since it is a final cause. It may be, then, that Aristotle's overall account of form (as we have it) is incoherent.

the animal kinds (*γένη*) that are presented to him by common parlance [cf. *Pa. An.* 643^{b}10–11], but does not try to group them further by finding similarities, . . . there is no classification scheme in the background, and all attempts to construct one for Aristotle have failed" (85).[15] In addition, according to Balme, of the 413 instances of the use of *genos* in Aristotle's strictly biological works, 354 of those instances denote a kind of animal, whereas of the 96 instances of *eidos* only 24 denote a kind of animal. "Thus *γένος*," says Balme, "is far the commoner word for a kind of animal" (85). Moreover, Aristotle, Balme (85) points out, uses the word *genos* not only for dog (*Pa. An.* 658^{a}29) but also for different breeds of dog (*Hist. An.* 574^{a}16). Also, according to Balme (86), for five different species of animal, each of those five is called a *genos* in one passage and an *eidos* in another. Thus Balme concludes that in the strictly biological works *genos* and *eidos* do not take on anything like the modern taxonomic meaning of "genus" and "species" but instead are taxonomically neutral and are normally used nontechnically to mean, respectively, kind and form.

It seems to me, however, that whether Aristotle presupposed a hierarchical taxonomic system in his biology (the issue does not concern me here), and given that his use of *genos* and *eidos* does not conform to our modern use of "genus" and "species," these matters must never be allowed to obscure the fact that Aristotle thought that biological species constitute a distinct level of being and are themselves objectively real—for that he clearly did, no matter whether he used the words *genos* or *eidos* or anything else to denote them (cf. Hankinson 1995: 110n3). Pellegrin's complaint (1982) that Aristotle often used the word *genos* "*below* the level of the species" (79) invites the rejoinder "So what?" since it merely reminds us (what Balme and Pellegrin are well aware of) that Aristotle often used that word in its common meaning of *kind*, or also *race* (cf. Urmson 1990: 65–66), which indeed Aristotle provides as one of the meanings of *genos*, *viz*. that of a "continuous generation of the same sort" (*Met.* 1024^{b}7).

Within the new orthodoxy we also find an attempt to deny the traditional view that Aristotelian species are essentialistic. Preus (1979: 342–343) provides a

---

[15]   Balme nevertheless concludes that Aristotle must have intended to apply in his biological works the technical distinction of *genos* and *eidos* developed in his logical and metaphysical works, "for it is incredible that he should have abandoned systematics in the very field where it has proved most fruitful, . . . In that case his biological work is incomplete" (98). Pellegrin (1982) takes Balme to task on this latter point, arguing extensively that Aristotle did not have a taxonomic goal, let alone one similar to the classical Linnaean system, pointing out (82) that in this Balme fell back into the errors of the traditional view and that Balme later agreed on this point (191n34). As for the passage in the *Politics* implying the contrary of Pellegrin's interpretation, which I quoted in full above, Pellegrin (1985) argues that this passage only indicates that "such a definition [of animal species] is possible," since "*this procedure was not put into operation.* This text, written as an unreal or contrary-to-fact conditional, does not correspond with anything in the biological corpus" (101). Pellegrin thus concludes that in this passage "Aristotle here calls attention to definitions of species not because of biological demands, but because of the nature of the political object to be explained" (102).

number of arguments against attributing to Aristotle what he calls "Noah's Ark essentialism." First, he refers to those passages in Aristotle's works (cf. chapter 2.1n2 and accompanying text) where Aristotle seems to subscribe to the ahistorical Great Chain of Being or *scala naturae*, the view that there is a smeary continuum between progressive forms of life. But I don't see this as necessarily contradicting Aristotelian essentialism, at least not in theory. The *appearance* of the continuum can be explained by ascribing it to variations in essential attributes between members of closely related though discrete essential forms (cf. n12 above), rather than to genuine smeariness between the forms themselves, or, what this amounts to, to no forms at all. It could also be explained by Aristotle's view of the lack of reproductive isolation between members of closely related forms, which I will discuss shortly.

Second, Preus refers to the problem of dualizing or equivocation in Aristotle's classificatory practice, in which some classes of animals do not fall under one higher class but instead straddle two. A classic example provided by Aristotle is that of the Libyan ostrich, which "has some of the characters of a bird, some of the characters of a quadruped" (*Pa. An.* 697^{b}14–26). But this is a problem for essentialism in supraspecific or higher taxa, and for Aristotle's method of division, not necessarily for his essentialism of forms at the level of the species category.

Third, Preus refers to those passages in Aristotle where Aristotle denies in many cases what today is called reproductive isolation. For Aristotle many pairs of closely related species are capable of producing fertile hybrids, for example "dogs and foxes," "wolves and jackals," and "partridges and hens," while out of all such matches "mules alone are sterile" (*Gen. An.* 746^{a}30–^{b}15). But again, this does not necessarily contradict strict essentialism. What needs to be remembered is that Aristotle believes that in all fertile hybrid crosses "as time goes on and one generation springs from another, the final result resembles the female in form" (*Gen. An.* 738^{b}31–32). So hybrid crosses are not really natural kinds on Aristotle's view, since they don't persist, not being a matter of necessity. Such crosses, then, I suppose, may be classed with what Aristotle calls deformities, such as a child that is said to have "the head of a ram or a bull" (*Gen. An.* 769^{b}15). A mule, then, would be a deformed horse *and* a deformed donkey. Indeed there is a passage in the *Metaphysics* where Aristotle strongly implies just that, that a mule is a type of deformity (1034^{b}3–4).[16]

Balme (1980) makes much of the fact that in Aristotle's theory of inheritance in the *Generation of Animals* (IV.3) what is passed on in the sperm is not only essential but also accidental characteristics and that the end of reproduction is to make the likeness of the father, but if that is not accomplished, then that of the mother, and if not that, then of ancestors, and if not that, then of the species itself, or then higher classes, this descent of reproductive effectiveness ending with monstrosities.[17] In his discussion, says Balme, "Aristotle makes no distinction

---

[16] Cf. my discussion on hybrid zones in chapter 3.3, which adds further weight to the above interpretation.

[17] All of this, incidentally, corresponds with Aristotle's ladder of life (*Pol.* 1256^{b}15–22,

here between substantial and accidental properties" (295). Each animal, then, has its own essence or form, which in males is both the efficient and final cause of reproduction. In this way Balme explicitly decouples Aristotle's essences from species. The biological treatises, he says, "make clear the difference between essence and species" (297). In sum, Balme's argument is that "in the *GA* Aristotle holds that the animal develops primarily toward the parental likeness, including even non-essential details, while the common form of the species is only a generality which 'accompanies' this likeness" (291). "Essence," says Balme, "picks out only those features for which a teleological explanation holds" (297), while each species, on the other hand, is "merely a universal obtained by generalization" (291), so that "species-membership is a consequential, not a primary cause, in animal reproduction and growth" (293).

Balme's thesis, however, does not seem to me at all satisfactory. The main problem is that, even though each female is according to Aristotle a "mutilated male" (*Gen. An.* 737^{a}28), females are nevertheless what he calls a "natural necessity" since in sexual species they are necessary for the preservation of those species (767^{b}9–10), which in turn are necessary for the ladder of life with its attendant teleology (cf. n11 and n17 above). Thus we can see why Aristotle would say in the *Generation of Animals* (in a passage entirely ignored by Balme, not to mention Pellegrin) that "the peculiar character of the species is the end of the generation in each individual" (736^{b}3–4). Contrary to Balme, then, in view of this teleological conception of species, as well as Aristotle's teleological conception of the ladder of life, Aristotelian species have a teleological role and are therefore essentialistic on Aristotelian principles. Aristotle's male chauvinism may not be entirely consistent with this essentialism—since natural masters (males), not a species in itself (*Met.* 1058^{a}29–^{b}25), are at the top of the ladder—but it cannot nevertheless be used to deny that essentialism.

Finally, as evidence that Aristotle was not really concerned with species, even in his biology, Pellegrin (1985) points out that "*he never gives us even a single example of a definition of any species of animal*" (99). Even his various definitions of man, according to Pellegrin (114n5), are only partial definitions. Instead, says Pellegrin, for Aristotle "the true objects of the science of living things are the 'parts,' their properties and variations" (106). Species, he says, "have only a verificational function" (108), as in the making of generalizations such as "The heart is of large size in the hare, the deer, the mouse, the hyena, the ass, the leopard, the weasel, and in pretty nearly all other animals that either are manifestly timorous, or betray their cowardice by their spitefulness" (*Pa. An.* 667^{a}20–22).[18]

De An. 414^{a}29–415^{a}1), in which the lower exists for the sake of the higher, with man (literally male humans, and natural masters at that) at the top of the ladder. Thus, the above theory of inheritance suggests a theory of recapitulation, in which the various levels of the ladder find expression in human progeny (Pellegrin 1982: 109), all of which is reminiscent of Plato's scale of degeneration in the *Timaeus* (cf. chapter 2.1n1).

[18] Pellegrin (106) gives a different example (*Pa. An.* 670^{b}32), but it does not support his point as well.

Contrary to Pellegrin, however, as we have seen, Aristotelian species have more than just a verificational role; they have a teleological role. And the fact that we do not have from Aristotle a full definition of any particular species need only be evidence of serious difficulties with his method of definitional division. Indeed right at the beginning of the *De Anima* he states that "To attain any knowledge about the soul is one of the most difficult things in the world" (402^{a}10–11), and in what immediately follows he expresses serious doubts about the efficacy of division in nailing down essences. None of this, however, should be used to deny the importance of or the ontology of species for Aristotle. What must never be forgotten is that in spite of the lack of real definition for any species Aristotle nevertheless repeatedly maintains the existence of what he calls *infimae* species,[19] and he explicitly calls these "substances" (*Pa. An.* 644^{a}24; cf. *Met.* 1034^{a}6–8). Moreover, from the examples which he gives, such as "sparrow, crane" (*Pa. An.* 644^{a}31–32), "horse, man, and dog" (*Met.* 1016^{a}26), it becomes clear that what he had in mind is basically what we today think of as species, though for Aristotle, I conclude, these are abstract and fixed essentialistic classes, whereas the modern biological view tends to think of them as concrete and evolving individuals.

Finally, when it comes to the traditional interpretation of Aristotelian species as essentialistic abstract classes, what must never be overlooked or underrated, it seems to me, is the sheer force and pervasiveness of the history of that tradition. While the Medieval schoolmen added to Aristotle's ontology an individual form for each human (Ayers 1981: 249–250) so as to allow for the immortality of individual souls, no one in fame and influence did as much as the Swedish botanist Carolus Linnaeus (1707–1778), the self-proclaimed "Prince of Botanists" (Leikola 1987: 45), to bring to fruition what may be called the Aristotelian paradigm.

Born the son of a Lutheran minister, Linnaeus was educated and later became a professor at the University of Uppsala in Sweden, one of the last strongholds of Aristotelian scholasticism in Europe (Leikola 1987: 45–47). As we have seen in chapter 2.3, Linnaeus held a fundamentally fourfold division of living beings, classes and orders being artificial, genera and species natural. As for species, following the Christianization of Aristotle by the later Medievals, with the attendant Genesis account of creation, Linnaeus stated in one his earliest works, *Fundamenta Botanica*, published in 1736, and in a number of other early works (cf. Larson 1968: 291n2), that "There are as many species as there were created different forms in the beginning"—"*Species* tot numeramus, quot diversæ formæ in principio sunt creatæ" (Leikola 1987: 49, 56n10). By "creatæ" Linnaeus not only meant creation by God in the biblical sense; he more interestingly believed that the first members of each species were either a single breeding pair (as with humans and other animals) or a single hermaphrodite (as with most, he believed, plants) (Larson 1968: 291–292, Leikola 1987: 50). In chapter 2.3 we already examined the kinds of characters that Linnaeus used to base varieties on. When it

---

[19] Cf. *Pos. An.* 96^{b}16, 97^{a}35–^{b}6, *De An.* 414^{b}27, *Pa. An.* 643^{a}17, 643^{b}17, 644^{a}4, 645^{b}25, *Met.* 1018^{b}5–6, and 1038^{a}20. It should be noted that, contrary to Balme (1962: 84), Pellegrin (1982: 146) includes *Pa. An.* Bk. I as a strictly biological work.

came to species and genera, however, at least with plants, Linnaeus based essential characters on generative and nutritive organs. As he stated in *Fundamenta
Botanica*, "There are as many genera as differently constructed organs of fructifi-
cation will bring forth of natural species of plants"—"*Genera* tot dicimus, quot
diversæ constructæ fructificationes proferunt plantarum species naturales" (Leikola 1987: 49, 56n10), which he expanded for species to include relatively constant nutritive parts such as roots, stems, and leaves (Leikola 1987: 49).

Given Linnaeus' view of species and genera on the one hand, and varieties on
the other, the former essentialistic and fixed, the latter varying and arbitrary, it is
all too easy to see the Aristotelian influence, with essentialistic characters determining species and genera, and accidental characters determining (or helping to
determine) varieties. As such, in spite of his creationism, it is important to see
that Linnaeus' species concept is truly an abstract class concept.[20] This is evident
from the fact that, although on Linnaeus' view individual members of a species
come and go in terms of existence, the species itself remains the same, a "unity in
generation" ("unitatem illam progeneratricem"), unchanged since the beginning
of creation. This of course could not be if its members were *constitutive* of the
species. As Linnaeus put it in his *Systema Naturae*, first published in 1735,

> As there are no new species; as similar always gives birth to similar; and as unity
> governs the order in every species, it is necessary that this unity in generation be
> attributed to some Omnipotent and Omniscient Entity, namely to God, whose
> work is called Creation. This is confirmed by the mechanisms, the laws, the prin-
> ciples, the constitutions and the sensations in every living individual. [Leikola
> 1987: 50][21]

---

[20] According to, for instance, Leikola (1987), a species was always for Linnaeus "the
totality of the individuals that had been produced from one original individual" (54). Although from the above discussion I am obviously in debt to Leikola for much of my understanding of Linnaeus' species concept, I find I must differ on this matter of interpretation. There is a vast difference, logically, between a group concept and an abstract class
concept (although, as we shall later see, the two are often confused), and it seems to me
that the evidence as examined above—the *form* concept in his definition from *Fundamenta
Botanica*—strongly supports the conclusion that Linnaeus' species concept falls within the
latter category.

[21] As Larson (1968: 293–298) amply documents, Linnaeus did not stick to his belief that
every species was created in the beginning. Beginning with his initial acceptance in the
mid-1740s of *Peloria* as a genuine hybrid species, and thus as a new species, one not in
existence since the beginning of creation, Linnaeus came in the early 1750s to believe in
the reality of what he called "mule species," such that in the beginning God only created
genera (or rather one species per genus), with intergeneric hybridizations later creating
further species (their generic membership depending on which of the male or female parts
of the parental species were employed). Later still, in the early 1760s, Linnaeus came to
believe that only orders were originally created. Not content to stop there, however, Linnaeus speculated further, just before his death, concluding that only classes were originally
created. As Larson points out, for all its empirical and theoretical defects, these final
speculations at least had the virtue of finally establishing, to Linnaeus' mind at least, a

The view that Linnean species are abstract classes receives only more support from Linnaeus' system for naming species. Although when we think of the modern binomial system of nomenclature in biology we are apt to think of the legacy of Linneaus, since it is from Linnaeus that our modern system is derived (including many of the actual names), it is important to realize that the binomial system was a later development in Linnaeus' system. Originally species identifications in his system were descriptions, beginning with the name of the genus followed by the essential characteristics which differentiated one species in a genus from other species in that genus (a practice inherited, obviously, from Aristotle's practice of defining species in terms of genus and differentiae, e.g., featherless biped for humankind). Later, however, beginning in the 1750s, Linnaeus adopted the much simpler binomial system (which he did not originate, but which he certainly standardized), for no other reason than as a shorthand to aid memory (Leikola 1987: 49), certainly not to mimic Christian proper names.

As we shall now see in the next section, the influence of Aristotle and Linnaeus in terms of species as essentialistic classes continues into the present day, albeit for the most part only in the heads of philosophers, and shorn of many of its Aristotelian and Christian trappings. It is to the modern view that we now turn.

3.2 Species as Elementary Classes

As we saw with Aristotle, there was a question as to whether his species essentialism was genotypic or phenotypic. With modern species essentialisms, however, where the essentialism is strict, there is no such question, since they all, in the spirit of reductionism, take them to be genotypic.

In fact these essentialisms are based on the same kind of essentialism that is now standard for chemical kinds. Looking to the periodic table, each element is defined in terms of its atomic (proton) number. For example, any atom that has seventy-nine protons in its nucleus is an atom of a particular kind (regardless of isotopes), in this case gold. (Isotopes are variations in the number of neutrons in the nuclei of an atomic element, each variation producing what is sometimes called a different "species" of that element.) But more than just defining the kind, the atomic number serves an explanatory function within chemical theory of why members of that kind have the properties which they indeed have (i.e., the number of protons in an atomic nucleus is equaled by its number of electrons,

---

natural hierarchy in the living world. Moreover, as Larson also points out, it is important to recognize that species on Linnaeus' later view were still essentialistic and fixed. But perhaps the most important point to recognize is that, in spite of Linnaeus' later speculations, it was his earlier views that gained him his initial fame and which remained in the minds of naturalists. This influence was best expressed by Charles Lyell (1832), as we saw earlier in chapter 2.3, when he wrote "the majority of naturalists agree with Linnaeus in supposing that all the individuals propagated from one stock have certain distinguishing characters in common which will never vary, and which have remained the same since the creation of each species" (3), as well as by the numerous authors (e.g., MacLeay, Darwin) who repeated the Linnean dictum *Natura non facit saltum*, which of course Linnaeus himself had come to reject.

which, along with variations in the number of neutrons, determines the chemical activity). Moreover, proton number has a predictive function, so that physicists can not only predict the properties of but can also synthesize new elements (all but two of the first ninety-four elements in the periodic table are naturally occurring on earth).

This form of essentialism, of course, is not confined to only chemical elements; it applies to compounds as well. H_2O, for example, defines what we call (pure) water. Any molecule that is H_2O is a molecule of water and no other, and any liquid, gas, or crystal that is completely made up of H_2O molecules is a liquid, gas, or crystal of pure water.

That this theory of essentialistic kinds is an abstract class theory ought to be clear, for not only is the *kind* not a physical or concrete entity, but the *kind* remains exactly the same even though particular physical members of it come and go. And that is the *quintessence* of an abstract class theory. Moreover, membership in the class is not determined by origin or descent. An atom of gold is an atom of gold whether it was produced by geological processes or by a philosopher's stone. All that matters is that it have the structure which is essential to that kind.

It is only natural, then, given the impressive developments in chemistry above,[22] to apply this successful form of essentialism to define kinds at higher or

---

[22]   The above is, of course, an oversimplified account. Nevertheless, it seems to me fundamentally sound. On the other hand there is much room for skepticism. Paul Churchland (1985), for example, gives voice to much of it. According to Churchland not only are most putative natural kinds really practical kinds, but all kinds may turn out to be practical kinds. Two of his points, in particular, seem to me worthy of serious consideration. First, "the possibility that, for any level of order discovered in the universe, there always exists a deeper taxonomy of kinds and a deeper level of order in terms of which the lawful order at the antecedent level can be explained. It is, as far as I can see, a wholly empirical question whether or not the universe is like this—like an 'explanatory onion' with an infinite number of concentric explanatory skins. If it is like this, . . . there are no natural kinds at all. All kinds are merely practical kinds" (14). Second, "Even if the world is not an onion of the relevant kind, there is still no guarantee that there exists a unique and final theory (= set of sentences) flawlessly adequate to its complete description. On the contrary, it may be that the cognitive medium of human natural language suffers certain fundamental structural limitations in its capacity for representing the intricacies of the universe" (15). Far be it to refute these two points in a footnote, their acceptance does not seem to me at present reasonable. First, a logical flaw: Dependence within a hierarchy does not necessarily entail ontological reduction, or rather elimination, for if it would, it would contradict the reality (the ontology) of the hierarchy. Second, as an eliminative materialist Churchland seems persistently insensitive to the possibility that some kinds, though they depend on lower-level kinds, supervene on those lower-level kinds and therefore cannot be reduced (in any sense of the term) to them. Finally, if one reads an authority no less than Steven Weinberg (1992), one gets a clear sense that in elementary particle physics a final theory is not only possible but possibly near at hand. As Weinberg puts it, "Some searches do come to an end. . . . If history is any guide at all, it seems to me to suggest that there *is* a final theory. In this century we have seen a convergence of the arrows of explanation, like the conver-

more complex levels of reality, namely, in the present case, biological species.

We begin with a look at Irving Copi (1954), a prime example. In his paper Copi defends a standard version of Aristotle's distinction between essence and accident. Using the example of salt, he says "The scientist singles out its being a compound of equal parts of sodium and chlorine as its essential nature" (189). Copi claims, moreover, that "Modern science seeks to know the *real* essences of things, and its increasing successes seem to be bringing it progressively nearer to that goal" (187). As for species and evolution, it seems to Copi that "the fixity of species is a casual rather than an integral part of the Aristotelian system, which in its broad outlines as a metaphysical framework can be retained and rendered adequate to the most contemporary of scientific developments" (179).[23]

Carl Hempel (1966), in reference not specifically to biological species but to "the thesis of *reducibility of biology to physics and chemistry*" (102), calls the thesis a "heuristic maxim, . . . a principle for the guidance of research" (106), where the reduction concerns both "the terms and laws of biology" (105). A little more specifically, he conceives of the reduction of biological terms in the sense of "*extensional definitions*" (103), where the extension of a biological term is characterized in terms of a common "molecular structure" (103).

Turning now to W.V.O. Quine. Quine is, of course, not only one of the premier logicians of the twentieth century, he is also, as generally recognized and as indicated by the quotation at the head of this chapter, the class realist *par excellence*. For Quine a complete description of the universe would include not only all physical objects but also all abstract objects, namely classes "over and above the physical objects" (1981: 15).[24] As for species, Quine (1965) tells us that

---

gence of meridians toward the North Pole. Our deepest principles, although not yet final, have become steadily more simple and economical" (231–232).

[23] My disagreement with Copi here could not be greater given my previous analysis. That analysis does indeed establish that the fixity of species (forms, essences, universals) is an integral part of Aristotelian ontology. That Aristotle had a fundamental insight here and basically got it right, that abstract classes cannot change or evolve, will be bolstered by a new argument in §5 of the present chapter, an argument inspired in the present author by his close reading of Aristotle.

[24] Quine (1981) thinks of his class realism as "a realism of universals" (182). Armstrong (1989), however, characterizes Quine's class realism as a "class nominalism" (xi, 10), moreover "an extreme form of Nominalism" (8), by which he means, of course, universals nominalism. The confusion here may perhaps be worth clearing up. Armstrong's point is that "because classes are not repeatables, and universals are repeatables" (10), classes are not universals, so that an ontology that includes only classes and physical objects must be a universals nominalism. I suppose this all depends on how one defines universals (it is generally held that universals have instances, not members, whereas classes have members, not instances), a debate that does not really interest me here. What does interest me here, and greatly, is Armstrong's point that classes are not repeatables. I think Armstrong here is right, although his point is in need of argument (which he does not provide). I will reserve my own argument to back up his point, and the implications that follow from it, for §5 of the present chapter, where I discuss the issue of abstractions and change, an issue that lies at the very heart of the many problems with the species-as-classes views.

"When we say that man is a zoological species, we mean that this abstract entity, the class of men, is a zoological species" (122–123). Quine (1982) also tells us that species words play a dual role, not only as predicates but as names of classes: "'Man' is true of each man and may also be said to name a class, mankind" (288). But the nature of Quine's class realism, and more specifically his attitude toward strict essentialism, is rarely properly understood.[25] With explicit reference to Aristotelian essentialism, and Aristotle's distinction between essence and accident, Quine tells us that it is "surely indefensible" (1960: 199–200), that it is "unreasonable" (1961: 156), that it is an "invidious distinction" and "uncongenial" (1976: 184), and that it is "loose talk" (1982: 289). Quine's dislike of Aristotelian essentialism, however, would seem directed only toward *individual* essentialism (a view properly Medieval rather than Aristotelian), not *kind* essentialism:

> Mathematicians may conceivably be said to be necessarily rational and not necessarily two-legged; and cyclists necessarily two-legged and not necessarily rational. But what of an individual who counts among his eccentricities both mathematics and cycling? Is this concrete individual necessarily rational and contingently two-legged or vice versa? Just insofar as we are talking referentially of the object, with no special bias toward a background grouping of mathematicians as against cyclists or vice versa, there is no semblance of sense in rating some of his attributes as necessary and others as contingent. Some of his attributes count as important and others as unimportant, yes; some as enduring and others as fleeting; but none as necessary or contingent. [1960: 199]

Whether concrete individuals require essences is a topic that will be further explored in chapter 4. But now what of abstract individuals, namely classes, and more specifically species? It is here that Quine, like many philosophers before and after him, subscribes to what may be called the genotypic version of the Aristotelian paradigm. In his celebrated essay "Natural Kinds" (1969a), Quine argues that a science reaches maturity only when its natural kinds cease to be defined in the loose terms of similarity and begin to be defined in the strict terms of microstructure on the example of chemistry, "and so by-pass the similarity" (135), with the similarity notion "finally disappearing altogether" (138), rendering the kinds based on similarity "respectable and, in principle, superfluous" (137). (Quine here displays that common reductive tendency to eliminate higher-level kinds in favor of lower-level ones. That similarity may indeed be an objective, irreducible relation, constitutive of higher-level kinds, will be examined in chapter 5.2/3.) Though Quine has very little to say in "Natural Kinds" about organismic similarity, he does suggest that a mature biology would reduce this concept "in terms of genes" (136). That Quine fully belongs to the class of philosophers here examined in this section, however, comes out just as clearly (if not more) in a roughly contemporaneous work of his:

---

[25] Sober (1980), for example, claims that Quine is a global antiessentialist, meaning that he (Quine) "wishes to banish essentialism from the whole of scientific discourse" (350). This is an understanding of Quine, common enough, that we will see is incorrect.

There is also in science a different and wholly respectable vestige of essential-ism, or of real definition, but it is tangential to the lexicographer's concerns. It consists in picking out those minimum distinctive traits of a chemical, or of a species, or whatever, that links it most directly to the central laws of the science. Such definition has little air of semantics and is of a piece rather with the chemi-cal or biological theory itself. Such definition conforms strikingly to the Aristo-telian ideal of real definition, the Aristotelian quest for the essences of things. This vestige of essentialism is of course a vestige to prize. [1976: 52]

So much for Quine. Saul Kripke and Hilary Putnam's view is also of great in-terest, especially since it is intimately connected with a highly controversial view of meaning and reference known as the *causal theory of meaning*. Their natural kind essentialism is but an extension of their application of a causal chain theory of reference concerning proper names. On the traditional theory of meaning, a proper name is but a shorthand for and is synonymous with a necessary and suffi-cient, and therefore definite, description. For example, "Aristotle" is the name of the most famous pupil of Plato, the tutor of Alexander the Great, etc. A serious difficulty with the traditional view is that it cannot accommodate identity in coun-terfactual and possible worlds. Had Aristotle not in fact tutored Alexander, he would still be Aristotle. Aristotle's identity does not depend on whether or not he tutored Alexander. His tutoring Alexander is a contingent rather than a necessary fact, so that it is not a necessary identification. On what, then, does Aristotle's identity depend? According to Kripke (1972), who is the principal founder of this new theory of reference, the identity of Aristotle depends on his parentage, more specifically the particular sperm and egg which gave rise to the particular zygote from which Aristotle began (112–113). Moreover, it is possible, on this view, for we today to correctly refer to Aristotle even if we were to be factually incorrect in all of our descriptions of him. This is possible only if our reference to Aristotle is part of a causal/historical chain which links back to the "initial baptism" in which he was originally named (96). What fixes the reference at the initial baptism is either an ostension or a description (97)—in the case of Aristotle, the description, I presume, would be "the son of Nicomachus and Phaestis." In this way the name "Aristotle" can function as what Kripke calls a *rigid designator*, such that "in every possible world it designates the same object" (48).

As for natural kinds, according to Kripke "terms for natural kinds are much closer to proper names than is ordinarily supposed" (127). As with proper names, what fixes the reference of a natural kind, on his view, is an initial baptism, either an ostension or a description. In the case of gold, the latter would include (and the former might imply) its yellow color, its malleability, etc., what Kripke calls "the initial identifying marks" (119). But as with facts about the life of Aristotle, we may turn out to be mistaken about these initial identifying marks, mistaken in the sense that they are not necessarily associated with what is in fact referred to. Thus, according to Kripke "the way the reference of a [natural kind] term is fixed should not be regarded as a synonym for the term" (135). Even if we were to be mistaken about *all* of the initial identifying marks—and for Kripke this is possible even in the case of gold (cf. 118–119, 124–125)—it would not follow, according to Kripke, that our original referent was not a natural kind. It would be a natural kind, on his view, if (in spite of our ignorance) it had an internal essence, which would make the name we initially gave it a rigid designator.[26] In the case of gold, "present scientific theory is such that it is part of the nature of gold as we have it to be an element with atomic number 79" (125), so that, if present scientific theory is correct and not mistaken (123), "a material object is (pure) gold if and only if the element contained therein is that with atomic number 79" (138). Or as Putnam (1975a) put it for water, "Once we have discovered that water (in the actual world) is H_2O, *nothing counts as a possible world in which water isn't H_2O*" (233).

Interestingly, what Putnam (1973) finds to be of great significance in the causal theory of reference is that it allows for terms to be *trans-theoretical*, i.e., "terms that have the same reference in different theories" (197), which in turn allows for a realist theory of meaning, such that "concepts in different theories may refer to the same thing" (197). One might immediately think here, as a classic example, of the species debate and the paradigm shift effected by Darwin (cf. chapter 2.3).[27] But both Kripke and Putnam, interestingly, rather than pay attention to Darwin and the biology that followed, clearly on the matter of biological species have attempted to revert to an essentialism of the pre-Darwinian kind, though now wedded to modern chemical theory. In other words, they explicitly treat biological species, the commonest example being tiger, in the same way they treat chemical kinds, typically gold and water. Though their essentialism regarding biological species is guarded, it is clearly part of their program. According to Putnam (1970), "An important class, philosophically as well as linguistically, is the class of general names associated with natural kinds—that is, with classes of things that we regard as of explanatory importance; classes whose normal distinguishing characteristics are 'held together' or even explained by deep-lying mechanisms. *Gold, lemon, tiger, acid*, are examples of such nouns" (139). What each of these has, according to Putnam, is an "'essential nature' which the thing shares with other members of the natural kind. What the essential nature is is not a matter of language analysis but of scientific theory construction; today we would say it was chromosome structure, in the case of lemons, and being a proton-donor, in the case of acids" (140–141). Putnam's essentialism here is guarded since he immediately continues by saying that "this is vague, and likely

---

[26] Although Kripke compares and identifies both proper names and natural kinds as rigid designators, his microstructural essentialism for natural kinds such as gold and water does not carry over to individuals such as Aristotle. In other words, Aristotle's DNA program is not what makes "Aristotle" a rigid designator (clearly it would fail in the case of identical twins). Instead, Kripke's essentialism for individuals is a *relational* essentialism based on origin and originating substance (cf. Kripke 1980: 114n57; Fales 1982: 85).

[27] What this example proves, it seems to me, is that realism implied by trans-theoretical terms need not involve strict essentialism, but that is a conclusion I will argue for later in this section.

to remain so." Later (1975a) he would add the caveat that we should not think of his account "as implying that the members of the extension of a natural-kind word necessarily *have* a common hidden structure" (240–241). He gives as an example disease names, some of which refer in fact to nothing in common except a cluster of symptoms. But there is nothing in his account to make us think that he suspected the same for names of biological species. His essentialism here remains what Hempel calls a heuristic maxim. So too for Kripke. Although his essentialism for biological species is likewise guarded—"whether a given kind is a species of animals is a matter for empirical investigation" (122–123)—he nevertheless thinks it probable that cows and tigers are natural kinds just as gold and water. Thus, "Even though we don't *know* the internal structure of tigers, we suppose—and let us suppose that we are right—that tigers form a certain species or natural kind" (120–121). Again, "Whether science can discover empirically that certain properties are *necessary* of cows, or of tigers, is another question, which I answer affirmatively" (128). Indeed for Kripke this is of a piece with science in general: "In general, science attempts, by investigating basic structural traits, to find the nature, and thus the essence (in the philosophical sense) of the kind" (138).[28]

Finally, according to the philosophers Kitts and Kitts (1979) "The search for essences is prompted by theoretical necessity. Speciation, which stands at the core of contemporary evolutionary theory, is a process by which groups of organisms become reproductively isolated from one another. Biologists seek a common explanation for the restriction of each of the members of a species to that species. . . . Since the discovery of the structure of genetic material it has been possible to get at this underlying trait not only through the manifest properties and the reproductive behavior of organisms, but more directly by means of chemical techniques" (621–622). But more than this, as we have seen earlier (chapter 2.3), Kitts and Kitts claim that "To suppose otherwise is not to give reason to change our view of species, but to give reason to abandon the concept of species altogether" (618).

Since no one has as yet proposed a strict molecular essence for even so much as one species, the question becomes whether this strict essentialism for species is even compatible with modern evolutionary theory. Whether modern evolutionary theory is compatible with species as abstract classes is a larger issue I will reserve

---

[28] Kripke and Putnam's particular brand of species essentialism has garnered them some rather unflattering remarks. For example, Michael Ruse (1987) makes the comment that these thinkers "at times show an almost proud ignorance of the organic world" (358n1). Jagdish Hattiangadi (1987) remarks that "I find it astonishing that in this paper Kripke purports to establish by analysis alone a certain metaphysical aspect of reality. He proposes a metaphysical doctrine of essentialism by analysing proper names. Not since Hegel established with the aid of Reason, that there can be only so many planets, in the year that a poor deluded astronomer discovered one more, has there been such a triumph of reason over mere empiricism. Kripke's timing is perhaps less perfect than Hegel's for his corollary that cats have an essence was demonstrated some time after Darwin had convinced us otherwise on mere empirical grounds, but his thesis is no less remarkable for all that" (74).

for §5. The more restricted problem will occupy us at present.

But first, why should morphological essentialism be unacceptable, even for philosophers, given modern evolutionary theory? No doubt one problem is that, given any such supposed essence, one would probably be able to find exceptions which are nevertheless conspecific. As Sokal and Sneath (1963) put it, "Every systematist knows of instances where a character previously considered to be diagnostic of a taxon is lacking in a newly discovered organism which clearly belongs to the taxon" (13). A deeper problem is that modern evolutionary theory entails species variability over varying geographies. But a deeper problem still is that, as Elliott Sober (1980) put it in a different context, "The key idea, I think, is that the membership condition [for essentialism] must be *explanatory. . . .* A species essence will be a causal mechanism which works on each member of the species, making it the kind of thing that it is" (354). Clearly a morphological essentialism fails most deeply at this point. But in reply, why should the membership condition be explanatory? The reason is that without it being explanatory there is nothing to distinguish any property from just happening to be common from one that is necessarily common. There is thus no way to determine whether a given property is truly essential or not.

On the other hand a molecular essentialism, more specifically one based on DNA, clearly would fulfill the above requirement. But is this theoretically compatible with the facts and fundamental theories of modern biology, philosophical considerations aside?

There are two perspectives here on which to base a sound judgment, the one synchronic, the other diachronic. Beginning with the former, one might begin by naively suggesting chromosome number as determining species essence. The human genome, for example, has forty-six chromosomes. But some humans, specifically those with Down's syndrome, have a genotype with an extra chromosome twenty-one, and therefore forty-seven chromosomes, but they are no less human for it. Moreover, a great many species share the same chromosome number. Chromosome number, therefore, cannot determine species essence. To find it in the genetic material one must consequently dig deeper into the chromosome, into the genetic code itself.

But here we only find deeper problems for essentialism. One problem, of course, is purely practical, and involves discoverability (cf. chapter 1.2). Although we now have gotten quite good at sequencing DNA, it would be a prohibitively enormous undertaking to sequence, let alone compare, the entire DNA of every, say, modern gorilla. We would have to assume, furthermore, that each DNA sample is representative of the genotype of its donor organism, an assumption that is quite weak given the fact of mutation.

A deeper theoretical problem is that, as John Dupré (1981) put it, "it is equally possible that there should be as much or more genetic variability as morphological variability" (84). This possibility is certainly worth exploring further. But here I will not do as Dupré, and appeal to the terms and concepts of classical genetics, such as gene locus and heterozygosity, for if element-like essentialism is to have any hope of success today it must cash out in terms of molecular genetics. But in the language of molecular genetics "genes are actually very complex . . . and exceedingly difficult to define" (Futuyma 1986: 47; cf. chapter 2.4n29 and accompanying text). Instead, then, I will turn directly to the more modern and reductive terms and concepts of molecular genetics itself.

Although variation caused by varying geographies does not necessarily preclude a fixed genetic essence, other considerations do. The synonymy of codons, discussed in chapter 2.2, entails the consequence that for any given phenotypic effect there are literally billions or trillions of different DNA sequences that may code for that self-same effect. A phenotypic effect consists of an enormous mass of protein molecules, each protein molecule consists of one or more polypeptides, each polypeptide consists of a chain of amino acids, and each amino acid is coded for by a codon, a triplet of nucleotides on mRNA, of which there are four nucleotides (cytosine, uracil, guanine, and adenine) and which pair in four possible ways on a corresponding DNA strand (double helix) as CG, GC, TA, or AT (in mRNA uracil takes the place of thymine in DNA). Since there are only twenty amino acids involved and sixty-four possible triplets of nucleotides given a four letter code, there is going to be considerable synonymy between codons (for a table see, for example, Futuyma 1986: 46).

Although this entails that literally billions or trillions of different DNA sequences may code for the self-same phenotypic effect, it does not follow, of course, that in any given population this tremendous variability will find even a low percentage of expression, since sequence segments get passed down by heredity and trace back to the parental population. The problem now, however, is that even if we were to have a candidate for a possible molecular essence, that sequence or set of sequences is contingent because of the synonymy of codons, and so does not fulfill the explanatory (necessity) criterion above, which is clearly satisfied in the case of chemical essentialism (elements). In other words, in chemical essentialism descent does not play a role, whereas in the above scenario it would have to, with the result that we no longer have a *molecular* essentialism.

Some, of course, may fail to appreciate the significance of this upshot. What will generally be found more impressive is the argument from mutations. Given any possible molecular (DNA) essence, it is bound sooner or later to undergo a change in at least some of its members due to the phenomenon of random mutation in base pairs, each of which is called a point mutation. According to Abercrombie *et al.* (1990), "Experiments suggest that an average of about one base-pair changes 'spontaneously' per 10^9 base-pair replications" (381). This may not seem like much, until we realize that not only can this rate be increased by mutagens, but also that, in the larger organisms with which we are more familiar, there are literally billions of base pairs comprising each organism's total genotype (in humans this amounts to roughly 100 billion in each somatic cell). This renders it highly unlikely that a species essence would remain the same in even one organism supposing it to have one. This situation becomes even more pronounced once we take into account what are called macromutations or chromosome mutations (e.g., inversions, translocations, and deletions), which involve larger sections of the chromosome and which are also subject to mutation rates and mutagens.

Turning now to the diachronic situation, David Hull (1965) makes the point that Darwin put an end to essentialistic thinking about species: "If species evolved so gradually, they cannot be delimited by means of a single property or set of properties" (203). In response to this view, Elliott Sober (1980) thinks that evolution does not of itself undermine species essentialism (although he rejects it on other grounds). "One often hears," says Sober, "that evolution undermined essentialism because the essentialist held that species are static, but from 1859 on we had conclusive evidence that species evolve. This comment makes a straw man of essentialism . . . . For one thing, notice that the discovery of the transmutation of elements has not in the slightest degree undermined the periodic table. The fact that nitrogen can be changed into oxygen does not in any way show that nitrogen and oxygen lack essences. . . . The mere fact of evolution does not show that species lack essences" (356; cf. Sober 1993: 146–147). But in this Sober seems to me deeply mistaken. His mistake is best expressed by Alexander Rosenberg (1985):

> It is not impossible, of course, to say that one quantity of matter has changed from being radium to being radon by decay, but there is no such thing as the element radium changing into the element radon. At most, all quantities of radium may decay into radon, and then there will be no samples of radium left. But its place in the periodic table will not be expunged. There is no difficulty in the notion that through transmutations or decay, quantities of another heavier element may become samples of radium. But the *kind* radium cannot change into the *kind* radon. The notion that species evolve is, in an essentialist view of the matter, to be understood in this latter sense, as the change not of organisms but of the kinds they belong to. As such, species can evolve no more than the kind radium can change into the kind radon. [189]

Sober's mistake, then, is to confuse change in the essence of an individual or group of concreta with change in the kind itself to which it belongs. The former kind of change certainly does not entail the latter, as Rosenberg has clearly shown. But perhaps modern chemistry is the wrong paradigm for comparison. I myself do not think it is. Essentialism, as we have examined it so far, and as we will examine it later in the following two sections, metaphysically entails an ontology with abstract classes, since the essence, which determines membership in the kind, remains the same even though members of the kind come and go. The essence *per se*, then, cannot be the set of the physical instances of it, otherwise it would necessarily change as that set changes. But it does not change as that set changes. It is therefore an abstraction. For species conceived as abstract classes to be possibly compatible with evolution, then, the classes must themselves be capable of change. This question, with my negative reply, will be examined in §5. For the present, the foregoing will suffice to have demonstrated on the grounds of modern molecular and evolutionary biology that strict element-like species essentialism, contrary to what many modern philosophers think, is an untenable view and should rightly be thought of, as it is among virtually all biologists today, as a dead issue.

3.3 Species as Cluster Classes

What is not a dead issue, biologists and philosophers included, is the view that species are cluster classes of some sort. Cluster classes are loosely essentialistic, for although there is a set of defining properties which defines the class, no one property from within that set is necessary or sufficient for membership in the class. Instead, all that is required for sufficiency of membership is possession of a certain minimum number of properties from within that set, a minimum quorum.[29]

As with strictly essentialistic classes, we can divide biological species conceived as cluster classes into basically two kinds, the one morphological or phenotypic, the other genotypic or molecular. The former, interestingly, has often been traced back to an eighteenth-century French botanist, a younger contemporary of Linnaeus by the name of Michel Adanson (1727–1806). As Simpson (1961) characterized his view, "It sufficed to form a taxon if each member had a *majority* of the *total* attributes of a taxon" (42). Thus, "A species consists of individuals with a maximum number of shared characters" (41). Although Adanson did believe that some characters are taxonomically more important than others, this he seems to have applied only in determining taxa higher than genera (cf. Mayr 1982: 194–195). At the level of species and genera, he seems to have believed that such taxa are determined visually by overall similarity, with no characters being diagnostic or essential.

Simpson's characterization seems to make out of Adanson's view a concept of species as physical entities of a sort. But any such interpretation seems to me incorrect. That cluster classes are abstractions should be even more apparent than in the case of strict essentialism, for with cluster classes none of the members of the class need instantiate all of the properties which define the class; indeed in many cases they can't (*viz.*, polytypic species). The reality of each cluster class, then, if it is to be real at all, cannot be physical but must be abstract. We have, then, a definite sense in which the class (in this case each species) remains entirely the same even though members of it come and go.

Indeed species conceived as cluster classes has an interesting history. Renford Bambrough (1960–61) connects cluster theory with Wittgenstein's remarks on the nature of ordinary language, specifically his remarks on games and family resemblances (cf. Wittgenstein 1958: §§65–67). Bambrough also contends that these remarks need not be closely restricted to concepts such as games and family portraits but that they can be applied as *the* solution to the problem of universals as a whole. For Bambrough cluster theory provides a realistic middle ground between nominalists on the one hand, for whom universals are real in name only because according to them there are no strict essences, and realists on the other hand, for whom universals are objectively real because according to them there

---

[29] The term is Latin and has its origin in Roman legal practice. In the Roman senate, passage of a law was not possible unless a certain minimum number of senators were in attendance. In this sense no one senator had a privileged position over any other, since no one senator had to be in attendance for the quorum to be met.

are strict essences. Strongly implicit near the end of Bambrough's account is that
species can accordingly be accounted for as cluster classes. Wittgenstein, interestingly, would probably have taken strong exception to this, since one of his gems
(Wittgenstein 1922) was that "Darwin's theory has no more to do with philosophy than any other hypothesis in natural science" (§4.1122). Wittgenstein, of
course, at least as modern philosophy of biology is concerned, couldn't have been
more wrong. At any rate, cluster theory, if it was indeed what Wittgenstein had in
mind, went on to have an interesting history of application to the species problem, whether inspired by Wittgenstein or in spite of him.

Of great interest is a book by Morton Beckner (1959), arguably the first comprehensive text on philosophy of biology. Favorably mentioning Wittgenstein and
his theory of family resemblances, Beckner gave to cluster theory perhaps its
most formal exposition to date. According to Beckner,

> A class is ordinarily defined by reference to a set of properties which are
> both necessary and sufficient (by stipulation) for membership in the class. It is
> possible, however, to define a group K in terms of a set G of properties f_1, f_2,
> . . . , f_n in a different manner. Suppose we have an aggregation of individuals (we
> shall not as yet call them a class) such that:
>
> 1) Each one possesses a large (but unspecified) number of properties in G
> 2) Each f in G is possessed by large numbers of these individuals; and
> 3) No f in G is possessed by every individual in the aggregate [22]

If only the first two conditions are met, the group is what Beckner idiosyncrati-
cally calls "polytypic," and if the third condition is also met, it is what he calls
"fully polytypic."

There is much more to Beckner's formalization of polytypic classes, which
includes an interesting comparison with what he calls monotypic classes (what I
call strictly essentialistic classes), but what interests me most is his application of
it to biology, specifically the species category. In this application Beckner does
not believe he is doing anything radical. Instead, he believes he is adding formal
clarity to the new view in taxonomy found in Julian Huxley's 1940 anthology
The New Systematics. Two points seem to me especially interesting.

First, although for the new systematists species are fully real, there is a basic
consensus, according to Beckner (68), that a certain degree of subjectivity is unavoidable in the delimitation of actual species. Beckner identifies this element of
subjectivity in the minimum quorums, which he fully admits are "an essential aspect of polytypic [cluster] classes" (24). As he expands on this elsewhere, "The
vague term 'large' enters twice into the definition of a polytypic concept. The
species-definition tells a good deal about the makeup of a species, but the systematist must decide just how large 'large' is in a particular case" (69).

Although Beckner (24) is correct in saying that the vagueness of the term
"large" necessarily allows for borderline cases in cluster class membership, and
although this may seem a theoretical plus because borderline cases clearly do
occur in nature, it is surely not true that every species, at least when perceived

horizontally, has borderline cases. Many species involve no fuzzy borders whatsoever. It ought to be the case, then, that many species can in principle be objectively delimited. But if cluster classes necessarily involve a degree of subjectivity, then this would, it seems to me, count against species as being cluster classes.

Furthermore, not only is a degree of subjectivity and arbitrariness necessarily involved with cluster classes, but they imply the Platonic concept of *degree of membership.* Certainly we often hear it said, for example, that so-and-so is more of a man than that other so-and-so. But biologists *qua* biologists seldom if ever talk like that. For biologists a sick or deformed man is no less a man than a healthy and anatomically correct one. In this sense too, then, cluster theory would not seem to conform to present biological theory and practice.

Of further interest is Beckner's claim that his formalization makes fully meaningful the view of the new systematics in which the population, rather than the individual specimen, is now the basic taxonomic unit. "In what sense," he asks (69), "is the population 'more basic'?" According to Beckner, the new systematists do not make this clear. Beckner's solution is to say that if we conceive of populations as cluster classes, then no one specimen can be typical, and a large sample of the population is needed for an adequate description of the population. The population, then, takes on a reality of its own, which is not afforded on type-specimen or strictly essentialistic concepts. The only role left for type specimens is therefore as name bearers and for fixing the names of newly found species, and this, according to Beckner (70), is precisely the view of the new systematics.

This is certainly a plausible answer. Nevertheless, for all the difficulties biologists have in defining the term "population" (as we have seen in chapter 2.4), it is clear that in the main they conceive of populations as fully physical entities, or at least as groups of organisms connected by certain interorganism relations. On Beckner's view, however, a population is not fully physical, but is constituted by physical organisms plus an abstract cluster class, and it is dubious that this is what biologists have in mind when they speak of populations.

This difficulty is especially highlighted when we consider Beckner's application of cluster theory to the species category. In Beckner's case (and for practically everyone else discussed in the present study) one must do to Beckner what Beckner does to certain biologists, and that is to attempt a greater degree of clarity on the topic in question. Beckner's most succinct statement on the ontology of species is as follows:

> On my definition, "species" is fully polytypic with respect to G, and, in addition, is both a functional and historical concept. [64]

Ignoring for a moment the "functional" and "historical" attributions, one might suppose that for Beckner a species *just is* G, or perhaps G plus a minimum quorum condition, which is to say an abstract cluster class. But this would be far from correct. It is clear that for Beckner *groups* of organisms are species, but not just any groups. For instance, he refers to "the generally acknowledged fact that no set of conditions is both necessary and sufficient for calling a group a species"

(59), and he claims that "ability to interbreed with other members of the group and geographical contiguity are not decisive criteria for ranking the group as a species" (64). What makes a group of organisms a species, then, if species are to be fully polytypic, is whether it meets the three conditions quoted above. Moreover, although Beckner is aware that the set G may include more than just morphological characters, it is clear that for Beckner morphological characters are more important and are minimally adequate, since they "are made the basis of the species-descriptions that occur in taxonomic journals" (63).

The group aspect of species comes out even more clearly when we consider what Beckner means when he says that species are both functional and historical. By "functional" he means that "it is logically impossible for a taxon to be a species unless its members contribute to the function of self-reproduction of its component populations" (77). And by "historical" he means that "it is logically impossible for a taxon to be a species unless it is monophyletic with respect to a single interbreeding population" (77).

In claiming that species are historical entities, and more specifically monophyletic, Beckner is only trying to keep in harmony with the new systematics, according to which all taxa of whatever rank should ideally be monophyletic and classification should be based on phylogenetic descent (cf. 58). But it seems to me that the greatest difficulty Beckner faces is that his concept of the ontology of species is hopelessly inconsistent with this desire. For Beckner, a species is constituted of organisms *plus* an abstract cluster class. But can a species thus constituted possibly *do* or *be* what Beckner hopes of it? I do not think so. In trying to combine abstract entities with physical ones, Beckner is trying to have his cake and eat it too. That he simply cannot have it both ways follows from the fact that they each are very different kinds of entities. (Indeed his position is somewhat reminiscent of the tradition of soul/body dualism and the problem of interaction.) Not only is a physical group distinguished from an abstraction by the very fact that the former can be pointed to while the latter cannot, but they both involve very different principles of change. A group changes every time there is a change in the membership that constitutes the group. On the other hand, it is part of the very nature of abstract classes, cluster classes included, that they remain the same while individual members come and go. At least Bambrough (1960–61) got this right when he said "There are an infinite number of actual and possible chairs. . . . and the word 'chair' . . . can be applied to an infinite number of instances without suffering any change of use" (276). Indeed it may be argued, as I shall argue in §5, that abstractions do not change at all. In the case of cluster classes, it seems to me that any change in the defining set of membership conditions, no matter how small, is not a change in the class but a new class altogether. But if any of this is so, then Beckner's abstract/physical dualism for species ontology is thoroughly incoherent. A species, then, if it is to be physically conceived, may entail an abstraction, but that abstraction cannot also be a part of what it is.

A more deeply theoretical problem is that what is required for abstract classes of any kind, cluster classes included, is a principle of identity and individuation. As Quine (1981) put it, "no entity without identity" (102). With physical objects such as organisms there really isn't a problem: diachronically we identify and individuate according to spatiotemporal continuity maintained by synchronic cohesion (cf. Locke 1700: II.xxvii.4), while synchronically (in addition to cohesion) we identify and individuate according to the principle that no two objects can occupy the same spatiotemporal position (cf. Haack 1978: 43, Denkel 1996: 23–26), which we typically accomplish simply by pointing. But one cannot point to an abstract class, since it is spatiotemporally unrestricted. How then can they possibly be individuated? According to Quine (1981), for whom as we've seen classes are abstract entities, "Classes are identical when their members are identical" (100), in other words two classes are numerically identical if and only if they are coextensive (cf. Quine 1960: 209). Simply, this means that if every member of A is numerically identical with B and *vice versa*, then we really have one class, not two. There are serious problems with this principle when applied to abstract classes in general, but I will reserve them for §5. With cluster classes we get a very unique problem, namely the logical possibility that two different sets of membership conditions could in fact be coextensive throughout the whole of their membership histories.[30] If this is true, then the principle of the identity of co-

---

[30] This can be illustrated using the type of schema commonly used when expounding cluster theory. Suppose for the sake of simplicity that we have a population in the wild consisting of only ten organisms, which we label from a to j. (If that number seems too small, we could have each letter represent not one organism but several, perhaps even hundreds or thousands.) Suppose furthermore, again merely for the sake of simplicity, that each organism has a total of nine characters and together they instantiate a total of twelve characters. Suppose now that two biologists, ardent followers of Beckner's scheme of classification, come along one after the other and that they each independently classify these organisms into a single cluster class, each biologist happening to observe only ten characters in total and each using a minimum quorum of seven characters and seven organisms to define both occurrences of the term "large" in Beckner's definition of cluster classes. A possible schema, then, is as follows:

Organisms:	Cluster class A:	ABCDEFGHIJ	Cluster class B:	CDEFGHIJKL
a: ABCDEFGHL		aaaaaaaa		aaaaaa a
b: BCDEFGIJK		bbbbbb bb		bbbb bbb
c: ACDEFGHIL		c ccccccc		ccccccc c
d: ABDEFGHIJ		dd dddddd		dddddd
e: ABCEFGHKL		eee eeee		e eeee ee
f: ABCDGHIJK		ffff ffff		ff fffff
g: ABCDEIJKL		ggggg gg		ggg gggg
h: ABCEFHJKL		hhh hh h h		h hh h hhh
i: DEFGHIJKL		iiiiiii		iiiiiiiii
j: CDFGHIJKL		jj jjjjj		jj jjjjjjj

It might seem odd that the first biologist overlooked characters K and L and that the second biologist overlooked characters A and B. Nevertheless there is nothing in Beckner's definition of cluster classes that stipulates that every character of the members of cluster classes must be taken into account. Indeed from the point of view of Beckner's definition each biologist has constructed a perfectly good cluster class. First, for each cluster class each member organism possesses a large number of the characters in G, determined by

extensives, a mainstay of modern set theory, proves useless for cluster classes. The principle of identity that we must fall back on for cluster classes, then, must not be extensional but intensional, namely the defining set of membership conditions.

I do not see that this is a problem for the theory of cluster classes itself, but it is a problem for Beckner's application of that theory to the species problem. According to Beckner, "every organism is a member of one and only one taxon of each rank" (55), and he immediately makes it clear that in this he includes the rank of species. But not only does this not follow from the nature of cluster classes, specifically when the minimum quorums are left vague, it also does not follow from the nature of nature. This is because some congeneric species, popular examples of which include North American oaks (*Quercus*) and Hawaiian *Drosophila*, form what are known as hybrid zones, an area of overlap in which two species produce fertile hybrids. Cluster theory allows for such an overlap, so that some organisms may indeed be members of two species, and in that sense it may conform to the facts. But that it allows for the logical possibility that *all* of the organisms in two species may be members of both species must surely count against it.

Even with these difficulties aside, is cluster theory really consistent with hybrid zones? Hybrid zones are often thought to be instances of incomplete or partial speciation (hence the two species are sometimes called "semispecies"). Futuyma (1986) notes that "hybrid zones sometimes contain rare alleles that are not found in either parental semispecies" (115). Would unique hybrid zone characters meet Beckner's second condition for a polytypic group? Although it is by no means clear, it would appear not to be the case. But a greater difficulty is whether cluster theory would entail that semispecies are two species or one. As Templeton (1989) has pointed out, species joined by a hybrid zone "are often real units in terms of morphology, ecology, genetics, and evolution. For example, the fossil record indicates that balsam poplars and cottonwoods (both from the genus *Populus*) have been distinct for at least 12 million years and have generated hybrids throughout this period" (10). According to Wiley (1981), "If the hybrid zone is narrow and there is evidence that it is old, then the forms are probably

---

each biologist to be at least seven of the ten characters in G. Second, for each cluster class each character in G is possessed by a large number of the member organisms, determined by each biologist to be at least seven of the ten organisms. And third, for each cluster class no character in G is possessed by every member organism. Each cluster class, then, is fully polytypic according to Beckner's definition of that term. And yet from the very same organisms two different cluster classes have been constructed (or if one wishes, inferred). And if we want we can make each of these two cluster classes conform to Beckner's definition of species, by making the group of ten organisms both functional and historical according to his definition of these terms (perhaps they are two generations of fertile offspring stemming from the cross of two now dead polyploids, the offspring generations subsequently dying without further issue due to, say, a fire). What we would have, then, are two perfectly good cluster class species coextensional throughout the whole of their membership histories.

species because they are successfully maintaining their identities in spite of gene flow. . . . If the zone is wide, then the two forms are probably geographic variants of the same evolutionary species" (29). But on Beckner's view, both the width and the age of hybrid zones must be irrelevant, and whether balsam poplars and cottonwoods are specific or subspecific must be determined by the subjective fiat of taxonomists.

There are further difficulties facing Beckner's thesis,[31] but I want now to consider the application of cluster theory to the species category by the school of taxonomy one would naturally expect to apply it, namely numerical (phenetic) taxonomy, a school that delimits and groups taxa based on overall similarity calculated by a high number of phenetic characters with the aid of computers. What is interesting is that not only do the authors of the first textbook on numerical taxonomy, Robert Sokal and Peter Sneath (1963), explicitly expound and endorse traditional cluster theory (14–15), but they explicitly refer to Adanson (16), Wittgenstein (14), and Beckner (14–15) as important forebears, going so far as calling their own system of taxonomy "Adansonian" (50).[32] Surprisingly, however, they never tell us in this book what they mean by "species."

According to Sneath and Sokal (1973), "It has taken some years for the lessons of phenetic taxonomy to penetrate to the species level" (362). They attribute this to several factors, namely (i) "the development of more powerful computers and computer programs," (ii) the growth of the view "that the biological species definition was difficult to apply and in practice was nonoperational," (iii) "the consistent application of phenetic principles to taxonomy at all levels," and (iv) "the realization that . . . the phenetic approach in population biology has considerable inherent interest for the evolutionist" (362–363).

In spite of these claims, however, it is still by no means clear in their later writings what they mean by "species." Sokal (1973) tells us that "The *phenetic species concept advocated by the numerical taxonomists* is based on the numerical evaluation of the boundaries of populations in a character hyperspace (A-space)" (361). This abstract, multidimensional (non-Euclidean) hyperspace is defined by Sneath and Sokal (1973) as follows: "A-space (*attribute space*) has formally n dimensions, one for each attribute or character, in which there are t points that represent the OTU's" (116). Depending on the study, OTUs (operational taxonomic units) will differ in rank when establishing higher taxa (though normally species will be used), but for establishing species themselves the OTUs are individual organisms (Sneath and Sokal 1973: 69). Thus, when establishing species a t point in an A-space will represent an organism (or possibly more than one organism, particularly in the case of clones).

Sokal and Crovello (1970) tell us that "Insistence on a phenetic species con-

---

[31] For mild criticisms (though no less interesting), cf. Simpson (1961: 19–23, 94–95).

[32] As Panchen (1992) points out, this changed once it became known that Adanson's aim was not to produce classifications based on overall similarity but only "to show the futility of artificial classifications based on one or a few characters (like Linnaeus's sexual system for plants)" (133).

cept leads inevitably to a conceptualization of species as dense regions within a hyperdimensional environmental space in the sense of Hutchinson" (52). (I will discuss Hutchinson in the next section.) A little more specifically, Sneath and Sokal (1973), although they waver between the following two possibilities, tell us that

> We may regard as a species (a) the smallest (most homogeneous) cluster that can be recognized upon some given criterion as being distinct from other clusters, or (b) a phenetic group of a given diversity somewhat below the subgenus category, whether or not it contains distinct subclusters. [365][33]

From these three definitions, then, it would appear that according to the phenetic species concept, unlike Beckner's view, species are not *dualistic* entities (half abstract, half physical) but instead are *monistic* entities (fully abstract), namely dense phenetic clusters inside an A-space, such that individual organisms *determine* the clusters but are not *constitutive* of them.[34] If one may use the anal-

---

[33] Although Sneath and Sokal (1973) speak of clusters as each a "condensation of points in certain regions of the A-space" and tell us that "current evolutionary theory (as supported by empirical observation) makes the random distribution of organisms in phenetic A-space extremely implausible" (194), they also admit a certain degree of vagueness in defining clusters (194). Nevertheless they tell us that "The *center* of a cluster can be represented in two general ways, as a point representing an actual organism, or as a point representing a hypothetical organism (such as the 'average man' who has 0.8 wives and 2.3 children!)" (195).

[34] Rosenberg (1985) describes the species concept of the numerical taxonomists as "a definition of species as sets of organisms sharing a set of observable or objectively detectable properties that satisfy a stipulated mathematical relation to one another" (184). Similarly Ereshefsky (1992b): "they [Sokal and Crovello] propose that species are simply those groups of organisms that have the most overall phenotypic similarity. Such phenetic groups are found by the methods of phenetic taxonomy" (5). Similarly Sober (1993): "the idea that species are groups of organisms with a great deal of overall similarity" (158). Similarly Ridley (1993): "Informally, the phenetic species concept defines a species as a set of organisms that look similar to each other and distinct from other sets. More formally, it would specify some exact degree of phenetic similarity, and similarity would be measured by a phenetic distance statistic" (386). Likewise the botanist Lennart Andersson (1990: 375) conflates the species concept of the numerical taxonomists with the overall similarity group concept of Cronquist (1978; cf. chapter 2.4) so that "Species may be visualized as clusters of individuals in a multidimensional space, where each dimension marks a character axis. . . . as the category of clusters that are inwardly continuous and outwardly discrete" (375). Interestingly Andersson (376)—who surprisingly does not even mention numerical taxonomy or any pheneticists—combines his own version of the phenetic species concept with the ecological species concept of Van Valen (1976; cf. §4 below) in order to account for both the patterns and processes of species and speciation. At any rate, all of the above interpretations are fundamentally *misconceived*. For the numerical taxonomists, as I have shown above, sets or groups of organisms are *not* constitutive of species. Indeed Andersson's view that an organism can literally occupy (rather than simply be represented by) a point in an A-space is logically untenable. To think otherwise is to vio-

ogy, although a king, *qua* king, is dependent on his subjects, his subjects are not part of his being, *qua* king. Clearly a king does not necessarily change, *qua* king, every time there is a numerical change in his subjects. Likewise a phenetic cluster in an A-space could theoretically remain exactly the same even though the organisms that determine it (particularly in the case of clones) come and go. Such an entity is therefore clearly an abstract class,[35] though quite different from the cluster class of Beckner.[36]

If my interpretation is correct, then it seems to me that the most serious problem for a cluster hypervolume in an A-space to be a species is that it must be able to change and evolve. Can such an abstract entity possibly do this? According to Sokal and Sneath (1963), "Natural taxonomic groups are formed by the restriction of evolution to certain regions of the phenetic hyperspace" (217). Similarly, Sneath and Sokal (1973) claim that "If evolution is predominantly by replacement of successive single mutations the total evolutionary pathway in A-space ... consists of successive displacements on each character axis in turn" (317). Again, whether abstract entities can be meaningfully said to evolve is an issue I will reserve for §5.

A related difficulty is that Sokal (1973), in a curious attempt to reach "a new synthesis of the species concept" (360), suggests that "Since it is very difficult to separate the phenetic characteristics of a species from its response surface to the

---

late the fundamental criterion of identity and individuation for physical objects, namely that no two physical objects can occupy the same spatiotemporal position. This is most evident in the case of clones (*ceteris paribus*), which cannot possibly be individuated in an A-space.

[35] Interestingly, in his update on the development of numerical taxonomy, Sokal (1986) tells us that "Accepting the philosophers' distinction between individuals and classes, the pheneticist clearly considers supraspecific taxa as classes, that is, constructs whose members are determined by a class concept" (424). Does this mean that the pheneticist considers *species* as *individuals*? Sokal does not tell us, but if that is his meaning it defies credulity. Sokal then immediately tells us that "In consequence, the reality of taxa is not a special concern of pheneticists. Taxa are human constructs, and natural taxa are those that are natural to humans" (424). In this the pheneticists explicitly follow Gilmour (1940). But such subjectivism, of course, is totally inconsistent with the certainty of biologists on the objective reality of evolution, let alone that of (horizontal) species (cf. chapter 2.4).

[36] Although they have great respect for Beckner's work, Sneath and Sokal (1973) tell us that "One of the difficulties of Beckner's definition is that in natural taxa we do commonly have *f*'s that are not possessed by large numbers of the class. Furthermore, we cannot test whether any given *f* is possessed by large numbers of the class before we have made the class, and therefore we cannot decide whether to admit this *f* into the set *G*. This difficulty can be avoided by defining class membership in terms of common (or shared) attributes" (22), what they earlier (Sokal and Sneath 1963) defined as "polythetic" classes (13–14). In other words, the problem of defining natural taxa and the problem of discoverability, both of which are serious difficulties for Beckner's view, are avoided by defining taxa not in terms of traditional cluster classes but instead as clusters or hypervolumes in an A-space. However, although the first problem might thus be avoided, the second problem returns for the numerical taxonomists in another way, as we shall soon see.

environment, the n-dimensional niche space of Whittaker et al. . . . and the niche-habitat relation lines and surfaces of Maguire . . . are really an added dimensionality of the A-space of the numerical taxonomist" (368–369). Unhappy with the *n*-dimensional realized niche of Hutchinson (to be discussed in the next section) because it "suggests a rather static species concept," he adds that "The niche of a species must be the union of the hypervolumes defining the various local populations in the A-space" (369), thus better approximating the fundamental niche of a species which "itself can and will change as the population changes" (369). However, in acknowledging that "the term species is involved in virtually all definitions of niches" (368), his suggestion of adding the fundamental niche dimensions of species as further dimensions of the A-space results in a very confusing species concept, since the dimensions of a species in an A-space will include species the A-space dimensions of which will include species and so on. The result is a hopelessly regressive abstract species concept.

Apart from the above difficulties, the species concept of numerical taxonomists faces more immediate difficulties. Sober (1993) points out that "the difference between species and taxa at lower and higher levels is left unclear" (158). A more interesting difficulty, suggested again by Sober (1993), is the case, for example, of "a sexual species of lizard and an asexual species that is descended from it. The asexual species consists of parthenogenic females. Perhaps from the point of view of overall similarity, we should group the sexual females together with the asexual females and treat both as distinct from the males. Almost no biologist would be willing to do this" (158). The more general difficulty, of course, is the issue of sibling species on the one hand, which by phenetic clustering would be conflated into single species, and highly polymorphic or polytypic species on the other hand, which by phenetic clustering might be broken into two or more species. As for sibling species, Sneath and Sokal (1973) admit that "sibling species will . . . constitute a single natural taxon, in the sense in which we use the term," but they immediately emphasize that "this is precisely what it [numerical taxonomy] is intended to do; a phenetic method should give phenetic groups" (62; cf. 423). In the case of radically differentiated cyclomorphic species such as Lepidoptera (butterflies and moths from caterpillars), they admit that "Larvae and adults of the same species will . . . frequently not cluster congruently," but they rest satisfied in the probability that "congruence will be greater the higher the rank" (98). Interestingly, in the case of radical sexual dimorphism they take the tack that "we really do not have any hard data measuring the similarity of . . . highly sexually dimorphic species by appropriate methods of phenetic analysis and until such measurements are made, a statement . . . [of criticism] is clearly misleading" (423). But consider the following description by Eldredge and Cracraft (1980) of a case of sexual dimorphism that is truly extreme, a species of deep-sea cephalopods of the genus *Argonauta:*

> The large female carries a "paper" shell as an egg case, while the tiny male (only about an inch long) is reduced essentially to a reproductive organ, and thus not immediately recognizable as an argonaut in terms of its superficial morphologi-

cal features. To confuse the issue even further, the male reproductive organ de-taches and moves about freely within the mantle cavity of the female. The de-tached organ was long regarded as a parasite. . . . Only when the full anatomical details of the males became known were there ample grounds for the hypothesis that these two quite dissimilar groups should be united into a single, albeit highly dimorphic, species. [98]

What should be clear from all of this is that there is a serious incongruence between numerical methods for delimiting species and the entities that biologists eventually recognize as individual species. No biologist worth the name would knowingly place the larval stage of an organism in a different species from that of its adult stage, or the males and females of breeding pairs into separate species (let alone the reproductive organ of the male). That numerical taxonomy entails such counterfactual consequences speaks highly against it.

As with strictly essentialistic classes, one may take a more reductive approach and conceive of species as genotypic or molecular cluster classes. Not surpris-ingly this move has its adherents among both philosopers and biologists. Among philosophers the most notable view is that of Arthur Caplan. With explicit refer-ence to and endorsement of Bambrough, Beckner, Sokal, Wittgenstein's theory of family resemblances, and traditional cluster theory, Caplan (1980) tells us that species have traditionally been thought of as morphological cluster classes. He has no problem with this except that it fails to explain "the cause of the particular similarity between organisms in classifying them" (161). This Caplan attributes to "the stability and permanence of certain genotypes and environments" (161).

In all of this Caplan finds no incompatibility with evolutionary theory. Thus he says "When the vast majority of class-defining properties are no longer in-stantiated by any organisms it is reasonable to assume that the class is either ex-tinct or has evolved into a different class. Thus, at some point Archaeopteryx crossed the class line dividing reptiles and birds. Reptilian properties gradually disappeared while avian characteristics emerged" (159–160).

What is confusing in Caplan's account is that it is by no means clear what he means when he speaks of species. In the passage quoted immediately above it might appear that what is being asserted is that species taxa are groups of organ-isms defined by abstract cluster classes. Indeed elsewhere in his paper Caplan tells us that "Species taxa are perhaps best understood as groups of individuals which share similar genotypes and similar environments" (161). Thus far it would seem, then, that species taxa are the *extensions* (the "groups of individuals") of abstract cluster classes and ecological niches, not the classes themselves. But immediately following this he tells us that "The hidden 'essences' of species taxa are the genotypes and environments which produce the similarities of traits we observe among organisms" (161–162), which might make one think that Caplan is groping toward some sort of complicated abstract concept of species. We get no help in trying to ascertain his view when he later (1981) tells us that "a species is a class of organisms which possess a common genotype" (139), since he seems to use the word "class" as a synonym for "set" or "group." He says, for instance, that "a set or class of organisms" may "interact with one another" (132) and that a "class" may be "spatiotemporally localized" (133).

Interestingly, this confusion is not all that uncommon. Those who believe that species are classes of one sort or another often conflate the classes with their extensions (members). Caplan is an especially good example of this. What is clear, the confusion aside, is that one cannot have it both ways, as we saw with Beckner.

It is interesting that in his reply to Caplan (1980, 1981), David Hull (1981a) clearly interprets Caplan as providing in the fully abstract sense a genotypic cluster theory for species. "Caplan maintains," says Hull, "that species are spatiotemporally unrestricted classes defined by 'clusters' of genotypes" (141). Perhaps one can assume, since Caplan is listed as a commentator to an earlier version of Hull's paper, that this clarification of Caplan's view is acceptable to Caplan. But if so, it still fails to meet the many objections we have seen so far.

On the other hand, why not say that a species *just is* the extension (the group taken horizontally, the total membership taken vertically) of a genotypic cluster class, that it is defined by this class but not constituted by it? (This sometimes sounds like Caplan's view.) If this approach is taken, there would seem to be no problem with accommodating evolution and extinction. Species could evolve by gradually changing their membership from one class to another, they could split by gradually splitting their membership into two related classes, they could instantly speciate as in polyploidy by jumping in membership from one class to another, they could remain in stasis as per the theory of punctuated equilibria by maintaining their membership in one class, and they could go extinct either immediately or gradually by either immediately or gradually ceasing to have membership in their class and by not acquiring it in any other.

Indeed one might be tempted to take this approach with all of the class theories examined in this chapter. However, serious problems attend this approach, not the least of which I shall call *the problem of reiterated membership.* Reiterated membership compels one to identify the species not with any particular group of organisms but with instead the class itself, namely the set of characteristics and membership conditions which defines the group. Taking the traditional cluster class view for our example, suppose that we have a group of organisms defined by a particular cluster class, so that on the view we are considering this group would be a species. Now suppose that at some point in time this species (group) goes extinct. Suppose further that at a later time another group of organisms comes to be defined by the same cluster class vacated by the previous group. On the view we are considering, this new group would be a different species than the previous group, even though both are defined by the numerically same cluster class. Now suppose that this course of events is repeated many times, and with very short intervals. Surely no biologist (at least of cluster class persuasion) would treat each successive group as a distinct and new species, either in name or in practice. Instead, all of the groups would be treated as conspecific, and rightly so it seems to me, not only for practical considerations but on account of Occam's Razor.[37]

Perhaps the most serious difficulty in taking the view that a species is defined though not constituted by an abstract class is that it would seem to destroy the dependency relation between the class and its members. On every view we have seen so far, at least with cluster classes, the abstraction is dependent for its reality on physical organisms. Destroy that dependency relation and one is left (if still a realist) with independent Platonic entities, which is clearly unacceptable to the modern biological way of thought. In the world of modern biology, there can be no kings without subjects (cf. chapter 4.2n21 for further discussion).

Much more interesting are the attempts by biologists to define species in terms of genotypic cluster classes. James Mallet (1995) is a good case in point. Although he does not provide a formal species definition, he gives a good description of what he means by his "genotypic cluster definition:"

> When we observe a group of individuals within an area, we intuitively recognize species by means of morphology if there are no or few intermediates between two morphological clusters, and because independent characters that distinguish these clusters are correlated with each other. Adding genetics to this definition, we see two species rather than one if there are two identifiable genotypic clusters. These clusters are recognized by a deficit of intermediates, both at single loci (heterozygote deficits) and at multiple loci (strong correlations or disequilibria between loci that are divergent between clusters). Mendelian variation is discrete; therefore we expect quantized differences between individuals. We use the patterns of the discrete genetic differences, rather than the discreteness itself, to reveal genotypic clusters. [296]

Mallet's species concept has a lot going for it. For one, Mallet clearly conceives of species as pattern entities rather than process entities. "The most important feature of the genotypic cluster definition," he tells us, "is that species can be affected by gene flow, selection, and history, rather than being defined by these processes" (296). Again, "By concentrating on genotypic clusters as opposed to an interbreeding concept, we are able to separate the causes of species distinctness from the observable distinctness itself" (296). As we have seen in the Introduction and will see more fully in chapter 5.2, a species concept that conceives of species as pattern rather than process entities is much more inclusive and provides a much better candidate for a viable species concept for the Modern Synthesis. This is clearly Mallet's intention, not only from the title of his paper but

---

[37] This argument has an analogue in repeated hybridization events, which is firmly established as a fact of nature. According to Mayr (1987a), "uniparental entities could theoretically originate several times independently, and yet be indistinguishable for all practical purposes. Such a case was actually discovered by M.J.D. White in Australia. The parthenogenetic species [grasshopper] *Warramaba virgo* consists of two allopatric populations that were derived from different strains of the two parental species" (164). In plants, of course, repeated hybridization is quite common, as in allopolyploidy (cf. chapter 5.2n15 and accompanying text).

from his claim that "Clearly, the definition could apply to eukaryotes as well as to prokaryotes. If we all adopted a genotypic cluster method, we would have a unified species definition . . . ." (297).

It is much less evident, though nevertheless implicit, that Mallet conceives of species as primarily horizontal rather than primarily vertical entities. He says, for example, that "This definition will, of course, best apply to populations in contact" (296). If his species concept is, then, a primarily horizontal one, it satisfies a second major desideratum that I have focused on periodically throughout this book and will return to in chapter 5.2. Nevertheless, Mallet's species concept falls short of an ideal one on a number of counts.

To begin, it is clear that Mallet's species concept is not a pure cluster concept, since he adds a further criterion, namely the inability of genotypic clusters to fuse in contact zones (as we have seen above, he allows for a plurality of explanatory processes). If his was a pure genotypic cluster concept, then geographically isolated races would have to count as separate species. As he put it, "it is obvious that we could, if we wanted, define allopatric geographic races as separate species, since they are often separate genotypic clusters" (296). But Mallet does not want this. Instead, he wants a polytypic species concept with subspecies. The movement to rid zoology of subspecies he considers a "pity," not only because "there are certainly many races that intergrade freely at their boundaries, but that are strongly differentiated and relatively constant in morphology, genetics, and ecology" (298), but more importantly because for conservation biology "it seems obvious that we should not ignore the huge amount of genetic biodiversity found in infraspecific taxa" (298). This desideratum is fine enough, but it is by no means clear that it is compatible with Mallet's species concept. This becomes evident when we ask what it means for speciation. "Evolutionarily," he tells us, "the definition [Karl Jordan's polytypic definition] means that speciation is the formation of a genotypic cluster that can overlap without fusing with its sibling" (296). Although Mallet allows for both internal (lack of gene flow, historical inertia) and external (selection) processes to account for this inability, the inability is not so much a matter of genotypes as it is other things, namely isolating mechanisms and historical constraints in the former case and environmental factors in the latter. Consequently, he allows species to go extinct not only in the normally thought of way but also by hybridization: "if the environment changes, the two clusters may again fuse; . . . We all agree that species are lost by extinction; it seems not unreasonable to have a definition under which species may be lost by hybridization as well" (297). The problem is that if two species, as he allows, can go extinct by introgression, where is the boundary that he wants between species and subspecies? Or between species and species? Genotypic clusters alone will not give us that boundary; that can only be decided arbitrarily. But neither will the other things listed above, given his scenarios.

A further problem with Mallet's attempt to add to his bare genotypic cluster definition is that it exposes him to a serious problem faced by the species concept he is most keen on usurping, namely Mayr's biological species concept. Mayr's concept is explicitly *relational* (as we shall see in chapter 4.1), which is to say a species cannot be real on his view unless there are other species, preferably sympatric, from which to distinguish it. In other words, it is impossible on his view to have a planet with only one species on it. This resolves itself into the logical difficulty of how species could ever have first evolved.

But there is a much more serious metaphysical problem. In defining species as genotypic cluster classes, even in spite of his added criterion, it should be clear that species on Mallet's view are abstract entities. They are not groups of organisms in any one of many possible senses but rather clusters of genotypes, each a loose essence as I have used the term above. This is evident from the problem of reiteration, outlined in my discussion on Caplan. This problem is even more evident in Mallet's case. That he allows for species to go extinct by introgression must mean that he also allows for species to re-evolve by repeated hybridization. Since *relations* are evidently not any part of what a species is on his view, his view resolves into a property view, albeit a cluster of properties, in this case genotypic/molecular properties. As such there is nothing in his concept to ground species in the physical world. They are instead abstract collections of genotypic clusters. In §5 below I will argue that abstract entities, clusters included, cannot evolve on any reasonable view of the world. Had Mallet included relations in the ontology of species, he could have grounded them in this world, as I shall argue in chapter 5.2. Since he did not, and since he obviously did not subscribe to any version of the species-as-individuals view, his view must be deemed unsuccessful from the perspective of evolutionary biology and the Modern Synthesis.

That a genotypic cluster concept of species is a class concept is explicitly recognized by M.H.V. Van Regenmortel (1997) in his application of it to virus taxonomy. Accepting the class/individual distinction of Ghiselin and Hull (cf. chapter 4.2), according to which classes are spatiotemporally unrestricted and cannot do anything such as evolve, Van Regenmortel argues that not all classes need be of the Aristotelian type. Explicitly following Beckner, he argues that classes can also be polythetic (Beckner's fully polytypic). To illustrate this, he provides a table and says "Suppose a species is defined by a set of five properties . . . . If these properties are distributed in the way shown . . . the class will be polythetic. This example represents a polythetic class because each individual possesses a large number of the properties (i.e., four out of five), each property is possessed by a large number of individuals and no property is possessed by all individuals" (21). "In this kind of class," he adds, "certain elements may evolve and there is no difficulty in reconciling class membership with phylogenetic change" (21). Again: "the concepts of polythetic class and fuzzy set [the latter explicitly allows for class membership to be a matter of degree] make it possible to reconcile phylogenetic change with class membership and this removes the rationale for considering species only as real individuals" (19).

Aside from confusing the terminology—"polythetic," as we saw above (n36), is the term introduced by Sokal and Sneath (1963) and it does not have quite the same meaning as Beckner's "polytypic"—Van Regenmortel confuses the members of a class with the class itself. This is a common metaphysical error, a theme that runs throughout this chapter. Van Regenmortel wants species to evolve, and the only way this can happen on his view is for polythetically similar genotypes of successive individual organisms to undergo directional change. On the other hand, notice that above he says "Suppose a species is defined by a set of five properties . . . ." This is the problem of reiterated membership all over again. In fact he admits that "in the case of structures as simple as some viruses it cannot be excluded that the same pathogenic entity might have arisen or evolved from a different parent more than once during biological evolution" (22). The problem is that the species is no longer merely *defined* by the set of properties but instead is *constituted* by them, fully polytypically if Beckner's minimum quorums are included and fuzzily if not. The species, then, on either account, remains a fully abstract entity and the problem of how an abstract entity can change or evolve, which I will elucidate in §5, presents a very serious, if not insurmountable difficulty for such a concept.

As if to muddy the waters only further, in the final section of his short article Van Regenmortel favorably appeals to Manfred Eigen's quasispecies concept to make the point that "it is no longer accepted that a virus species can be defined by a single genome sequence" (23). If the implication is that Eigen's species concept is in accordance with his own, that implication must be dismissed. Eigen's species concept, as we shall now see, is quite unique, even though it falls within the category of genotypic cluster class concepts of species.

Manfred Eigen, 1967 Nobel Prize co-recipient for Chemistry and a leader in origin of life research (cf. Eigen 1992), provides what is arguably the most fascinating view within the paradigm of species as cluster classes. In collaboration with a number of colleagues, Eigen *et al.* (1988) call their view *molecular quasispecies*, and they tell us that they use the term *quasispecies* "to point out the analogy with the species concept in biology" (6882).

Immediately, however, I have a difficulty with this, for not only does it seem to me clearly false that biology has a single species concept—it is clear from Eigen (1993: 44) that the species concept referred to is the reproductive isolation concept of Mayr—but I fail to see what analogy there can be between what we shall see is a fully abstract species concept and an avowedly (Mayr, to be discussed in §4 and chapter 4.1) concrete species concept.

At any rate, although their focus is on viral species, it is explicitly clear in reading Eigen *et al.* (1988) that their species concept is meant to apply to all biological species from viruses to eukaryotic species such as *Homo sapiens*. "The present theory," they tell us, "relates to asexually replicating organisms, but this restriction is not essential" (6881). Although their exposition is highly technical and I will not for a moment pretend to understand the technical aspects of it, the basic idea behind it is relatively simple, especially as it is expounded in Eigen (1993), and can be briefly described as follows.

To begin, every population, on this view, involves what biologists call a "wild type," as Eigen (1993) put it "the form that predominates in a population and that is particularly well-suited to the environment in which it lives" (44). At the genetic level, however, less so with clones but especially with sexual species, it is extremely unlikely that any one total organismic genotype is going to predominate, so that "The wild type of a quasispecies refers to an average for all the members, not to a particularly fit individual. . . . Its wild type is the consensus sequence that represents an average for all the mutants, weighted to reflect their individual frequency" (45). Eigen (1983: 110) provides the following hypothetical example (unrealistically short, as it must be) of nucleotide sequences for five organisms to show how a consensus sequence is constructed:

$$
\begin{array}{l}
\text{UCGUCCA} \\
\text{AAUUACG} \\
\text{ACAAAUG} \\
\text{ACGUGCG} \\
\underline{\text{ACGCACG}} \\
\text{ACGUACG} \qquad\qquad \text{(consensus sequence)}
\end{array}
$$

Next, the concept of a quasispecies requires what is known as a Hamming sequence space, an "abstract point space, the sequence space originally introduced by Hamming" (Eigen *et al.* 1988: 6883). In this non-Euclidean multidimensional nucleotide hyperspace the genotypic wild type "would occupy a single point" (Eigen 1993:44) and "each point represents a unique sequence" (44), so that "mutational neighbors [are placed] adjacently" (Eigen *et al.* 1988: 6883) and "the degree of separation between points reflects their degree of dissimilarity" (Eigen 1993: 44).

Mutation, of course, is an expansive force on the genotypic wild type, while natural selection, on the other hand, is a compressive force. Where "the expansive force of mutation would strike a balance with the compressive force of selection" (45), one gets a particular cloud or cluster of genotypes within the Hamming sequence space. As Eigen (1993) describes it, "Like a real cloud, it need not be symmetric, and its protrusions can reach far from the center because some mutations are more likely than others or may have higher survival values that allow them to produce more offspring. That cloud is a quasispecies" (45).[38]

While it is the error rate that "directly determines the integrity of a quasispecies" (45), it is the equilibrium between mutation and selection that determines its "collective identity over time" (45). Disequilibrium in either direction results eventually in extinction, in the contraction direction by a loss of variability, in the expansion direction by a loss of integrity.

What I find most fascinating about this view is the practical implications Eigen draws from it for biological control, specifically in the case of eliminating

---

[38] Consonant with this view is the empirically validated idea of mutational constraints. Although mutation is an expansive force, it would not be like an expanding balloon and be uniformly expansive in a sequence space. Templeton (1989) provides a brief discussion on mutational constraints, citing the example that "no picture-winged *Drosophila* has ever produced a clear-winged mutant, nor has a clear-winged species produced a picture-winged mutant. . . . [so that] at the species level there is a block to certain types of mutations. This is simply another way of stating that constraints exist that make certain types of mutations impossible or highly improbable" (18).

viruses such as HIV. As Eigen (1993) suggests, "If the error rates of viruses can be increased moderately, just enough to cross the critical threshold that defines their quasispecies, they would experience a catastrophic loss of information. The viral quasispecies would fall apart because it would be producing too many non-viable mutants" (49).[39]

Now, although Eigen and his colleagues do not make this altogether clear, I suggest that on a close reading of their writings it should become sufficiently clear that a quasispecies on their view is neither a physical entity defined by an abstract entity nor a dualistic entity part physical and part abstract, but instead a *fully abstract* entity (much like the species concept of the numerical taxonomists), as Eigen *et al.* (1988) put it, "a nonsymmetric cloud of mutants" (6882). Or as Eigen (1993) put it, "That cloud is a quasispecies. . . . the quasispecies is a localized distribution in sequence space" (45).

What tends to mislead is that Eigen and his colleagues describe a quasispecies as having a population and as being the object of natural selection, which in itself suggests either of the two interpretations that I rejected above. For instance, Eigen (1993) tells us that "Biologically, the quasispecies is the true target of selection. All the members of a quasispecies—not just the consensus sequence—help to perpetuate the stable population. The fitness of the entire population is what matters, not the fitness of individual members" (45). What is confusing is that we normally think of a population in biology as being constituted of *physical* organisms. But it is clear that a quasispecies population is not so constituted. For example, Eigen *et al.* (1988) tell us that "The *population of mutant states* in the quasi-species is strongly modulated by the fitness distribution" (6890). Similarly, Eigen (1993) tells us that "This region in sequence space can be visualized as a cloud . . . . It is a self-sustaining population of sequences" (45). Thus, by a quasispecies population Eigen and his colleagues mean a population of *points* in sequence space, which is clearly an abstract population.

As for natural selection, we normally think of selection as acting on physical organisms, so that when Eigen (1993) tells us that "Biologically, the quasispecies is the true target of selection" (45), we naturally get the impression that a quasispecies is at least partly physical. But as we have just seen, the quasispecies is a population of abstract points in a sequence space. Can such a population, in whole or in part, be the target of natural selection? I suppose it depends on one's concept of natural selection. If, like Sober (1984a), one thinks of natural selection as a physical force, then the answer must be no, since a physical force could not possibly act on abstract entities. Eigen, however, does not think of natural selection as a physical force. According to Eigen (1995), natural selection is a "physical consequence of error-prone self-reproduction far from equilibrium" (10). As a

---

[39] This could only be accomplished, of course, by the use of mutagens. The problem with this approach, it seems to me, is that since viruses can multiply only within the cells of their host organisms, it would appear impossible to find a mutagen that would act only on the HIV viruses. Presumably any mutagen would also affect the invaded cells in the host organism, most likely causing cancer and thus defeating its original purpose.

consequence, it is not a physical force, so that it may be compatible with the abstract entities that are quasispecies.

One might still feel, however, that a quasispecies is merely intended as an abstract *representation* of a "real" species, so that the concept of the former is not intended to replace the concept of the latter. But this apparently is not the case. For example, Eigen *et al.* (1988) mention "the replacement of single species by quasispecies" (6882). Moreover Eigen (1993) tells us that "The substitution of 'quasispecies' for 'species' is not merely semantic" (42). There really is, then, an attempt at a conceptual revolution here, motivated in part, it would seem, by a desire to include asexual species (in particular viruses) as real species.

As a fully abstract entity, it is important to notice that a quasispecies also fulfills the role of an abstract class. What determines the class (i.e., the cluster of points or hypervolume in sequence space) is a physical population with a wild type and an equilibrium between mutation and selection. Within this hypervolume each point, including the wild type, may represent one or more organisms, or it may represent none. Moreover, any organism with a genotype represented within that hypervolume would be a member of that quasispecies, regardless of its origin, just as an aberrant offspring with a genotype falling outside of that hypervolume would not be a member of that quasispecies, in spite of its origin. Fundamentally, then, descent is irrelevant.

What this means (keeping in mind the fundamental criterion of identity and individuation for physical objects) is that although a *physical* population determines a quasispecies, and each organism of that physical population will probably also be a *member* of that quasispecies, neither that physical population itself nor any of its organisms are *constitutive* of the quasispecies. And this, after all, is not such a radical idea, since much the same relation holds for chemical kinds. An atom of gold, for example, is a *member* of its kind, namely gold, but it is not *constitutive* of its kind. This is evident from the simple fact that its kind remains the same even though its member atoms come and go. But if each atom of gold were constitutive of its kind, this could not be so.

The problem now is whether a quasispecies as an entity in its own right can meaningfully be said to change or evolve. Eigen (1983) tells us that "The process of evolution can . . . be regarded as a route in this space [sequence space]" (108). Similarly, Eigen *et al.* (1988) tell us that "Evolutionary optimization proceeds along defined pathways in sequence space" (6891). But do we really want to take this literally? Again, this is an issue I will reserve for §5.

For the moment, it is interesting to note a serious difficulty with quasispecies evolution even if we grant it for the sake of argument. Eigen *et al.* (1988) define evolution as "destabilization of the local quasi-species upon arrival of an advantageous mutant that establishes a new quasi-species" (6890). Eigen (1995) tells us that "If a better adapted mutant is found, the previous distribution is no longer below the error threshold. It becomes unstable and its information content vaporizes only to condense in the vicinity of a new wild type. Despite the continuity of the underlying molecular process, we see that evolution proceeds via discrete jumps" (13). But if every discrete jump "establishes a new quasi-species," then in a sense we've regressed back to the early days of the twentieth century, to the sal-tationism of De Vries and the early Mendelians, for whom "the theory of muta-tion assumes that new species and varieties are produced from existing forms by sudden leaps" (De Vries 1912: vii), a view that was supposed to have been put to rest by the Modern Synthesis (cf. Mayr 1982: 546–567). Indeed, as Mayr (1991) put it, "Sudden new origins, . . . are not evolution. The diagnostic criterion of evolutionary transformation is gradualness" (18). Thus, what price quasispecies?

A further difficulty for the quasispecies theory is that it depends on a common but somewhat questionable concept, namely the concept of population (cf. chap-ters 2.4 and 4.4). One problem is that for the majority of biologists the concept of population does not apply to asexual but only to sexual organisms. It is axiomatic in classical population genetics, for example, that the populations under consid-eration are sexual. This is clear from a simple glimpse of its foundation, namely the Hardy-Weinberg law.

Even when confined to sexual organisms, however, the concept of population, as Mayr (1987a) has aptly pointed out, is not homogeneous but heterogeneous, ranging "from the local population (deme) [a 'deme' is normally defined as a local interbreeding population] up to the species ('the largest Mendelian popula-tion' of Dobzhansky)" (163). If one takes the deme approach, the problem is that, as Wilson (1992) put it, "Few such objectively definable populations exist in na-ture. Most that do look like textbook examples are endangered species, with so few organisms left that there is no doubt as to the boundaries of the population they compose" (64). Of course, even if the deme approach should prove to have objective application in the majority of cases, the problem now is that every local population, no matter if conspecific with another, will have its own optimal en-semble and hence its own wild type, so that each population will have its own quasispecies. But this is contrary to the beliefs and practices of most modern bi-ologists, for whom species are often composed of more than one population, and represents a regression back to the days when every local variety was thought of as a distinct species.

On the other hand, if one takes the wider, Dobzhansky approach to pop-ulations, it is difficult to see how one could generate a quasispecies for such populations. Often in biology a species will comprise a number of geographical populations, each involving geographical variants. In such cases, then, there will be more than one wild type to consider when generating a quasispecies, and it is by no means clear, since it is nowhere discussed by Eigen and his colleagues, how this scenario would be accommodated in their theory. Eigen *et al.* (1988) tell us that "if mutations happen much more frequently than it is commonly assumed in population genetics, then not a single fittest type but an optimal ensemble—a *master sequence* together with its frequent mutants—will survive" (6882), that the quasispecies "consists of a *master sequence* . . . which is the most frequent se-quence and commonly has the maximum selective value, and a mutant distribu-tion centered around the master" (6885), and that "The *wild-type* is not a single individual but rather a distribution having a defined consensus sequence that usu-ally coincides with the master sequence" (6890). Similarly, Eigen (1995) tells us that "the large number of mutants do indeed cluster about the best adapted type, so that the mean 'consensus' sequence does indeed represent the entire population" (13). But can a Dobzhansky population with geographical variants have a consensus genotype? Originally in genetics the wild type was thought to be the most common genotype (or phenotype) of a species. But as the great amount of genetic mutants in species became better known, the concept became highly strained. As Futuyma (1986) put it, "The revelation of all this variation was something of a shock in the 1920s and 1930s, when genetic uniformity was taken for granted. It resulted in a swell of new opinion among many geneticists, led by Theodosius Dobzhansky: a population is an immensely diverse assortment of genotypes, and there is no such thing as a wild-type, or normal, genotype. Rather, the norm is diversity. The words 'normal' and 'abnormal' begin to lose their meaning" (96). As a consequence, rather than being discarded the concept of wild type was appropriated to refer to the most common allele at a particular gene locus. The problem for quasispecies theory, then, is that if one takes as one's physical population a Dobzhansky population, the existence of geographical variants will involve alleles not found in other geographical subpopulations of the same Dobzhansky population, so that any given gene locus may not, and some gene loci will not, have a wild type for the Dobzhansky population. In the language of molecular biology, a consensus genotype for the wider inclusive population, the Dobzhansky population, will cease to have meaning, and with it the concept of quasispecies. For endangered species and for laboratory populations, then, the concept of quasispecies might retain some meaning, but for many populations in the wild it would seem a theoretical *cul-de-sac*.

Technical incompetence prevents me from providing a more insightful analysis, but it seems to me that the problems outlined above are very serious for what is otherwise a truly fascinating species concept.

3.4 Species as Ecological Niches

The final view that we will examine in this chapter is similar in a number of respects to the cluster view, not the least of which is that it is not a single view but rather a family of related views. But first a little history. According to the *Oxford English Reference Dictionary* (1995), the word "niche" means "a shallow recess, esp. in a wall to contain a statue etc." (979). Such nonbiological uses of the term go back at least 250 years. According to T.W. Schoener (1989), however, "Etymology and prior usage help little to understand the present ecological meaning" (79). Instead Schoener (85) tells us that the first ecological use of the term was in 1910, by a naturalist named R.H. Johnson. But one can certainly find much earlier uses of the idea (minus only the term). Lyell (1832), for example, used the phrase "the economy of Nature" (42), and after quoting from an 1820 essay by Augustin Pyramus de Candolle, in which Candolle distinguished "stations" from "habitations," Lyell wrote:

> we may remind the geological reader that station indicates the peculiar nature of the locality where each species is accustomed to grow, and has reference to cli-

mate, soil, humidity, light, elevation above the sea, and other analogous circumstances; whereas by habitation is meant a general indication of the country where
a plant grows wild. Thus the *station* of a plant may be a salt-marsh, in a temper-
ate climate, a hill-side, the bed of the sea, or a stagnant pool. Its *habitation* may
be Europe, North America, or New Holland between the tropics. . . . The terms
thus defined, express each a distinct class of ideas, which have been often confounded together, and which are equally applicable in zoology. [69]

Not much later the most famous heir to Lyell's uniformitarianism, namely Darwin (1859), would write of the "somewhat different stations" of primroses and
cowslips (49) and more generally of "each place in the economy of nature," some
of which are "unoccupied" and which natural selection tries to "fill up" (102; cf.
104), a concept he put to profound use in his theory of competition, diversifica-
tion, and speciation. In turn Darwin's follower Henry Bates (1862), in his classic
paper on mimicry in butterflies, would write of "The two sets of forms [that]
seem to agree, . . . in habits, and apparently occupy the same sphere in the economy of nature in their respective countries" (498).

At any rate, I will carefully examine modern definitions of ecological niches a
little later. For the present I will begin by examining the views of recent theorists
each of whom conceives of a species as being in some way identifiable with an
ecological niche. Though with some hesitation, I include this view as an abstract
class view for basically five reasons. First, it is commonly though not universally
held that ecological niches are abstract entities (cf. Rosenberg 1985: 199). Second, in common with other class views, a niche is thought to have *members* (as
opposed to parts), whether these members are lineages of organisms or just organisms themselves. The main difference here, of course, is that on many conceptions of ecological niches there may be empty (unoccupied) niches, whereas in
modern set theory there are not many but only one empty or null class (cf. Quine
1969c: 30). This latter view follows, as we have seen in the previous section,
from the criterion of the identity of coextensives. That I find fault with this criterion will be discussed in the following section. If my analysis in that section is
correct, then there may be more than one null class, which would remove at least
one obstacle to there being empty niches and to niches being classes. Third, the
ecological view shares with the other class views the possibility that a species
may change in membership and yet still remain the same. Fourth, in common
with at least some of the other class views, it is possible (on at least some conceptions of niche) that a niche could have only one member organism and thus be a
class of one. And finally, on at least some interpretations the ecological view
shares with the other class views (or at least some of them) the conclusion that
extinction is not necessarily forever.

As the ecological view of species is not homogeneous but highly heterogeneous, it is a view that cannot be adequately analyzed without first examining its
heterogeneity. The view most often referred to as an ecological view of species is
that of the ecologist Leigh Van Valen (1976). Motivated mainly by the phenomenon of semispecies (Van Valen uses the term "multispecies") as well as a desire

to base his species concept on actual causal processes in evolution (69), Van Valen tentatively put forward the following *ecological species concept:*

> A species is a lineage (or a closely related set of lineages) which occupies an adaptive zone minimally different from that of any other lineage in its range and which evolves separately from all lineages outside its range.
>
> An adaptive zone . . . is some part of the resource space together with whatever predation and parasitism occurs on the group considered. It is a part of the environment, as distinct from the way of life of a taxon that may occupy it, and exists independently of any inhabitants it may have. . . . The boundaries of an adaptive zone may be fixed and if so will remain the same whatever species are present, like apartments in an apartment house or a surface with basins separated by ridges. [70]

Although Van Valen does not explicitly use the word "niche," the concept is clearly there. The concept of adaptive zone is simply a wider and more inclusive concept than that of niche. According to Abercrombie *et al.* (1990), an adaptive zone is "A more or less distinctive set of ecological niches established and occupied by an evolutionary lineage with time" (9). But Van Valen uses the concept in a somewhat different (or perhaps additional) sense, including higher taxa as well. As he tells us, the word "minimally" in his definition above "delimits species from higher taxa," and it is this, he tells us further, that makes species "less arbitrary taxa than are those in higher categories" (72). Thus, species occupy niches, while higher taxa occupy ever wider adaptive zones.

It is not difficult to find similar views in other biologists who for whatever reason use the concept of niche though not the term. Ehrlich and Raven (1969), for example, are impressed by the phenetic cohesion in "good species" in spite of the weakness of gene flow in some cases and the sheer lack of it in others (including especially asexual species). Accordingly they conclude that the primary cohesive force for each species is what they call "similar selective regimes" (61). Speciation on this view, then, occurs either when the selective regime itself changes or when populations within a species become subjected to "differing selective regimes," which is "often reinforced by selection operating against hybrids" (63).

What is unclear in the above two views (among other issues to be dealt with later) is whether species are to be thought of as primarily horizontal or as primarily vertical entities (cf. chapter 2.3/4). There is no such confusion in George Gaylord Simpson's *evolutionary species concept.* Motivated primarily by the lack of an evolutionary criterion in Mayr's biological species concept (cf. chapter 4.1), by its exclusion of asexual species, by a desire to base his own species concept on actual causal processes in evolution, and by paleontological desiderata (Simpson was a paleontologist), Simpson (1961) proposed the following definition:

> *An evolutionary species is a lineage (an ancestral-descendant sequence of populations) evolving separately from others and with its own unitary evolutionary role and tendencies.* [153]

That evolutionary species on Simpson's species concept are meant to be vertical entities is quite clear. "The evolutionary species," he tells us, "implicitly brings in the element of time" (154). Moreover an evolutionary species on his view is clearly meant to have both a beginning and an ending in time (165), which is evident alone from his diagrams (164 and 168).

What is of particular interest, however, is the role played by ecological niches in the above definition. Although Simpson admits that the main difficulty with his species concept "is the definition and recognition of roles," he immediately goes on to tell us that "Roles are definable by their equivalence to niches, using 'niche' for the whole way of life or relationship to the environment of a population of animals and not for its microgeographic situation" (154).

Although Van Valen's ecological species concept also employs the concept of lineage in the ontology of species, he allows for more than one lineage in niche occupation, as well as polyphyly or multiple origins (cf. Van Valen 1976: 73), so that his species concept is fundamentally a class concept. Simpson's species concept, on the other hand, since it involves only a single lineage, is not so much a class concept as it is a physicalist concept (although Simpson himself denied this when later interviewed by Hull; cf. Hull 1988: 213n2), and so therefore more properly belongs somewhere in chapter 4. I introduce it in the present section, however, since Simpson defines evolutionary lineages in terms of niches. His species concept is thus connected with all the problems of the niche concept, which we will examine later in this section, so that it is subject to a similar fate. In defining species as physically vertical entities (at least arguably in effect) it should be kept in mind that Simpson's species concept is also subject to many of the problems that we shall encounter in chapter 4.4/5.

For the present, it should be noticed that Simpson's use of niches involves a major difficulty. Since Simpson explicitly tells us that "The evolutionary definition given above omits the criterion of interbreeding" (153), and since he explicitly includes asexual organisms (161–163), we might infer from Simpson's species definition and his discussion on niches that a lineage is divided into different species only according to different niches, so that if a lineage changes its niche then it is no longer numerically the same species. However, when he comes to discuss the evolutionary aspect of species, Simpson tells us that "The definition neither states nor implies that the unitary role of any one species is necessarily unchanging. Evolution (apart from quite exceptional saltations) could not occur if roles did not change *within* species. The concept involves a species having a unified role at any one time (not necessarily the same role at all times)" (155). But if not defined and delimited by any one niche, then what is it, we must ask, that delimits the temporal ends of a species on Simpson's view? Although Simpson employs "the approximate mean horizontal difference" between contemporaneous species taxa as his yardstick (167; cf. 165), he nevertheless admits that "Certainly the lineage must be chopped into segments for purposes of classification, and this must be done arbitrarily . . . because there is no non-arbitrary way to subdivide a continuous lineage" (165; cf. 117–119). Thus, in taking the vertical dimension as primary contra Mayr, the curious result is that most species on Simpson's view turn out to be arbitrary, a matter of "personal judgment," as he says, so that "classification is an art" (165). As we shall see in chapter 4, others have suggested nonarbitrary ways of delimiting vertical species, but with equally undesirable results.

Interestingly, it is the role played by niches in Simpson's species concept that motivated Van Valen (1976) to attempt an improvement with his own species definition. "I will modify Simpson's concept," he tells us, "first stating the revised version formally and then justifying it" (69). One important difference between them is in terms of what niches are supposed to be a property of. From Simpson's description quoted above, it would appear that a niche is a relational property of a lineage. From Van Valen's description, on the other hand, a niche is clearly a property of an environment. This is a fundamental difference, being rooted in the modern history of the concept itself, as we shall see later in this section, with each side having its full share of problems.

Van Valen's is not the only attempt to improve on Simpson's species definition. In an explicit attempt to "resurrect and defend the species concept of Simpson . . . as best suited for dealing with the species and its origins" (1978: 79), Edward Wiley (1981) provided the following definition:

An evolutionary species is a single lineage of ancestor-descendant populations which maintains its identity from other such lineages and which has its own evolutionary tendencies and historical fate. [25; cf. 1978: 80]

There are a number of peculiarities with Wiley's species concept which make it especially curious. First, to say that a species has a "historical fate" sounds like orthogenesis, the view that a species has a developmental program analogous to the genetic program in organism ontogeny. Orthogenesis, of course, is a view now thoroughly discredited (cf. Abercrombie *et al.* 1990: 410–411).

Second, like all ecological views, Wiley includes asexual organisms within his species definition. In a very interesting argument (1978), in which he maintains that asexual organisms "do represent independent lineages . . . and thus do correspond to evolutionary species" (85), he suggests that "asexual species can be accounted for under the evolutionary species concept in the same way we place allopatric demes into a single sexually reproducing species" (86), namely by "lack of differentiation" based on "real genetic, epigenetic, ecological, or other bases" (87). What is so curious about this is that Wiley (1981) defines "lineage" in his species definition above as "one or a series of demes that share a common history of descent not shared by other demes" (25). The concept of deme, of course, ubiquitously defined as a local interbreeding population, has no application whatsoever to asexual organisms, thus excluding them in species.

But the curiosities do not end there. Wiley (1978) tells us that the principal advantage of his species concept over that of Simpson is that of "not implying that species must change (evolve)" (90n1), which makes his view consistent, he believes, with the stasis stage of species in Eldredge and Gould's theory of punctuated equilibria (cf. chapter 4.3). On the other hand, although Wiley finds much in Van Valen's species definition "and Van Valen's subsequent discussion that I can agree with" (1978: 87), he finds it deficient on a number of counts. For one thing, "My criticism with the adaptive zone as used by Van Valen stems from the fact that I do not think supraspecific taxa have a niche *per se*" (1981: 254). But more importantly, Wiley does not think that a *species* has a niche *per se* either, or at least that it can be defined by one. Although Wiley (1981) says that "In either sexual or asexual species identity may be manifested in distinctive ecological roles" (25), he nevertheless maintains that "It is possible for two species to share essentially the same niche within the same range." Moreover, "given the original definition of Van Valen (1976) one might argue that a species forced to extinction through interspecific competition was not a species at all" (1978: 87). Even more distancing from both Van Valen and Simpson is Wiley's view that speciation, rather than being caused by niche change or niche differentiation, actually *precedes* niche change or niche differentiation: "So far as I am aware," says Wiley (1981), "potential empty niches (if such things exist) or the subdividing of an existing niche by a number of species must await speciation" (253).

Finally, Wiley (1978: 80, 1981: 21–23) does not think that species are classes of ecological similars at all, but instead that they are *concrete individuals* following the view of Ghiselin and Hull (cf. chapter 4.2). The curiosity here is that unlike Wiley, Ghiselin and Hull, as we shall see in chapter 4, adamantly refuse to accept asexual organisms as members of genuine species. Wiley, then, it seems to me, is trying to have it both ways, but in doing so it ought to be clear that rather than slightly improving on Simpson's and Van Valen's species definitions or following the individuality thesis of Ghiselin and Hull, he is really doing neither. His species concept, therefore, remains somewhat of an enigma. At any rate, like Simpson, Wiley's species concept more properly belongs in chapter 4, and we shall examine problems with his physicalist ontology for species specifically in chapter 4.4/5.

Wiley and Simpson, of course, are not the only theoretical biologists to only partly employ niches in a species concept. Perhaps the most famous employment is Ernst Mayr's (1982) ecological version of his *biological species concept:*

> A species is a reproductive community of populations (reproductively isolated from others) that occupies a specific niche in nature. [273]

This definition is interesting on a number of counts. First, in the long history of the development of Mayr's species definition, which spans some fifty years, this is the first time that the concept of niche plays a defining role. (For a brief recapitulation of Mayr's species concept from 1942 to 1991, see the end of my chapter 4.1.) Prior to 1982, Mayr always defined species wholly in terms of reproductive isolation. Now (Mayr 1982), he tells us, "Reproductive isolation, however, is only one of the two major characteristics of the species. Even the earliest naturalists had observed that species are restricted to certain habitats, and that each species fits into a particular niche" (274).

Second, Mayr's species concept has always been that of a horizontal concept, or what he normally calls a nondimensional concept (cf. chapter 2.3n15). As El-dredge (1985b) so interestingly put it, "Allochronically, . . . Mayr believes that species give up the ghost ontologically" (52). But if so, then Mayr's niche con-cept in his 1982 definition, unlike the niche concept in Simpson and others, is also fundamentally a horizontal concept.

Third, "as one of the two major characteristics of the species," one has to seri-ously wonder about the nature of niche occupation and therefore of niches in his species concept. Hengeveld (1988), in his critique of Mayr's (1982) niche crite-rion, claims that niche concepts "should be considered as essentialistic reasoning in ecology" (50), which of course is a serious criticism given Mayr's long and repeated rejection of essentialistic thinking in biology. If niche concepts are es-sentialistic, however, they do not appear to be of the strict variety. I would think they are more like cluster concepts. But even so, not everyone thinks of them as abstract. Mayr (1987a) tells us that "My own species concept (1942–1982) has clearly always been that of a concrete entity" (151). But if this is true, since niche occupation is clearly a property of species in Mayr's 1982 ontological definition, then niches must also be concrete entities on his view. Otherwise it would be meaningless to say that a concrete entity could "occupy" or "fill" an abstract en-tity. In other words, then, niches on Mayr's view must be physical entities, which alone is suggested by his use of the word "habitat" in the quotation from 274 above. (As we shall see, a physical conception of ecological niches is not that uncommon.)

Fourth, until 1982 Mayr's biological species concept, since it was based on sexual reproduction and the corresponding concept of a closed gene pool, exclud-ed asexual species as real species. Now (Mayr 1982), however, he tells us that "It is only where the criterion of reproductive isolation breaks down, as in the case of asexual clones, that one makes [exclusive] use of the criterion of niche occupa-tion" (275).

But now something very interesting happens. First, Mayr's 1982 species defi-nition quoted above now becomes somewhat misleading, since a community of asexual clones can in no meaningful sense be reproductively isolated from others. But more importantly, in including asexual species as genuine species based ex-clusively on niche occupation, all species in the species category now have one of fundamentally two characteristics: for sexual species it is not simply reproductive isolation but reproductive isolation *plus* niche occupation, while for asexual spe-cies it is simply niche occupation. Mayr's species concept is thus disjunctive, and either of "the two major characteristics of the species" is a property of each and every real and genuine species. But stated in this way, niche occupation obvious-ly becomes the more fundamental of the two, and reproductive isolation becomes somewhat unnecessary.

But Mayr's troubles caused by trying to include asexual species do not end there. Elsewhere (Mayr 1987a) he tells us that

> The fact remains that there are entities in nature that do not qualify as biological
> species, but which fill the same place in the ecosystem as do biological species.

> They cannot be ignored in any study of niche occupation, utilization of resources, and competition. An ecologist will say that some of them occupy the same place in nature as do genuine species, and that some way of dealing with them must be found. Although by descent they have continuity in time within the clone, they lack the internal cohesion characterizing a gene pool. Therefore they are classes and not individuals-populations. [166]

Mayr, of course, is well acquainted with the philosophical literature on the species problem, particularly that of David Hull, so he is well aware that the use of the word "class" in biology is usually quite different from that in philosophy (cf. Mayr 1982: 56–57, 254). Moreover, Mayr more than anyone has repeatedly reminded his readers of the distinction between species as a category and species as a taxon. As Mayr (1987a) put it, "That the species category is a class is not disputed by anyone. What is at issue is the ontological status of the species taxon" (146). As he says at the top of the same page, "Are they classes, are they individuals, or what are they if they are neither classes nor individuals?" (cf. Hull 1981a: 143). Unquestionably, then, since Mayr clearly thinks of classes in the philosophical sense as abstract rather than physical entities,[40] it follows according to his 1982 species definition and what he says above about asexual species that not all species are exclusively concrete entities but rather some are concrete and some are abstract entities. But then, if on his view ecological niches are physical or concrete entities, how can asexual species as abstract entities possibly occupy or fill them? There is thus not a little problem of incoherence in Mayr's 1982 view. Mayr (1987b) tells us that when in 1982 he added the niche concept to his species definition he did not do so with the purpose of including asexual species, but that instead it was an afterthought, a further application. His main purpose, he tells us, was a "toughening it up" rather than a "watering down" of his biological species concept, to add "one more constraint." "Time will show," he says, "whether this additional qualification is useful or confusing" (214). I suggest that, in his own usage, it has proved more confusing than useful.

One more species concept that cannot be ignored among those views which partially use the concept of niche is Alan Templeton's *cohesion species concept*. Although Templeton (1989) tells us that "The cohesion concept of species defines a species as an evolutionary lineage" (20), he formally defines his species concept as follows:

> The cohesion species concept is the most inclusive population of individuals having the potential for phenotypic cohesion through intrinsic cohesion mechanisms. [12]

---

[40] Cf. Mayr (1970): "A category, thus, is an abstract term, a class name" (13). Cf. Mayr (1987a): "both individuals and classes can be real" (146). Cf. Mayr (1987a): "By far the most important defining property of a class is its constancy, a necessary correlate of its being based on an essence. At the same time class membership is not spatiotemporally restricted" (148).

As a definition, of course, the above is perhaps more vague than any other we've examined so far. But unlike most definitions we've seen, Templeton goes to perhaps the greatest lengths to unravel and explain it. His view turns out to be very detailed, and it is by no means easy to extract from his discussion the precise nature of species according to his concept. In attempting to figure this out, however, I will try to be as brief and clear as possible.

But first, some of the motives behind Templeton's species concept have to be addressed. As with other authors in this chapter, Templeton is impressed by the fact that asexual organisms often have a high degree of phenotypic cohesion, which is to say that at the horizontal level they do not normally provide us with the specter of a smeary continuum. Instead, "the asexual world is for the most part just as well (or even better) subdivided into easily defined biological taxa as is the sexual world. This biological reality should not be ignored" (8). Second (4), asexual species display evolutionary "fates" through time much the same as sexual species. Third, Templeton (9) points out that many sexual organisms also fall outside of the logical domain of the traditional biological species concept, namely all obligate self-mating organisms (obligate hermaphrodites) as well as organisms with mandatory sib mating (e.g., many species of wasps). Fourth (10–11), as we have seen in my discussion on Beckner, in both the plant and animal worlds there is the phenomenon of hybrid zones between what are otherwise biologically distinguishable species.

In an attempt to unify a number of species concepts so as to provide a better fit with the biological world, Templeton "defines species in terms of the mechanisms yielding cohesion rather than the manifestation of cohesion over evolutionary time" (12). Although Templeton is not altogether clear on this point, he later makes it clear that on his view the causal mechanisms responsible for species cohesion are actually *intrinsic properties*—"traits" (Templeton 1998: 34)—of the species themselves. This implies that by definition there cannot be any external causes of species cohesion, a problem for his view that we shall see shortly.

But now what are these intrinsic mechanisms? This is where Templeton's discussion gets very detailed. He tells us that "The basic task is to identify those cohesion mechanisms that help maintain a group as an evolutionary lineage" (13). Without getting into the details, Templeton divides these mechanisms into two basic classifications headed by what he calls "genetic exchangeability" and "demographic exchangeability." The former "simply refers to the ability to exchange genes via sexual reproduction" (14). Demographic exchangeability, on the other hand, unlike the exchange of genes, refers to the exchange of whole organisms, and is where ecological niches come in: "To the extent that individuals share the same fundamental niche, they are interchangeable with one another with respect to the factors that control and regulate population growth and other demographic attributes" (15). In employing Hutchinson's distinction (to be better discussed later) between the fundamental niche and the realized niche (the latter, the actual niche, is a subset of the former, the potential niche), Templeton tells us that "members of a demographically exchangeable population share the same fundamental niche . . . although they need not be identical in their abilities to exploit that niche. The fundamental niche is defined by the intrinsic (i.e., genetic) tolerances of the individuals to various environmental factors that determine the range of environments in which the individuals are potentially capable of surviving and reproducing" (14–15). On this view, then, just as genetic exchangeability is clearly a property of sexual species, these "various environmental factors" that partly define the fundamental niche must also be themselves, as part of the defining cohesion mechanisms, among the properties of a species. (As we shall see, niches on Hutchinson's view are properties of species, not of environments, so that there are no empty niches on Hutchinson's view. Thus my interpretation of Templeton's species concept is in keeping with his use of Hutchinson's niche concept.) On Templeton's view, then, a niche is but one of a number of properties of a species.[41] I will reserve critical comments on Templeton's species concept for a little later.

Finally in my survey, it should be pointed out that among philosophers the ecological view of species has not been popular. Most philosophers who have written on the species problem subscribe to either the nominalist position, the strictly essentialist position, one or other of the cluster views, or (most popularly) the individuals view. Among adherents to the ecological view, I have only been able to find David Johnson (1990). Johnson, however, is concerned primarily to defend the Kripke-Putnam view of natural kinds by showing how abstractions can be causes. As Johnson put it, "the presumed fact that a natural kind is an abstract and repeatable nature, category, or property seems to make it impossible for it to be the initiating cause for a chain of events that results in someone's employing a word or phrase to refer to that kind now, because we commonly suppose abstractions cannot be causes of anything" (66). Although Johnson is a pluralist with regard to the objective reality of species, in that "no 'correct,' 'God-given' conception of biological species always is going to prove maximally efficient and explanatory" (67), and although he thinks that the species problem can only be answered in practical rather than metaphysical or objective terms (67), he nevertheless suggests that "sometimes it is useful to identify a species with an ecological niche itself, rather than the group of genetically related organisms that happens to fill this niche at a given time" (67–68). In identifying a species with a niche itself, even by implication an empty niche, Johnson is at least following, if not in a biological tradition, in a metaphysical as well as a physics/chemistry tradition, where natural kinds (e.g., gold) are identified with abstractions rather than their physical members. But Johnson makes a further claim. Although he rejects

---

[41] On page 13 in his paper Templeton provides a basic list of the properties a species may have. Among these are offspring viability, reproductive isolating mechanisms, genetic drift, mutational constraints, as well as historical constraints. Later he tells us that "Not all species will be maintained by the same cohesion mechanism or mixture of cohesion mechanisms, just as proponents of the isolation concept acknowledge that not all isolating mechanisms are equally important in every case" (20). Does this mean that on Templeton's view not every species will have a fundamental niche as one of its properties? Templeton does not tell us, but I strongly suspect that on his view this is not the case.

the conception of niches as classes (which he conceives as "lists of properties") (68), he clearly wants to go beyond his relativistic policy and to "suppose that they [niches] have some sort of objective existence in nature—and therefore are describable but not definable—despite the fact that they are repeatables, and at any given moment may or may not have instances" (68). In fact Johnson suggests that "it might be a factual, empirical—perhaps even partly experimental—question which ecological niches are elements of this present universe, and which others (although equally conceivable with those in the first group) are not" (69). Finally, it is Johnson's contention that if natural kinds are thought of in "the proper way" (66), that is, as niches, then it becomes apparent how they can have causal force: "the fact that ecological niches predictively determine (some of) the properties of the things which occupy them proves . . . that repeatable kinds can have causal force" (69). As to the type of causal force, Johnson tells us that it is not "the 'billiard ball' variety discussed by Hume and his followers" but rather "a type of 'pull' causation," different, however, from "final causality in the Aristotelian sense," but which is nevertheless a "real causality" (70).[42]

Again, I will reserve critical comments for a little later. What I would like to do now is to take this discussion to its next stage. What should become abundantly obvious from all of the foregoing discussion is that the concept of ecological niche in biology and its philosophy is very far from being monolithic. Consequently, any attempted solution to the species problem that involves niches cannot take for granted its particular concept of niche. Accordingly, the species problem expands into the niche problem. If the niche problem admits of a solution, it would clearly exert a sort of selection pressure on the various species concepts that employ niches. On the other hand, if the niche problem does not admit of a solution, then this must be seen as a fundamental flaw in any species concept that employs niches.

What, then, is the niche problem? To understand the niche problem is to understand its history, which is fairly recent. Again, I will keep my discussion as brief as possible, and I include it only out of necessity.

At the beginning of this section I briefly examined the history of the nonbiological use of the term "niche," followed by a brief examination of some of the early uses of the concept (though not the term) in biology. It is now time to examine the history of the use of both the term and the concept in biology. The following short history is based on the papers by Schoener (1989), Griesemer (1992), and Colwell (1992). (For references to the names that follow see any one of these papers.)

Although (as pointed out at the beginning of this section) the first ecological use of the term "niche" was by R.H. Johnson in 1910, his use was isolated to merely one passage in his writings. Instead the first true development of the eco-

---

[42] Johnson's ascription of causal force to abstractions (in particular, niches) has sparked somewhat of a debate within the pages of *Biology & Philosophy*. Cf. Hogan (1992) for a reply to Johnson, and Miller (1994) for a reply to Hogan. My own reply will be reserved for the end of this section.

logical use of the term was by Joseph Grinnell beginning in 1913, but most importantly in 1917. The second important development was by Charles Elton in 1927. Whereas Grinnell's niches included both biotic and abiotic factors so as to account for evolution and speciation, Elton's niches included basically predator and prey factors so as to establish food chains. What is of fundamental importance is that both thought of niches as *properties of the environment* and both accordingly allowed for (or rather necessitated) the existence of empty niches. Moreover, both allowed for the existence of "equivalent niches," niches not only similar but identical (i.e., repeatable). Otherwise there are some interesting differences. Grinnell's niche concept is more of a *place* concept, whereas Elton's is more of a *role* concept, although both recognized each in their respective niche concepts. Furthermore, employing a concept of competitive exclusion presaged in Darwin (cf. Darwin 1859: 109–111), Grinnell held that there can be only one species per niche, that competition over the very same resources would force one or the other species into extinction. Elton, on the other hand, thought that two species could in principle occupy the same niche without competitive exclusion forcing one or the other into extinction.

The next important development in the concept of niche, which Schoener calls "revolutionary" (an assessment with which both Griesemer and Colwell agree), was by G.E. Hutchinson beginning in 1944. Rather than define niches as properties of the environment, Hutchinson defined niches as *properties of populations or species* (not of individual organisms *per se*). Moreover, Hutchinson's concept is purely abstract, a non-Euclidean multidimensional space reminiscent of the A-space and Hamming sequence space that we encountered in the previous section. As Hutchinson himself defined it in 1957, a niche is "an *n*-dimensional hypervolume . . . every point in which corresponds to a state of the environment which would permit the species S_1 to exist indefinitely. For any species S_1, this hypervolume N_1 will be called the *fundamental niche* of S_1. . . . all X_n variables, both physical and biological, being considered, the fundamental niche of any species will completely define its ecological properties" (Schoener, 89). Since few if any species actually utilize all of the environmental dimensions that would allow them to exist indefinitely, Hutchinson complemented the fundamental niche with the *realized niche*, the actual multidimensional ecological property of the species at any one time, a subvolume of its fundamental niche. Competitive exclusion, then, entailed on this view that no two species have identical realized niches. It also entailed as a matter of definition that there could be no empty niches, only unutilized resources.

Schoener provides some interesting conjectures on the motivation behind Hutchinson's revolutionary revision of the Grinnell/Elton niche concept. Most importantly, not only are niches on the latter view fundamentally vague in the sense that they cannot be defined with any precision, a lack of precision remedied on Hutchinson's species-property view, but "If competition between species commonly caused their niches to change, then observation of an occupant's characteristics cannot be used in any simple way to determine characteristics of the 'recess'" (90).

As Colwell puts it, Hutchinson's "hyperspatial niche, whatever its complexities, was at base simply an ecological description of the phenotype [phenome in my usage] of some particular population or species" (241). However, a number of operational and theoretical difficulties were seen to attend Hutchinson's view. One major difficulty involves the determination of the fundamental niche given the realized niche (Schoener, 93). The fundamental niche, as a potential niche, would seem just as conjectural and vague and indeterminable as niches defined environmentally. A further difficulty, as Griesemer puts it, is that "Hutchinson's abstraction [specifically his fundamental niche] is static, atemporal, and does not immediately suggest how to represent variability of organisms in their utilization of the environment over time or variability of populations or species in space and time" (238–239).

These problems soon led to a further development in the niche concept, what is known as the *utilization distribution* view. This view, as Griesemer puts it, "shifted the focus from the permissive range of environmental conditions to the actual resource utilization distribution of species . . . [and] represents a final shift in the meaning of niche" (239). Still an abstract hyperspace, the niche on this view, as Schoener describes it, is "nothing more than a frequency histogram of resource use by some population" (91). It is thus still a property only of a population or species, but now the main emphasis is on an operational definition of niche dominated by a practical concern over actual resource competition between species. (It should be noticed that, unlike earlier views but as with Hutchinson, competitors and predators are not here part of the niche concept.) Of course, as one might expect, serious problems were also found to attend this view. One difficulty is that many plant species would seem to have the same resource utilization. The most serious problem, however, as Schoener points out, is the relation between competition and niche overlap and the inability to determine this relation unequivocally. In short, competition hypotheses are extremely difficult to test, and the diversity in real ecologies has resulted in a diversity in conclusions by ecologists.[43] The result was a questioning of their theoretical assumptions, including a questioning of the value of a niche concept.

All of this has resulted in a wide variety of responses within biology to the value of the niche concept. Some, as both Schoener and Griesemer point out, would have the concept expunged altogether. Some, such as Erhlich and Holm (1962), include the niche concept with other concepts (including species) that they consider to have "low information content and little or no operational mean-

---

[43] As Schoener relates (104), in a study on two species of iguanid lizards in Texas it was found that during dry years there was competition even though the niche overlap was lower than during wet years in which there was no competition, the conclusion being that "competition is associated with lower niche overlap." On the other hand, in a study on two species of iguanid lizards in the Lesser Antilles it was found that when the species were of similar size there was competition and the niche overlap was higher than when the species were not of similar size in which there was no competition, the conclusion being that "competition is associated with higher niche overlap."

ing" (653). Others, such as Wiley (1981), believe that it is a useful concept and should be retained, but that "The word 'niche,' like the word 'character,' defies adequate definition, probably because it is a primitive term" (254). Generally, however, the current scene in biology, both within and outside of ecology, is constituted by a pluralism in niche meaning, with all four principal niche concepts enjoying a wide currency. This richness is seen as an embarrassment by many ecologists.

It is only with this background that the various uses of niches in defining species, and the various criticisms against them, can be fully understood and assessed. For example, if a niche is thought to be only one among other properties of a species, then clearly it would be a mistake to identify a species with a niche, for a species must be *something else* to *have* this property.

In the case of Templeton, where a Hutchinsonian niche is among a number of species properties, his species concept suffers from the difficulty that he uses fundamental instead of realized niches. But more importantly, on Templeton's view one property of a species (the fundamental niche) entails another property of a species (demographic exchangeability), and whatever are the causes of these properties (which according to Templeton is anything and everything that accounts for the phenotypic cohesion of a species) are themselves *also* properties of the species. But this is to conflate causes with their effect (more on this later).

This problem is equally acute when Templeton discusses speciation. On Mayr's biological species concept, the causes of speciation are extrinsic to the species. Templeton, however, thinks of this as a theoretical defect. "Under the cohesion concept," he says, "the evolutionary significance of a species can arise directly out of its defining attributes" (24). But if one is going to conflate causes with their effect, why not then also include the causes of *those* causes, and so on *ad infinitum?* It is not clear how such a *reductio* can be avoided, except, of course, if one thinks of niches in the environmental sense, along with other causes of speciation, as external to the species themselves.

A further problem with Templeton's species concept is a problem of internal coherence similar to what we've already seen with both Simpson's (1961) and Mayr's (1982) species definitions. Quite simply, both "genetic exchangeability" and "demographic exchangeability" are abstract class concepts, whereas "evolutionary lineage" implies a concrete entity, the consequence being that cohesion species are both abstract and concrete. Evidently Templeton takes the concrete aspect as primary, but what (to use a problem that I shall amplify in chapters 4 and 5) would he say in the case of two separate lineages, contemporaneous and yet separately evolved (as in the case of repeated polyploidy), which are both genetically and/or demographically exchangeable with each other? In taking the concrete aspect as primary, he would have to say that they are two separate species. And yet they have the same cohesion mechanisms. Consequent to his species definition, then, they could be taken as one species or two.

At any rate, when species are identified in some way with niches, the niche concept normally used is the environmental or role view, which accordingly has generated the majority of criticisms. For example, Ghiselin (1987a) argues that those who identify species with niches "confound professions with organizations" (138),[44] that "an individual remains the same irrespective of its activities" (139), that "closely related organisms can have very different niches, distantly related ones virtually identical niches" (140), that "Two species can easily occupy identical niches when living in different places, and even when living in the same place" (140–141), and that "a species may itself remain unchanged, while the environment around it changes. Thus extinction of one species can readily advance another to the position of 'top predator'" (142).[45] Similarly, Wiley (1978)

---

[44]   The point is that an organization, which is a historical, economic individual, for example IBM, is not to be confused with the role (or roles) that it plays in an economy. More specifically, Ghiselin's point is that those who identify species with niches necessarily employ an incoherent ontology. A role is a class, which may be filled by an indefinite number of organizations, whereas an organization is an individual which fills a role. To identify a species with a niche is to think of it as both a class and an individual at the same time, which is incoherent: "On the one hand they want species to be individuals," he says, "so they can evolve. On the other hand they want them to be classes of 'ecologically similar' organisms" (138). Ghiselin then immediately proceeds to point out that only individuals can evolve and that classes cannot. I find it interesting that Ghiselin shares my interpretation of the niche view of species as a class view, and I shall agree in the next section that abstractions cannot change or evolve. But I shall argue in the next chapter that species do not need to be individuals in order to evolve, and indeed that this (individuals) is an inappropriate category for species. For the moment I would like to suggest that Ghiselin's organizations example better displays a serious flaw when we substitute employees for organizations. Clearly a job is also a sort of class, since indefinitely many people may have the same job. What is interesting is that we often equate or categorize people according to the type of job they have and that this allows us more often than not to make correct predictions or hypotheses about a number of their personal qualities (e.g., math skills, honesty, physical strength, etc.). Of course many people like to think of their identity as separate from their job, but perhaps this is more a matter of vanity than anything else. We are, after all, fundamentally social creatures. The stress upon one's individuality as distinct from one's social role is a relatively recent phenomenon, a distinctly 1960s thing born of the rebelliousness of the time, a trend that seems mostly to have died following the end of the Vietnam war, as the hippies of yesterday eventually settled down and became the professionals of today. To think of species as individuals, then, and moreover as distinct from their ecological roles, may indeed be a product of the same social trend, and if so it may eventually suffer the same fate. At any rate, any criticism of the niche view of species which presupposes that species are individuals simply begs the issue, and with the greatest respect for Michael Ghiselin I suggest that his professions-and-organizations criticism, as well as many of his criticisms that follow, does precisely just that.

[45]   This is an interesting argument. However, it would appear to suffer a fatal difficulty. As Templeton (1989) points out, "Speciation is generally a process, not an event" (22). The same is also true of extinction (cf. Damuth 1992: 110–111). Thus, should a species vying for the role of top predator be going through the process of extinction, its main competitor for that role, even if it is itself the main cause of the other's extinction, is surely not all the while remaining unchanged. As an important dimension of its niche changes, it itself changes, and by the time it has become top predator it may have changed considerably. Of course, one might reply to this by appealing to the example of what are called *introduced species*, foreign species that are introduced into a new environment, whether

claims that "species do not have to occupy minimally different niches or adaptive zones from other species within their ranges to be considered species. It is possible for two species to share essentially the same niche within the same range" (87). It is apparent that the criticisms of both Ghiselin and Wiley presuppose the environmental view of niches, more specifically that of Elton as opposed to Grinnell. Should Grinnell's niche concept someday prove superior, many or perhaps even all of their criticisms may be found more than a little wanting.[46]

What ought to be clear, as Colwell (245) points out, is that the issue of which niche concept is correct cannot be decided simply by fiat. While conceptual considerations may take us a long way, empirical considerations cannot be ignored. The competitive exclusion principle, as an important example, is by no means *a priori* true. Logically there is no reason why two different species may not indefinitely share the very same environmental dimensions and role in the very same environment, say for example that of top predator. Any denial of this possibility is an empirical hypothesis subject to testing. The problem is that competition is extremely difficult to test, as well as extremely difficult to define. Indeed the concept of competition has quite a varied history in itself, although it seems to have achieved in recent years a much improved degree of consensus (cf. McIntosh 1992). With this in mind, what is the current state of expert opinion on the empirical support for competitive exclusion? As Law and Watkinson (1989: 248) point out, the really rigorous work began with G.F. Gause, whose experiments eventually led to his famous 1934 formulation of the competitive exclusion principle, often later incorrectly referred to as Gause's axiom or Gause's law. Since then ecologists have done much further work, with the consequence that, according to Law and Watkinson, "Much of the later work has served to confirm or elaborate the various observations made by Gause" (260). Nevertheless, Law and Watkinson warn that "the empirical foundation of community ecology is not altogether secure" (275; cf. chapter 2.3n16). Any preference for Grinnell's over Elton's niche concept, then, when it comes to competitive exclusion, must likewise (at least at present) be "not altogether secure."

A similar issue is the question of whether there can be empty niches. Once again, the issue cannot be decided by fiat. There are important empirical considerations that must be taken into account, as well as certain conceptual constraints. Colwell provides a short and useful discussion here. According to Colwell there

---

naturally or by man (in the latter case one naturally thinks of the havoc wreaked by species introduced to Hawaii and New Zealand). An introduced species could conceivably drive the top predator into extinction in a very short time and without undergoing any significant change in itself except for an increase in the number of its member organisms. But of course Ghiselin cannot use this example since he thinks of species as *individuals*. For this example to work the *entire* species would have to be transplanted, not a part of it. But then this is to return to the previous scenario.

[46]   In this case the criticism of Van Valen's view that we saw earlier in Wiley (1978)— "But, given the original definition of Van Valen (1976) one might argue that a species forced to extinction through interspecific competition was not a species at all" (87)— should be viewed as suffering from an improper time scale.

are basically two sorts of evidence that have been adduced in favor of empty niches. One sort of evidence one often finds is the claim that a species or group of species having a place in a particular environment have no equal in a similar environment somewhere else, so that the latter environment provides an empty niche that could be filled if only there were the appropriate organisms. Mayr (1988a) provides an interesting example: "The tropical forests of Borneo and Sumatra, for instance, provide resources for 28 species of woodpeckers. By contrast there are no woodpeckers what-so-ever in the exceedingly similar forests of New Guinea, and on that island hardly any other species of birds make use of the woodpecker niche" (136). Colwell makes the interesting claim that such a hypothesis is in principle verifiable but not falsifiable: "If an introduction fails to become established, or becomes established but displaces or replaces one or more established species, the argument can always be made that the wrong species was chosen, or even that the appropriate species has not yet evolved" (244). Interestingly, if we take the Popperian view of science, in which only in principle falsifiable hypotheses can be genuinely scientific, then the hypothesis that there are empty niches is not scientific but metaphysical. I have argued elsewhere (Stamos 1996b) against this criterion in favor of a disjunctive criterion, disjunctive in the sense that if a hypothesis is either in principle verifiable or falsifiable then it is in principle genuinely scientific.[47] On my view, then, which I believe better accords with actual practice in science, the hypothesis of empty niches is a genuine scientific hypothesis.

The second sort of empirical support for empty niches comes from actual species introductions. What is needed is accurate observations indicating that the introduced species does not disturb in any ecologically significant way any of the residents. So far, however, as Colwell (245) points out, there is fundamental disa-

---

[47] Briefly, the empirical content of a theory, on Popper's view, is not what it says about the world, but instead what it prohibits. On this view, then, the empirical content of a theory "increases with its degree of falsifiability" (Popper 1934: 113). In connection with this, Popper claims that strictly existential statements, such as "There are white ravens," have no empirical content because they are in principle unfalsifiable, the reason being that "We cannot search the whole world in order to establish that something does not exist, has never existed, and will never exist" (Popper 1934: 70). The basic problem here is that genuine science clearly includes hypotheses that are strictly existential. For example, in 1928 Dirac reluctantly conjectured that there must be positively charged electrons (positrons), a conjecture that was experimentally verified only a few years later in 1932. On Popper's view, however, we should have to conclude that Dirac's conjecture was empirically empty and not genuinely scientific. Such a view is clearly wrong. (It will do no good to point out that Dirac's theory was not an isolated one, since as Popper [1934: 31–32] himself tells us, a theory's status has nothing to do with its origin. Thus, whether it is isolated or connected is irrelevant.) In my paper I therefore suggested a disjunctive criterion for genuine science, such that if a hypothesis is in principle either verifiable or falsifiable, then it is genuinely scientific. Since Popper's view also includes the restriction that genuine scientific hypotheses must be putative laws of nature (in addition to being falsifiable), my criterion has the added advantage of rejecting this and thereby admitting evolutionary biology (contra Popper) into the circle of genuine sciences.

greement between ecologists on this very matter, some claiming that the needed observations have been attained, others denying it. At any rate, it should be clear that one cannot argue *a priori* based on the assumption that any introduction will necessarily disturb the ecology of the residents, or on any other assumption involving organism interrelations—as Rosenberg (1985), for example, does with two species of butterflies and two species of birds, to the Hutchinsonian conclusion that niches "do not exist independently of their inhabitants" (199). Such assumptions are empirical, not logical or metaphysical, and can only be decided by rigorous observation and experiment.

The most reasonable conclusion, then, it seems to me, is that the empty niche hypothesis is genuinely scientific but that the evidence in support of it is "not altogether secure."[48]

Further conceptual difficulties, however, remain to be discussed. Some theorists write of niches as if they could exist over vast stretches of time, in other words as if there could be vertical niches correspondingly delimiting the beginning and end of a vertical species. Simpson (1961), for example, seems to me to imply this when he states that his concept of an evolutionary species involves "having a unified role at any one time (not necessarily the same role at all times)" (155). It is the "not necessarily" that leaves open the possibility of a vertical niche. But the existence of any such vertical niche seems unlikely in the extreme, particularly if predators and prey and other organisms are considered as part of a niche, since these other species are possibly evolving over time. Thus the concept of niche would seem to be meaningful only in a strictly horizontal sense. As the corresponding niches are extended over time, each tends to lose both its distinctiveness and theoretical meaningfulness.

At any rate, as with species, one might make a distinction between horizontal and vertical niches and argue for the priority of the former over the latter, and therefore for the priority of a horizontal ecological species concept over a vertical one. However, the distinction, if there really is one, is of no account, since however one conceives of environmental niches, the interesting thing about them is that, as Ridley (1993: 394) points out, they necessarily imply gaps between them, such that organisms which are not well adapted to a niche will be maladapted and selected against by virtue of a gap. It is the gaps, then, that causally speak against the ahistorical Great Chain of Being.

But it is these very gaps that pose a serious problem for any identification of species with niches. For example, males and females in markedly sexually dimorphic species occupy somewhat different niches. As Futuyma (1986) points out,

---

[48]   Nevertheless, conceptual difficulties always seem to get in the way. According to Griesemer, "'empty niches' can clearly exist in the sense that the species serving as resources might occur without the presence of a given species" (238). Colwell, however, is not so accommodating: "In itself, the fact that any imaginative naturalist can describe an unlimited number of unfilled niches for which plausible organisms might exist casts serious doubt on the operational utility of the environmental niche concept in its broadest sense" (245). Perhaps empty niches must remain forever hypothetical after all.

"the males of many woodpeckers have larger beaks than the females and feed in somewhat different parts of trees" (32). A much more dramatic example of dimorphism is in what might be called ontogenetic dimorphism, as in Lepidoptera (butterflies and moths). Mayr (1982) points out that "many species (for example, the caterpillar-butterfly) occupy very different niches at different stages of their life cycle and in different portions of their geographical range" (275). Surely a single organism is not a member of one species when it is a caterpillar and a member of another species when it is a butterfly. And it will not do to take the adult stage as primary and defining, for as Medawar and Medawar (1983) point out, in examples such as the mayfly "the ephemeral adult mayfly lives only a few hours and is in effect a reproductive organ of the larva" (119).

One possible response in such cases is to simply extend each species' niche. One often finds this move subtly undertaken in the literature. For example, in the case of woodpeckers Malte Andersson (1994) remarks that "the sexes have evolved extreme dimorphism in bill morphology. Coupled with foraging differences, this should reduce food competition and expand the food niche of a pair" (15). Similarly, in the case of Lepidoptera, perhaps one might use the distinction employed by Eldredge (1985a: 163) in distinguishing between ecologically generalized and ecologically specialized species, the former "broad-niched," the latter "narrow-niched." Not an exactly perfect application of what Eldredge had in mind, but perhaps we could say that Lepidoptera are broad-niched. Ridley (1993), on the other hand, would call each of our woodpecker and Lepidoptera cases an example of what he calls "multiple niche polymorphism" (115). Each of these different views depend on the group one starts with, and are therefore perspectival. Who is right and how can we decide? The issue is really of little account. Either way, what ought to be clear by now is that in each and every one of these cases the species is being *presupposed* and its niche or niches subsequently *inferred*. Of course, on the view that species are to be identified with niches, this direction in the arrow of inference is the opposite of what it should be. But how else can one go about it? Certainly to define or to describe a niche by one or a few ecological dimensions, as one often encounters in niche talk, is unacceptable. With the possible exception of organisms such as autotrophic bacteria (the resource utilization of which is completely abiotic), the reality is that the niches of higher (multicellular) organisms, and the ecological dimensions on which their ecological roles depend, are *highly* dimensional.[49] There may be empty niches, and that there must be seems to me a legitimate view, but no one can ever with any certainty define or describe one *a priori*. One can only do this *a posteriori*, after examining living organisms and the ecological dimensions they actually occupy. (It makes no sense to examine possible organisms.) The problem is thus epistemological, not ontological. But as such it makes the identification of species with niches fundamentally nonoperational, since to be operational one would

---

[49] The biotic dimensions include a variety (often large) of dietary, symbiotic, predator, and parasite dimensions, as well as temperature, moisture, elevation, sunlight, atmosphere, and whatever other abiotic dimensions.

have to be able to identify species *a priori*, by niches alone. The species-as-niches view, then, is in a certain sense incoherent.

In bringing this section to a close I want finally to make some comments on the issue of niches as abstract repeatables and the related issue of niches as causes. For niches to be abstract repeatables, they must be *types* capable of having *tokens*. It makes no difference whether the tokens are concrete or abstract entities; if they are tokens of a type, then the type is an abstraction and is, in a sense, repeatable, having an identity in plurality. But can the tokens themselves be abstract entities? What sense can be made of saying that a particular abstraction can be a token of another particular abstraction? One could certainly make sense here in the realm of mathematics and language, where the number two and the word "man," for example, are themselves each a type with tokens, and which in turn are tokens of higher-order types, namely even numbers and nouns respectively. But in the case of niches, what sense can be made of supposing that each token is an abstraction? Real organisms do not eat and interrelate with, in a word "occupy," abstract entities. Bishop Berkeley thought it rather strange that his metaphysics entailed that we literally eat and drink ideas, though in recognizing this paradox he enjoyed but a brief wakening (cf. Cowley 1991: 17, 97). Ghiselin (pers. comm.) and others, then, would seem correct in construing niches as physical and not abstract entities and in chiding Mayr by calling it a "truism" that species occupy niches (Ghiselin 1987a: 139). But if niches are capable of being repeatable, so that there are types with tokens, then Johnson would seem correct in construing niches as abstractions. Both Ghiselin and Johnson are correct insofar as the one is referring to niches as tokens and the other to niches as types. Indeed this seems to me what Johnson was getting at when on the causal role of niches he wrote that niches "are not *merely* abstract entities" (69).

But if niche tokens are physical entities, how is it that they can be repeated? Or in other words what does it mean to say that they are tokens of the same type? One possible answer is that the ecological dimensions of two niche tokens must be absolutely identical for them to be tokens of the same type. But then that would make it almost certain that niches are never repeated in actual history. I suggest a more reasonable view is that ecological *roles* are repeatable insofar as they supervene on a multiple base of physical ecological dimensions. This, of course, returns us to a view of niche tokens as abstract entities, but only in a sense, and in doing so it avoids the problem above. The ecological dimensions that organisms (and by implication species) *occupy* are indeed physical, as they would have to be, and accordingly may be thought of as physical niches, while the ecological *roles* they *play* supervene on those physical dimensions and are abstract, and accordingly may be thought of (both types and tokens) as abstract niches. If niches are to be causes, then, it is the physical ecological dimensions that must be the causes. The abstractions they entail, namely the roles and the niche types, must be causal only in a derivative sense.

This accords with Armstrong's (1989: 28) claim that abstract classes are not causal agents, that the causal work done is by the members of the classes in virtue of their physical properties and not their class membership. But efficient causality is plausibly not the only kind of causality. In fact ever since Hume, causality in the sense of force has been a matter of great debate. Causality as such is always inferred, never observed. But the same is not the case if we think of causality not as forces but simply as relations (cf. chapter 1.2n3 and accompanying text). Causal relations are something scientists and the rest of us discover every day. This opens up the way for what is the widest sense of causality, which may be traceable back to Aristotle. (Seneca certainly interpreted Aristotle in the following way, but it was so many years ago that I read him that I can't remember where.) The widest sense of causality is *sine qua non* causality (that without which not). In this sense of causality, niches, both types and tokens, are clearly causal. But they are clearly more than mere *sine qua non* causality, since unlike mere *sine qua non* causality they provide *direction*.

But beyond this I don't think anything stronger can or ought to be said about the nature of niche causality. Wiley (1981), it seems to me, rightly chides Mayr and others who "conceive of the niche as the causal agent which stimulates the adaptation":

> If species had some mechanism to help them identify how they should become adapted, then I could believe that "empty niches" were causal agents in the evolutionary process and I could believe that the invasion of empty niches was an evolutionary phenomena. But if daughter species are randomly adapted with respect to their ancestral species . . . then daughter species find their ways into new niches by chance, or they are pushed into new niches by necessity; they do not actively seek them out. [254]

Wiley's criticism here, of course, as we've seen earlier with Ghiselin, takes for granted that species are individuals so that they are therefore not to be identified with niches. But his argument still stands if we replace species with organisms. To think otherwise is to import a metaphysical conception of causality that really has no place in modern biology.

By the way, Wiley's argument has equal application even if we should substitute the word "niches" above with "roles." The point is not only conceptual but empirical. It need only be recognized that uncountable niches, such as the woodpecker niche in New Guinea, have remained empty for many millions of years in spite of the presence of suitable ancestors, and may remain empty forever. Causal force, of course, implies a sort of necessity, but clearly in the case of niches this necessity is lacking. Grinnell, for instance, in conceiving of niches as causes, explicitly employed the Scholastic dictum "Nature abhors a vacuum" (cf. Griesemer 1992: 235), but he was just as mistaken as the same principle is in physics.[50]

---

[50] For an interesting discussion of this principle including its refutation in seventeenth-century physics by Torricelli and Pascal, cf. Agassi (1968: 115–117). Interestingly, one can still today find biologists who subscribe to this concept for niches, as for example Lennart Andersson (1990): "The only generalization is that when an 'empty' niche is available, it will be filled by that one out of a multitude of initial competitors who finds the least costly (complicated) way to specialize" (378).

Granted that a given niche may allow one to predict some of the properties of the organisms that will inhabit it, nevertheless the nature of evolution is such that it does not allow one to predict that that niche will ever be filled, or if it is filled when and for how long.

What needs to be remembered is that the selection pressures that come from or that constitute niches are not the only causal factors that are relevant. As Darwin (1859) repeatedly stressed, for instance in the case of a species of woodpecker that doesn't peck wood and that preys on insects on the ground (184), many of the characters that characterize a species are not adaptations at all but instead are simply inherited from ancestral forms, providing what Templeton (1989: 19) calls historical constraints on evolution and adaptation. Sexual selection, moreover, provides selection pressures not from without but actually from within the species, and in some cases may even account for both speciation and extinction (cf. Andersson 1994: 46–47, 248–249).

Another purely internal factor that plays an important role in adaptation and evolution is the causal role played by genes, for instance mutational constraints. Ecologists, however, tend to downplay or even ignore the role of genes. According to Van Valen (1976), for instance, "genes are of minor importance in evolution and should ordinarily be considered there in nearly the same degree (if often not for the same reasons) as other molecules" (69). The same view, of course, is presupposed in Grinnell's view that "Nature abhors a vacuum." One can even see this view trickling into the public consciousness. In, for instance, the movie *Jurassic Park*, specifically in the scene where the scientist tells the visitors that all the dinosaurs were engineered to be females and therefore couldn't breed, the chaos mathematician played by Jeff Goldblum replies with the warning that "nature, uh, finds a way." All of this, of course, ignores the constraints involved with mutations. As Templeton (1989) pointed out, "constraints exist that make certain types of mutations impossible or highly improbable" (18), the result being that certain lines of evolutionary development are allowed while others are obstructed or even precluded. In the public consciousness, of course, with few exceptions such as the Teenage Mutant Ninja Turtles, mutations are generally thought of as bad things. Every biologist, of course, knows that mutations are a necessary part of evolution and that if it were not for mutations none of us would be here. Mutations and mutational constraints, then, must surely also count as causes in addition to external ones.[51]

---

[51]   David Raup (1991), one of the foremost experts on extinction, argues that ultimately most extinction is due to bad luck rather than bad genes. The only genetic cause of extinction that he lists (cf. 124–125) is dangerously low genetic variability which results when a species for whatever reason is reduced to a population level below its *minimum viable population* (MVP), above which it is "virtually immune to extinction." It seems to me, however, if I may be allowed to make a purely biological hypothesis, that some extinctions may be partly caused by mutational constraints (a specific change in selection pressure of the right kind, of course, is also required), which would give us a genetic cause of extinction that does not at all depend on population level, entailing the chilling consequence that perhaps no species is genetically immune to extinction when above its MVP.

So what about identifying species with niches? Hogan's (1992) complaint that "if ecological niches causally explain properties of species they *cannot* be identical with them" (207) ignores what even Wiley (1978) has to concede, namely the fact that "Species cannot be divorced from their environment any more than they can be divorced from their gene pools or their morphologies" (87). But does that make it legitimate to identify effects with their causes (recall Templeton)? Whatever the answer, in identifying species with niches one is necessarily including some causes and excluding others, namely one is including only the "opportunity" or "directional" causes. Whether this approach is ultimately viable, however, would seem to require, as I think I have sufficiently demonstrated, a lot more conceptual and empirical work.

3.5 Problems with Species as Classes

In this chapter we have examined a wide variety of species concepts, all of which involve abstract objects normally called classes. Classes, of course, are but a subset of abstract objects in general, which are problematic in themselves. Talk of abstract objects as objectively real, mind-independent entities makes many scientists uneasy. Many philosophers likewise do not like the thought of objectively existing abstract objects, and attempt to reduce them to something concrete. I believe that in chapter 1.2, I have made a plausible case for the conclusion that in addition to facts about concrete objects there are further facts about abstract objects, facts both particular and general, implying the existence of objective abstractions. But abstract classes are another matter. There may be abstract facts, both particular and general, but at the same time there might not be abstract classes. Clearly if classes are not objectively real, then virtually all of the species concepts examined in this chapter are doomed to failure right from the start.

In this final section on species as classes I will develop a theory of abstract class realism, not to defend the species concepts examined in this chapter (although it will initially have that effect, including the view that niches are classes and may be both empty and yet numerically distinct), but to defend species category realism, which many if not most biologists and philosophers (this author included) take to be an abstract class intensionally defined and a fitting subject for laws of nature. The relevance of this exercise for this chapter, on the other hand, is that it will show that abstract classes, and indeed abstract objects in general, should not be thought of as being capable of changing or evolving, a consequence acceptable to virtually all species category realists but obviously fatal to the species taxa realists examined in this chapter (specifically those who conceive of species taxa as abstract classes).

As a start, it seems to me worthwhile to briefly examine some of the classic modern objections to abstract classes. One of the earliest of these is that of John Stuart Mill. According to Mill (1881), the meaning of the word "class" is embodied in the *dictum de omni et nullo*, which means "That whatever can be affirmed (or denied) of a class, may be affirmed (or denied) of everything included in the class" (113). On Mill's view, then, "a class, a universal, a genus or species, is not an entity *per se*, but neither more nor less than the individual substances themselves which are placed in the class, and that there is nothing real in the matter except those objects, a common name given to them, and common attributes indicated by the name, . . . The class *is* nothing but the objects contained in it" (113–114).

Immediately one might reply that the *dictum de omni et nullo* is false, or at least arbitrary, since we often do affirm (or deny) of a class what we do not of its members. For example, the class of gold atoms is generally thought to be spatiotemporally unrestricted, but this affirmation is not also extended to any or all of its members. All atoms of gold are spatiotemporally restricted, but the class of gold atoms is not. Moreover, the class of gold atoms is the subject of laws of nature, and statements of these laws are true even if there never were any atoms of gold in existence.

Mill faces the further problem that in spite of his adherence to the *dictum de omni et nullo* he nevertheless wants to retain a distinction between natural and artificial classes. Natural classifications on his view are achieved in basically two ways: (i) by being formed according to "those [properties] which are causes of many other properties, or, at any rate, which are sure marks of them" (301); or (ii) by being formed "so that the objects composing each may have the greatest number of properties in common" (303–304). This twofold approach allows for classes which Mill calls "indefinite" or "unenumerated," which are important since "definite classes, as such, are almost useless" (70n2).

But there is surely a fundamental incoherence in all of this, an incoherence that generally attends all those approaches which attempt to reduce classes to nothing more than their members but which nevertheless want indefinite or open classes. In Mill's sense of definite or closed classes, any change in membership is clearly a change in the class. In his sense of indefinite or open classes, however, where he likewise wants a class to be *nothing more* than its members, there is a clear sense in which a class remains the same (e.g., the *kind* gold, the class of gold atoms) even though its members come and go. This can only be possible (providing that the class is to be objective) if the class is an abstract object. Moreover, in employing properties in common Mill is reverting to classes of properties, or to properties in the abstract. This is clear given the hypothetical case where, for instance, all the atoms of gold in the universe cease to exist (as well as anyone to classify them), and then at some time later atoms with the exact same structure come into existence (but still with no one to classify them). According to Mill's *dictum de omni et nullo*, the new atoms are not members of the class of gold atoms. But clearly no physicist would agree, and rightly so.

An equally interesting attempt to get rid of abstract classes is to be found in the writings of Bertrand Russell. Russell, in his ever-increasing application of Occam's Razor, argued early and throughout the rest of his career that supposing classes to be objectively real is not necessary since they can be reduced to something else. Occam's Razor for Russell meant that reduction implies elimination. But Russell was not dogmatic about the nonreality of eliminated entities. As he says (1919), "when we refuse to assert that there are classes, we must not be supposed to be asserting dogmatically that there are none. We are merely agnostic as regards them" (184). What was it that Russell thought that classes could be reduced to without losing any communicative power? It should be obvious, it seems to me, that they cannot be reduced to their members, as collections or aggregates, for the simple reason that classes remain the same throughout the coming and going of their members and also because their membership might be indefinite. But Russell (183) gives different reasons. Classes cannot, on his view, be reduced to "heaps or conglomerations" because class realists hold that (i) there is a null class, and (ii) there are classes with only one member but which are nevertheless not identical with that member. A better approach, then, according to Russell, is to suppose that classes are reducible to propositional functions, open sentences such as "x is a man" or "x is mortal," propositions that are neither true nor false until the variable (or variables) in them is given a particular value. Accordingly on this view a class of one is merely a propositional function with only one true substitution instance, while a null class is a propositional function for which there are no true substitution instances.

The problem with this overall approach, however, as Russell points out, is that "if a class can be defined by one propositional function, it can equally well be defined by any other which is true whenever the first is true and false whenever the first is false. For this reason the class cannot be identified with any one such propositional function rather than with any other—and given a propositional function, there are always many others which are true when it is true and false when it is false" (183). In other words, granted that different linguistic expressions of a given propositional function have the same meaning, in other words are logically equivalent, it follows that a given class cannot be reduced to any one explicit expression of that propositional function. However, once this is understood, he says, "it becomes very difficult to see what they [classes] can be, if they are to be more than symbolic fictions" (184). The basic point is that all the linguistic work that is done with explicit mention of classes can equally be done with propositional functions without any mention of classes, thereby reducing classes to what he also calls "logical fictions, or (as we say) 'incomplete symbols'" (182).

Quine's (1961) reply to Russell is basically twofold: First, "Russell's method eliminates classes, but only by appeal to another realm of equally abstract or universal entities—so-called propositional functions" (122). Insofar as the propositional functions in favor of which Russell would eliminate classes are admitted by Russell not to be explicit propositional functions but instead logically equivalent propositional functions, Quine's point might be expanded and strengthened by pointing out that, since there is an infinite possibility of languages (at least when semantically defined; cf. chapter 2.2n8), any given class on Russell's view logically reduces to a *class* of propositional functions, each member of which is a particular linguistic expression.[52] (Any attempt to accordingly reduce an individ-

---

[52] Given Quine's strong skepticism on the notion of synonymy in natural languages, however, it is unlikely that he would ever make such an argument. In reply to his many attacks, I will only say here that like many others I find no serious difficulty with the no-

ual propositional function class to another propositional function, of course, would only involve an infinite regress.) Thus, Russell can't escape classes after all. He can escape the *use* of the word, but not the *thing*, namely a spatiotemporally unrestricted abstract entity with spatiotemporal members.[53]

Quine's second point, however, instead of providing a powerful argument against Russell, returns to give Quine's own class realism serious problems. According to Quine,

> The phrase 'propositional function' is used ambiguously in *Principia Mathematica*; sometimes it means an open sentence and sometimes it means an attribute. Russell's no-class theory uses propositional functions in this second sense as values of bound variables; so nothing can be claimed for the theory beyond a reduction of certain universals to others, classes to attributes. Such reduction comes to seem pretty idle when we reflect that the underlying theory of attributes itself might better have been interpreted as a theory of classes all along, in conformity with the policy of identifying indiscernibles. [122–123]

As we have seen in our examination of Beckner's cluster class theory, what is required for classes is a principle of identity and individuation. For Quine, as we saw, this requirement can only be satisfied by the identity of coextensives. A recurrent theme throughout Quine's writings, however, is that attributes (properties) fail to satisfy this criterion, and failing to find any other, classes are then to be preferred to attributes. Quine's favorite example is hearts and kidneys, and his perhaps clearest expression of the problem is in Quine (1987):

> If a thing has this property and not that, then certainly this property and that are different properties. But what if everything that has this property has that one as well, and vice versa? Should we then say that they are the same property? If so, well and good; no problem. But people do not take that line. I am told that every creature with a heart has kidneys, and vice versa; but who will say that the property of having a heart is the same as that of having kidneys?
>
> In short, coextensiveness of properties is not seen as sufficient for their identity. But then what is? If an answer is given, it is apt to be that they are identical if they do not just happen to be extensive, but are necessarily coextensive. But

---

tion of synonymy. But unlike others, my conviction is grounded in the synonymy that obtains in the supervenience of biological entities on physical entities, as in the clear case of codons but also in the not so clear case of properties such as fitness (cf. chapters 2.2 and 5.3). In spite of the problem of determination in the latter case, biologists and philosophers of biology as a whole have not been skeptical of the objective existence of the synonymy. I suggest that philosophers of language, in spite of the difficulties that many have raised, should be no more skeptical of synonymy in natural languages. The positivistic impulse to translate epistemological indetermination into ontological nonexistence, along with the preoccupation with messy situations, is no more legitimate than in the case of species nominalism (cf. chapter 2.4).

[53] This is not to exclude, of course, higher-order classes, namely classes of classes and the relation of class inclusion. But higher-order classes ultimately rest upon classes with physical members (cf. the quotation from Quine at the head of this chapter).

NECESSITY, *q.v.*, is too hazy a notion to rest with.

We have been able to go on blithely all these years without making sense of identity between properties, simply because the utility of the notion of property does not hinge on identifying or distinguishing them. That being the case, why not clean up our act by just declaring coextensive properties identical? Only because it would be a disturbing breach of usage, as seen in the case of the heart and kidneys. To ease that shock, we change the word: we no longer speak of properties, but of *classes*. [22–23]

One may feel that this is an attempt to fool by use of the proverbial smoke and mirrors. For after all, whoever confuses hearts with kidneys? Any surgeon who would do so, as unlikely as that is, would face a heavy lawsuit. What allows any force to Quine's argument is that properties, or rather physical instances of properties, normally cannot exist on their own but have to be a property *of* something. Thus it may turn out that some properties can only coexist with other properties. But does that make them, as types, coextensive? Every creature with a heart may also be a creature with a kidney (suppose that's always the case), but that does not mean that hearts are kidneys and kidneys are hearts. Their principle of individuation is in their *structure* and *function*, and that's what makes the one different from the other and what makes it reprehensible should a surgeon confuse the two. In one sense, hearts and kidneys may be coextensive, insofar as they must be properties *of* something, but in another sense, in the sense of structure and function, they are not coextensive at all.

Quine's use of the principle of the identity of coextensives is problematic in another way. According to Quine (1969b), classes "enjoy, unlike attributes, a crystal-clear identity concept" (21). That concept, of course, as Quine (1981) put it, is that "Classes are identical when their members are identical; such is the principle of the individuation of classes" (100). What this means, again, is that if every member in class A is numerically identical with one member in class B and *vice versa*, then A and B are not numerically distinct classes but instead are numerically identical.

The identity of coextensives would be a crystal-clear identity concept for *sets*, as I defined them in chapter 1.3, but will it do for *classes?* I suggest not. First, as we have seen in §3 above, cluster classes may have the very same members and yet not be identical. But perhaps one could eliminate this problem by being more inclusive, namely by taking into account all possible members (all actual members, of course, are by necessity also possible members, for if they would not be possible they would not be actual). The problem now is that nonactual possible members are abstractions, just like possible worlds (cf. Kripke 1972: 19), and it defeats the generation of cluster classes to use members that have never existed.

One might reply that this only gives us yet another reason to reject the objective reality of cluster classes. Already in §3 we saw that cluster classes (with the possible exception of Eigen's view) involve an inherent degree of arbitrariness and subjectivity. Surely, it might be argued, the stricter kind of classes, such as gold, are not subject to the same objections.

The identity of coextensives, however, is also deeply problematic for the stricter kind of classes. It will be remembered that classes are abstract objects for class realists such as Quine, not physical objects. Quine (1981) makes at least this much crystal clear against those who would misconstrue him. Disavowing such "weasel words like 'aggregates' and 'collections' and 'mere,' said of classes" (184), he tells us that classes are "abstract and immaterial" (107) and that "I am a Predicate and Class Realist, now as of yore; a deep-dyed realist of abstract universals" (184; cf. §2n24). What, then, does it mean to say that "a class is 'determined' by its members" (106)? Certainly not, says Quine, that "to find the class you must find its members." Instead, "All it really means for classes to be determined by their members is that classes are the same that have the same members" (106).[54] But now here is the problem. If members are not constitutive of a class, which for Quine they clearly are not, then the principle of the identity of coextensives can only operate at best as *evidence* of class individuation. Armstrong (1989: 25–27) has a nice little discussion on the coextension problem, involving David Lewis' attempt to solve it by using necessity and possible worlds and Elliott Sober's reply that being an equilateral triangle and being an equiangular triangle are necessarily coextensive in all possible worlds but are nevertheless distinctive. It seems to me, however, that the real problem is when classes are indefinite and/or contingent as to their membership. In the case of the class of gold atoms, for example, if strict determinism is false (and there are good reasons to believe it is), then the total membership of this class is intrinsically indefinite. Moreover, if strict determinism is false, then it must always be contingent whether this class will have a member at any particular spatiotemporal coordinate. Therefore, even if one could know (*per impossibile*) all the members of the class of gold atoms throughout the whole history of the universe, that knowledge would be a product of contingent events, not necessary ones. In other words the total membership could have been otherwise. Moreover, some classes may never even have any members, *ever*, although it is possible that they could. A number of chemical elements have never been synthesized and never will, and yet they

---

[54] Armstrong (1989) clearly misconstrues Quine on this point. According to Armstrong, "A class, as Quine says, is determined by its members. Change its membership and it is automatically a different class" (27). What Armstrong says here is true of *sets* (extensionals, groups, collections, aggregates, heaps, etc.), but not of *classes*. Quine's classes are clearly not sets but instead are abstractions. To think otherwise is to seriously misconstrue the nature of Quine's class realism. Such a realism entails, it seems to me, the view that a class remains the same even though its members come and go. It is, however, a difficult matter of interpretation whether Quine himself would accept this consequence. On the one hand, he does not accept possible worlds as real (cf. Quine 1961: 4, 1981: 173–174), so that the members of a class would not on his view be all the members in all possible worlds. On the other hand, he prefers a symbolic notation without tense, in which the material world is to be thought of as a four-dimensional space-time world (cf. Quine 1960: 170–174). But in this Quine neither denies the reality of time or change nor affirms determinism (cf. Quine 1981: 10–11). Instead his motive is merely notational elegance (Quine 1960: 170).

each are in a sense real and are clearly individuated.[55] Appealing in this case to possible but nonactual members, of course, will no more do than in the case of cluster classes. In appealing to merely possible members, all that one is really doing (if one is not presupposing strict determinism) is appealing to the membership conditions. But then one is not individuating classes by members but by something else.

It would seem, then, that not only (per Quine) can a class not be defined by its members, but it cannot (contra Quine) be individuated by them either. According to Armstrong (1989), "Contemporary philosophers all accept these identity conditions for classes" (25). But if this is the case, and if they also think of classes as abstract, then I suggest they think again, for their view is incoherent.

How, then, are classes to be individuated if not by their members? Quine (1981) says that "classes, like attributes, are epiphrastic. You specify a class not by its members but by its membership condition, its open sentence" (107). That, of course, is not to say, like Russell, that a class is a sort of linguistic entity. But it is not to say that a class is its membership conditions either. Quine says you *specify* a class by its membership conditions, not that it *is* its membership conditions. But then what *is* a class?

According to Armstrong (1989), "Classes have turned out to be rather mysterious entities in modern metaphysics; nobody seems to have got a firm grip on their nature" (36). On the other hand, Max Black (1971), in his often quoted exposé on the concept of set in set theory (Black uses "set" and "class" synonymously), claims that

> We are not forced to choose between thinking of sets as mysterious aggregations of distinct things into "unified" wholes and thinking of them as unknown things shared in common by certain coextensive properties. For, in ordinary language, we do manage to identify sets and reason about them, without the inconveniences of a superfluous mythology. If we can become sufficiently clear about how we *talk* about sets, we shall have all the clarity about "the concept of set" that we need for a start. . . . the task of exhibiting the underlying rules of use should not be insuperably difficult. Let us then see how we manage to talk about *several things at once.* [628]

Such a Wittgensteinian solution, however—"Don't ask for the meaning, ask for the use" (cf. Wittgenstein 1958: §§40–43)—fails to satisfy from a metaphysical point of view. We still want to know if there is anything more to a term than its linguistic function. This is especially true if not all of the metaphysical possibilities have been explored. Until that happens, Wittgensteinian solutions should be put off until the very last, as a sort of default position.

According to Black (1971), in the case of classes as abstract entities "Nobody

---

[55] I believe that the highest atomic number synthesized so far is 109 (cf. Lide *et al.* 1995: 4:2). Clearly, of course, there must be a physical upper limit to how many protons an atomic nucleus may have. Although I believe that physicists do not yet know what that is, I would argue that any class above that limit (whatever it is) is not objectively real.

knows what the supposedly 'abstract entities' are intended to be" (634). It seems to me, however, that there is a possibility which (unless I am mistaken) has yet to be explored and which makes clear what the abstract entities are (or at least could be). It seems to me that much would be gained in clarity if we suppose that a class *just is* its membership conditions. The advantages of this view are many.

For a start, it shows how classes can meaningfully be conceived as abstract entities. That membership conditions are abstract should be evident from a number of considerations, such as the fact that they can have instances, and also from the fact that the reality of membership conditions does not depend on the reality of their members. This latter point should be especially evident in the case of classes which are the subjects of laws of nature. A law about the boiling point of water, for example, would be true even if there were no water in the universe, ever, so that the truth of the law does not depend on members of the class of water molecules (although of course our way of knowing the truth of the law is another matter) but on something abstract about the universe. Similarly, in the case of the species category, what Richard Dawkins (1983) has aptly referred to as Universal Darwinism—that only Darwinian natural selection can in principle account for the adaptive complexity of organisms, which Dawkins likens to "the great laws of physics" (423), and that Darwinian natural selection must be the main mode for species evolution wherever there might be species (cf. Dawkins 1986: 288)—was presumably true even before the evolution of life anywhere in the universe.

My solution would also account for what I have repeatedly said is the *quintessence* of classes, not only that they are abstractions, but that they remain the same even though their members may come and go (cf. Hull n56 below and Armstrong 1989: 27). It would also entail a rejection of the identity of coextensives. Not only is this a principle that does not work for classes (although it should be retained for sets), but it goes against actual practice in science by entailing that there can be only one empty class.

It would also solve the problem of the one and the many, namely how a class can be both a one and a many, a problem which Russell wrestled with in his early work and which contributed to his settled view that classes are convenient fictions (cf. Russell 1959: 58). On my view it is the membership conditions which constitute the unity of the class as a one, while the set of its members constitutes the many. Each member, then, however concrete, is a member of an abstract unity, a unity which is especially evident, once again, in the case of classes that are the subjects of laws of nature: these classes no more change as their members come and go than the laws themselves.

The above solution also solves the problem of spatiotemporality, which in the literature on the modern species problem pervades much of the discussion on classes and individuals (cf. this chapter and chapter 4). Classes as membership conditions are spatiotemporally unrestricted, while their members are either spatiotemporally unrestricted (when clearly abstract) or spatiotemporally restricted (when clearly concrete).

One might still wonder, however, in what sense membership conditions can be thought to be objectively real. This is an issue that will seem problematic for many, so that it deserves a closer look. It seems to me that if classes are membership conditions, then classes have the same nature as conditionals. The class of gold atoms, for example, can be expressed as "If x is an atom with seventy-nine protons in its nucleus, then x is an atom of gold." But equally important, in accordance with actual practice in science this approach accords objective reality to memberless classes the reality of which scientists claim to *discover* rather than *create* (cf. chapter 1.2). For example, all but two of the first ninety-four elements in the periodic table occur naturally on earth. Elements 43, 61, and 95 to 109 (except 108) have been synthesized in laboratories. In spite of Elements 95 to 109 having presumably no concrete members at any previous time or place in the universe, when scientists synthesized an atom with, say, atomic number 104, they presumed (if the textbooks are to be believed) that they were *discovering* an objectively real class of atoms, not that they were *creating* one (cf. Lide *et al.* 1995: 4:1–34). Discovery, of course, presupposes preexistence. Supposing classes to be membership conditions conforms to this view. It supposes that the membership conditions for Element 104 have always existed as part of the fabric of the universe, allowing atoms of that class to be synthesized. Conversely, if those membership conditions would not be part of the fabric of the universe, all attempts to synthesize those atoms would necessarily prove futile.[56] Modern set theory, on the other hand, in presupposing that all memberless classes are not distinct but are really only one class, the null class, would preclude any search for atoms such as Element 104, since it is logically contradictory to search for members in a class that by definition cannot have members. Fortunately, modern physicists, either in ignorance or defiance of modern set theory, chose to follow their scientific instincts.

As conditionals, this would mean that some classes are in the same ontological category with laws of nature (since laws of nature are fundamentally conditional in form), some with laws of logic and mathematics (since laws of inference in logic and mathematics are conditional in form), while other classes are merely stipulative and therefore conceptual, being created rather than discovered. While each of these three categories corresponds to a different kind of class, the first to natural or empirical classes, the second to logical and mathematical classes, the third to artificial classes, each involves a different modality: the first involves physical necessity, the second involves logical or mathematical necessity, while the third involves no necessity at all. Thinking of classes as membership conditions, then, gives us a taxonomy of classes that corresponds to actual class usage.

---

[56] The view I am suggesting here conforms somewhat closely with, for example, Hull's (1983) characterization of the traditional philosophical view that "natural kinds are somehow built into the framework of the universe. At any one time, a particular natural kind might not be exemplified. At one stage in the history of the universe, perhaps no gold atoms existed. However, when atoms of the right sort become formed, gold becomes exemplified. Hence, gold can come and go with respect to exemplification, but given the fundamental makeup of the universe, gold is a permanent possibility" (154–155).

Natural and logical/mathematical classes are objectively real, since they both are in principle discoverable and (at least insofar as they are the subjects of laws) would have to be included in a complete description of the universe (cf. chapter 1.2), while artificial classes are only subjectively real.

Nevertheless there are problems with my theory of classes, although I do not think that any of them are serious. One problem is with the cardinality of classes. Mathematicians, for example, speak of the class of whole numbers as being larger than the class of positive integers. But how can a membership condition be larger than another? On my view this problem is easily solved simply by having cardinality refer to members rather than membership conditions. In mathematics the members of classes are mathematically necessary, whereas in physics, for example, they are contingent. But in neither case is there irrevocably a problem of cardinality for my theory of classes.

Another problem is with class inclusion. It is common in set theory to say that one class is included in another, for example that the class of squares is included in the class of rectangles. But once again I don't see a problem here. Membership conditions can also be members of classes, and a class member can be a member of more than one class, both of which allow for class inclusion. Once again the class of chemical elements is perfectly instructive here. The class of gold atoms, for example, conceived of as the membership condition of having seventy-nine protons in an atomic nucleus, is a member of the class of chemical elements, since it is itself a kind of element. A particular atom of gold, on the other hand, is not a member of the class of chemical elements, since it is not an element but an atom, although it is a member of more inclusive classes such as the class of atoms.

The topic of class inclusion, of course, raises the issue of *Russell's paradox*, discovered by Russell in 1901, a discovery so revolutionary for subsequent mathematics and logic that no theory of classes can afford to ignore it (although Black 1971: 633 explicitly does). Accordingly, I shall examine it in some detail, with the hope of showing that thinking of classes as membership conditions is less arbitrary and more metaphysically satisfactory than other solutions to Russell's paradox.

There are basically two ways of stating Russell's paradox, one using only classes, the other using properties which determine classes.

Beginning with the first, although some classes are clearly members of themselves—the class of all classes with more than five members, for example, is a member of itself, since it has more than five members (incidentally, the class of all classes, another example, generates an infinite regress, for to be truly the class of all classes it must include itself, but when it is included a new class of all classes results, and so on *ad infinitum*)—it is obvious that most classes are not members of themselves. The class of all books, for example, is not itself a book, and so is therefore not a member of itself.

Granting this, a paradox arises. Is the class of all classes that are not members of themselves a member of itself? If it is a member of itself, then it is not a member of itself. And if it is not a member of itself, then it is a member of itself. Taking R to stand for the class of all classes that are not members of themselves, we get the following formalism:

$$(R \in R \subset R \notin R) \cdot (R \notin R \subset R \notin R)$$

Thus,

$$R \in R \equiv R \notin R$$

This, of course, is logically false, or in other words a contradiction, since if the left side of the biconditional is true, the right side must be false, and *vice versa.*

Another way to generate this paradox is by using the most basic axiom of what is now called *naive set theory.* According to this axiom, which is highly intuitive, any property determines a class. For example, if something has the property green (G), then it is a member of the class of all green things (G_c). Thus,

$$(Gx \subset x \in G_c) \cdot (x \in G_c \subset Gx)$$

Thus,

$$Gx \equiv x \in G_c$$

But what of the class property of not being a member of itself? If we use P for this property and P_c for the class of all things that have P, we get a formalism identical to the previous one:

$$(Px \subset x \in P_c) \cdot (x \in P_c \subset Px)$$

Thus,

$$Px \equiv x \in P_c$$

A contradiction arises, however, once we use P_c as the value of x. Thus,

$$(PP_c \subset P_c \in P_c) \cdot (P_c \in P_c \subset PP_c)$$

Thus,

$$PP_c \equiv P_c \in P_c$$

This, however, is logically false, because if P_c has the property of not being a member of itself, then it is a member of itself, since it is the class of all classes that are not members of themselves; but if it is a member of itself, then it cannot be a member of itself for the very same reason. In other words, PP_c entails that P_c $\notin P_c$, which in turns entails that $P_c \in P_c$, which is exactly the same result we saw above for R.

The great and revolutionary value of Russell's paradox was that it showed that the most basic axiom of naive set theory, the axiom which states that every property determines a class, must be entirely rejected, since, as every undergraduate who has taken a course in deductive logic knows, any proposition whatsoever can be validly inferred from a contradiction (in this case an axiom that entails a contradiction).

There have been a number of attempts to repair set theory so as to avoid Russell's paradox, but they all seem more or less arbitrary, as stipulations rather than natural solutions (i.e., solutions that solve more than just the problem or problems at hand). Russell's own solution was his theory of types, a theory of class hierarchy according to which (to put it all too briefly) individuals are of the lowest type, classes of individuals the second lowest type, classes of these classes the third lowest type, and so on, with the prohibition added that a class cannot have any members from its own or higher type, thus making it nonsense to ask whether *R* is a member of itself (cf. Russell 1908). Quine's favorite solution is to distinguish between "sets" (classes that are capable of being members of classes) and "ultimate classes" (classes that are not capable of being members of classes) and to stipulate that *R* is an ultimate class rather than a set (cf. Quine 1981: 108). A more interesting solution is that afforded by what is called the *iterative conception of set*. On this view, which has gained some popularity, the members of a set *precede* or are in some sense *prior* to the set itself. Thus, as Charles Parsons (1975) put it, "If we suppose that the elements [members] of a set must be 'given' before the set, then no set can be an element of itself, and there can be no universal set. The reasoning leading to the Russell and Cantor paradoxes is cut off" (505).

On my own view of classes, a class is *not* determined by its membership conditions (for otherwise my view would be defeated by Russell's paradox), but instead *just is* its membership conditions. Thus Russell's paradox is avoided. A class on my view can no more be a member of itself than parents can be their own children, and *R* is no more real than the class defined by round/square.

But there is a further advantage to my theory. As we've seen earlier in the case of Mill, and as we'll see later in the case of Kitcher, for some philosophers a class *just is* its members and nothing more. For others, as we've just seen with the iterative conception, a class is in some sense *posterior* to its members. My own view resides at the opposite end of the spectrum, since on my view of classes as membership conditions a class is logically *prior* to its members. The advantage of this view over the other two is that it allows for a most important desideratum, namely that there must be an objective sense in which a class can remain exactly the same even though its members come and go. On the view of Mill and Kitcher, as well as on the iterative conception, this desideratum cannot possibly be met, since for both views any change in membership entails a change in the class. That this failing has generally not been seen to be important is probably because it does not come up in the very foundation of set theory, namely mathematics (i.e., numbers don't "come and go"). But set theorists rarely restrict set theory to mathematics. They also apply it to entities such as chemical kinds and biological species. But it is precisely here that the above desideratum becomes extremely important, since chemical kinds don't change or evolve, and while biological species change and evolve and are capable of stasis, the species category itself (the subject of each realist species concept and definition) remains the same while its members (species taxa) come and go. These facts call out for a radical revision of set theory, a revision I have attempted in the present section.

Interestingly, if we take this revision and then return our gaze to the various class views of species, we can meaningfully ask which ontological kinds of species may be said to be objectively real. As membership conditions, classes can be thought of as types, but the reality of them as types is now no longer dependent on whether they actually have tokens. (If there can be no kings without subjects, then types without tokens would correspond to princes.) But if they are to have tokens, it should be clear that if two tokens are to be tokens of the same type, they must satisfy the same membership conditions.

Classes of huge macromolecules, therefore, may in a sense be real even though they might lack physical members.[57] Species as elementary classes, then, may be real, but as we have seen in §2, biological species do not conform to that view.

Species as cluster classes would appear to suffer an even worse fate, since (with the possible exception of Eigen's view) a degree of subjectivity and arbitrariness is necessarily involved in their membership conditions. They are like stipulated definitions, their being purely conceptual.

Niches present a more interesting case. For niches as abstractions, membership requires an intermediary, namely physical niches, the niche tokens. But how can we ever know if a particular ecological role, which is based on one particular niche token, is identical with another ecological role, which is based on another particular niche token, so that there is only one ecological role supervening on these two niche tokens, and so that the organisms which partake in these two identical roles are not members of two species but one? The two physical niche tokens, of course, need not be physically identical or even highly similar, for such is the nature of subvenient bases. But if they need not be identical or even highly similar, the identity (if there is to be identity) must be in the roles. There is only one way, it seems to me, for two numerically different roles to be type identical. To be type identical, members fulfilling the one role must be exchangeable (*sensu* Templeton 1989) with members fulfilling the other role, so that this exchangeability would constitute the common membership conditions if they are to be members of one and only one species. Let's take an illustrative example. The marsupial wolves of Australia, along with the placental wolves of Europe and North America, are often cited as a striking case of convergent evolution (cf. Wilson 1992: 95). For the two species to be ecologically conspecific they would have to be demographically exchangeable. In other words, if the entire population of marsupial wolves could be simultaneously exchanged with the entire population of placental wolves (with the restriction that both populations would have to be roughly the same size), the two populations would then be ecologically conspecific only if the exchange resulted in no significant change in either population as well as the other species with which they interact.[58] The problem, of course, is

---

[57] Just like chemical elements, however, there must be physical limits to which macromolecules (genotypes) can actually be formed. Any macromolecular classes outside of those limits, then, just like chemical elements beyond a certain upper limit in proton number (cf. n55 above), must likewise be counted as not objectively real.

[58] Sokal (1973) claims that "Surely convergence is unlikely to produce similar and inter-

how one could ever go about testing such a thesis. However, one might possibly test the thesis in its general sense by using much simpler organisms and exchanging, for example, the hosts of parasites, say fleas (though not such a test, cf. Mayr 1988a: 154 for convergence in fleas). But for higher organisms, organisms with greater complexity, it is difficult to see how one could test it. This epistemological difficulty, of course, does not preclude (except for a Logical Positivist) the ontological possibility. But what it does preclude is an operational taxonomy based on ecology.

There are deeper theoretical problems, however. If classes are membership conditions, then this accords with what I've repeatedly said is the quintessence of abstract classes, namely that their members may come and go while the class itself remains exactly the same. To deny this basic feature of abstract classes is to take away the abstract nature of the class, making each of its members into a part and the class into a collection, so that any change in membership (parts) is a change in the whole. But thinking of classes as membership conditions has a much more significant consequence, namely that any change in membership conditions is not really a change in the class but instead a new class altogether.

We are thus led to the issue of abstractions and change. It will be recalled that according to Aristotle change is only possible in things that have matter as a constituent. When Socrates, for example, grew from a baby to an old man the change was in his matter, not in his form. On Aristotle's view the forms (species, essences, universals) are incapable of change. In making this claim Aristotle seems to me to have had a fundamental insight, namely that abstractions, since they do not have matter as a constituent, cannot change. But Aristotle's claim, though it seems to me intuitively true, lacks sufficient argument by modern standards. In what follows I suggest a simple argument to help remedy that deficiency.

Let's return to my giraffe example from chapter 1.2. You go to the zoo (hopefully it is a humane zoo) and you see a group of giraffes of various sizes and ages. What strikes you most is the lengths of their necks. Now let's suppose for the sake of argument that the average length of their necks is not instantiated by any one of the giraffes. This average length is clearly an abstraction, a further fact about the group of giraffes that we can argue about, that we can have tested, and that we can have settled to our mutual satisfaction. Let's suppose that the average length is in fact five feet. Now suppose that we return a year later and the same giraffes are still there, only now we are told that the average length of their necks is five and one-half feet. We are tempted to think that over the year the average length of their necks has changed. But has it really? Certainly over the year at least some of the lengths of the physical necks of the giraffes have changed. But in addition to those changes, was there also an *additional* change, a change in the *average length* of their necks? If we are guided by parsimony in our ontology, we

---

fertile populations originally belonging to two different reproductive communities" (366). Although correct, it needs to be stressed that genetic exchangeability is quite irrelevant for ecological species; it is irrelevant just as it is for species of asexual organisms. Instead, all that matters is demographic exchangeability.

will conclude in the negative. Any change in the lengths of the physical necks of the giraffes entailed a *new* average length, a *new* abstraction, not a change in that abstraction *per se*. Granted, the changes in the physical necks are infinitesimally gradual, but that is insufficient to conclude that the abstract length has infinitesimal continuity just like the physical necks. There is not a further entity that changes along with the physical necks. To think that way is not only to think luxuriously but to hypostatize that abstraction in an excessive way, to give it a kind of reality that it really does not deserve. The abstraction is simply a derivative entity, not a substantial entity.

Another way of looking at this is through the distinction between constitutive and further facts. The average length of the necks of the giraffes is not a constitutive fact of that group; it is only a further fact, a derived fact. The constitutive facts are each length of the actual necks of the giraffes. To think of the average length of their necks as a constitutive fact is to think of it as a substantial entity, not a derived entity. It leads to the conclusion that the abstraction can maintain its identity *in spite of any change whatsoever* in the conditions on which it depends. But then that is to say that it really does not depend on those conditions, which is not only contradictory but outright otherworldly—in a word, "Platonic."

This analysis, of course, easily extends to the abstractions we have examined in this chapter (not to mention abstractions *per se*). But now a serious problem arises. If the analysis above is correct, then any change in an elementary class is not a change in that class but is a new class, any change in a cluster class is not a change in that class but is a new class, any change in a cluster hypervolume in an A-space or a Hamming sequence space is not a change in that hypervolume but is a new hypervolume, and any change in any of the dimensions of a niche is not a change in that niche but is a new niche. In other words none of these abstractions can change or evolve. Species, on the other hand, are normally thought by biologists to be units of evolution, if not *the* units of evolution. In short, on the modern view species are supposed to evolve. But if a species is an abstraction, or if it has an abstraction as one of its properties, then it cannot evolve. This is a serious problem for all of the views examined in this chapter. A species, of course, just like the group of giraffes at the zoo, may *entail* an abstraction, indeed at any particular instant, but if it is in any way *constituted* by one then it cannot evolve.

Why then have so many biologists and philosophers who have entertained these views casually assumed that their conceptions of species as abstractions involved no barrier to evolution? I suggest (somewhat like Nietzsche argued)[59] that we are easily misled by the metaphysics of language. There is a *name* for the abstraction—whether it is "class," "cluster class," "hypervolume," "niche," or whatever else—and because there is a name for the abstraction we subconsciously suppose that there must also be a *thing out there* to which the name corresponds, a thing moreover that may itself change just as other things for which we have names may change. But in this (as so often happens in metaphysics) we are sim-

---

[59] Cf. Nietzsche, *The Will to Power* §484, and *Twilight of the Idols*, "Reason" in Philosophy §5.

ply being misled by language.

There is a further problem that concerns me, again a linguistic one, but this time very different, namely the lack of coherence between the abstract species concepts examined in this chapter and the linguistic practices typical of biologists in biological literature. Darwin, for example, thought that varieties are incipient species. But if a species is an abstraction, then what is a variety and how can it be an incipient species? Was Darwin simply wrong? Likewise when we turn to today, we find that biologists often tell us that this particular species is mostly to be found in this particular area and that that particular species is partly to be found in that area and partly in that other area. They also often tell us that most of this species has this particular character and that only in a small part of that species is this other character to be found. In addition, with the aid of computers they sometimes plot for a given species multiple character distributions over a given geography by a method known as multivariate analysis, resulting in maps with continuous and varying distributions reminiscent of weather maps.[60] But if species are abstractions, then biologists in general are wrong and none of the above types of statements and practices should be made of species *per se* but only of their members. This, of course, does not prove that the views examined in this chapter are wrong, but their violence to the basic intuitions of practicing biologists should make them somewhat suspect.

Where the views examined in this chapter do gain their greatest appeal, it seems to me, is in that they allow for the reality of asexual (uniparental) species. Many biologists and philosophers cannot accept that asexual organisms are not members of species. As we have seen earlier, so-called asexual species display a degree of phenotypic cohesion generally on a par with sexual species and are thus taxonomically "good" species; moreover they equally seem to inhabit niches, to play roles in food chains, to be subject to biological control, and to evolve. To include them among genuine species, then, seems not only natural and legitimate but it also gives direction to and apparent justification for the view that species are classes of some sort. What only adds to this is the view, common among philosophers and lay people, though not quite so common anymore among biologists, that extinction is not necessarily forever, that in principle there could once again be dodos and dinosaurs.

As we shall see in the next chapter, however, there is a major view of species which denies this, a view which denies asexual organisms any species status whatsoever and which affirms that extinction is necessarily forever. It is to this view that we now turn.

---

[60] For a brief discussion on multivariate analysis, cf. Gould (1977: 234–236). Gould makes the interesting point (among others) that multivariate analysis makes subspecific designations in most cases arbitrary (cf. chapter 2.3).

Chapter 4

Species as Individuals

It is only when the Biologist and the Philosopher join hands that they can begin to see the subject [of biological individuality] in its entirety.

—Julian Huxley (1912: 2)

4.1 Precursors from Hegel to Mayr

The view which conceives of species as individuals, as concrete, cohesive wholes, bounded in space and time, so that species names are proper names, is a radical and revolutionary departure from the traditional view, in which species are conceived of as abstractions of some sort. From reading as much as I can of the vast literature on the species problem, particularly from recent years, I think I can safely say that the species-as-individuals view is now the dominant paradigm on the ontological status of species, although its position as a majority view is not overwhelming when all of the other views are taken together in opposition. The inception date of the species-as-individuals view is normally attributed to Michael Ghiselin's 1966 paper, although it was his 1974 paper that really got the controversy started. Ghiselin, of course, claims for himself no great originality; instead his claim has always been that he has but made explicit what the majority of biologists and naturalists have held implicitly or unconsciously for over the past two centuries. In the next section I will devote close attention to the arguments of Ghiselin, and also of David Hull, the principal proponent among philosophers of Ghiselin's view, as well as the arguments of a number of other biologists and philosophers who more or less follow in Ghiselin and Hull's foot-steps. Although critical assessment of their view will be found interspersed throughout the first three sections of the present chapter, culminating in the fourth section, in the next section I will also show that their view is not as unified and uniform as one might naturally suspect. In the present section I will examine Ernst Mayr's biological species concept for its position as the foundation of the modern species-as-individuals view, as well as Mayr's reservations for attributing individuality to species and Hugh Paterson's variation on the theme. Indeed a number of earlier precursors of the individuality thesis are also worth examining. Much of the modern debate, it seems to me, suffers precisely from the failure to take account of precursors. Accordingly I will briefly examine the views of G.W.F. Hegel and Karl Popper among philosophers, and Julian Huxley among biologists, none of whom are discussed let alone mentioned by Ghiselin or Hull. This examination should prove not only interesting but fruitful.

From the above short list of precursors many will be surprised to see that I do not include Buffon. They will be surprised because there has been a trend in recent decades toward crediting Buffon with shifting the history of the species concept away from conceiving of species as abstractions and toward conceiving of them as historical/physical individuals. Interestingly, it may even have been Ghiselin who began this trend. Based apparently on Buffon's famous sterility criterion, quoted in chapter 2.3, Ghiselin (1966) claimed that Buffon was "an early adherent of a biological species concept" (208), so that although on his view Buffon was a species nominalist in the traditional sense of species as universals, Buffon was a species realist in the modern sense of species as individuals. Without committing himself either way on the issue of universals realism, Ghiselin accordingly claimed that "Thus nominalism is fully consistent with the real existence of biological species" (208). He would later (Ghiselin 1974) reaffirm that the species-as-individuals view "goes back at least to Buffon" (279; cf. Ghiselin 1969: 53, 1987b: 208, 1997: 14). More recently Phillip Sloan (1987) has devoted a paper to defending this thesis. Based largely on a few highly suggestive passages in the *Histoire Naturelle*—such as in volume II (1749), in which Buffon says "this chain of successive individuals, which constitutes the real existence of the species " (Lyon and Sloan 1981: 170), or in volume IV (1753), in which Buffon says "one can always draw a line of separation between two species, that is, between two successions of individuals that cannot reproduce with each other" (Farber 1972: 267)—Sloan argues that on Buffon's mature view "species are not to be conceived of as classes of individuals, but as spatiotemporal individuals" (124), both "physical and real" (121), in other words "concrete historical lineages" (126). Sloan claims moreover that this move on Buffon's part was "revolutionary" (118), "novel" (121), and "marks a critical innovation" (123). More recently still, Jean Gayon (1996), based mainly on a passage found in volume XIII (1765) of the *Histoire*, quoted in chapter 2.3 but quoted here more fully—"Species are the only entities of nature—perduring entities, as ancient, as permanent, as Nature herself. In order to understand them better, we shall no longer consider species as merely collections or series of similar individuals, but as a whole independent of number, independent of time; a whole always living, always the same; a whole which was counted as a single unit among the works of the creation, and which consequently makes only a single unit in nature" (Lovejoy 1959: 101)— argues that either in this passage "Buffon presents species as individuals" (220), that in the three-stage development of his species concept "It is while talking of this third concept that Buffon sees species as individuals" (224), or that "Buffon came close to qualifying a living species as *individuals* [sic]" (228), that he came "close to that formulation" (228).

It seems to me, however, that when all things are considered Buffon did not even come close to a conception of species as individuals. Beginning with the passage that Gayon makes so much of, although the part where Buffon says a species is "a whole always living" might seem to support Gayon's view, the rest of the passage seems to go against it. In that passage Buffon also says a species is "a whole independent of number, independent of time" and "always the same," which strongly suggests that he thought of species as anything but concrete individuals. An individual in any *biologically meaningful* sense of the term would not be "independent of time," let alone "a whole independent of number." This is the language of abstract classes (where individual organisms belong to but are not constitutive of species), not concrete individuals (where individual organisms are parts of species).

What only further supports this interpretation is Buffon's theory of the *internal mold* (*moule intérieur*), a theory virtually ignored in Gayon's discussion (cf. his 226) and passed over much too lightly in Sloan's—on Sloan's interpretation, species on Buffon's view are "immanent material lineages created by the process of like begetting like through the intermediacy of a formal causal agency [the internal mold] which is passed on perpetually" (121). Although Buffon is somewhat obscure in his explanation of what an internal mold is and does, he does tell us in volume II (1749) of the *Histoire* that it is "an organic matter perpetually active, and always ready to mold itself, assimilate, and produce beings similar to those which receive it" (Lyon and Sloan 1981: 207). More importantly for Buffon's species concept, he tells us that it is the internal mold in which "the essence of the unity and continuity of the species consists, and will so continue while the great Creator permits their existence" (Lyon and Sloan 1981: 202). And later, in volume XIII (1765), in the same preface in which Gayon claims to find the species-as-individuals view, Buffon states that "The type of each species is cast in a mold of which the principle features are inneffaceable and forever permanent, while all the accessory touches vary" (Lovejoy 1959: 101). And later still, in volume XXXIV (1778), he claims that "The type of each species has not changed; the internal mold has kept its shape without variation. However long the succession of time may be conceived to have been, however numerous the generations that have come and gone, the individuals of each kind represent today the forms of those of the earliest ages" (Lovejoy 1959: 103).

As such, Buffon's concept of the internal mold is largely reminiscent of Aristotle's genotype-like concept of the *eidos*, the main differences being the role of the Creator and that both the male and the female contribute the genetic material. At any rate, it should be clear that Buffon's theory of the internal mold is fundamental to his species concept. Without internal molds there would be no similar organisms in nature, or successions of similar organisms via reproduction, or sterility barriers. All of these are merely the visible effects of that which is invisible to the human eye but which constitutes the real reality of a species, namely its internal mold (a "whole always living"?). Thus conceived Buffon's concept of species logically fails to fulfill what is (as we shall see) one of the defining characteristics of concrete individuality, namely that the end (death, extinction) of a concrete individual is necessarily forever.

What this analysis further suggests, interestingly, is that Buffon's famous sterility criterion should not be taken as constituting a precursor of the modern biological species concept, characterized by reproductive isolation on the one hand and genetic integrity on the other, and so should not therefore (contra Ghiselin) be taken as a precursor of the modern species-as-individuals view. According to Mayr (1982), Buffon's famous sterility criterion "does not truly define a concept but only provides a method for testing whether two individuals belong to the same or to different species" (334–335). This is the conclusion to which we are here led. But the real reason why sterility on Buffon's view is only evidence of, rather than constitutive of, species distinctness is not, as Mayr supposes, because Buffon was a species fixist; rather it is because of Buffon's theory of the internal mold. What fundamentally constitutes a species is its specific internal mold, which like Aristotle's *eidos* is invisible to the eye, and the sterility between individuals of two different species is merely the external evidence that there are two internal molds instead of one. Had Buffon held that internal molds are not fixed but evolve gradually over time, his sterility criterion would still function in his theory as an evidential test (horizontal), so the fact that Buffon thought that species are fixed has nothing to do with why he was not really a precursor of either the modern biological species concept or the modern species-as-individuals view. (Cf. Stamos 1998 for a fuller discussion on the above issues.)

Of greater interest, it seems to me, is the view of Hegel (1770–1831). According to Ralph Eaton (1931), "The logic of the second half of the nineteenth century is shot through with controversy over the *law of the inverse variation of extension and intension*" (265). According to this law, the intension of a universal or class varies inversely with its extension and *vice versa*: i.e., if one increases the number of characteristics which define a universal or class, its extension decreases, while if one decreases the number of characteristics which define a universal or class, its extension increases; conversely, if one increases the number of members of a universal or class, its intension decreases, while if one decreases the number of members of a class, its intension increases. (Cf. Eaton [266–269] for a brief discussion on the various problems with this law.) Hegel's rejection of this law, which may have inaugurated the later controversy, is codified in his theory of the *concrete universal* (Hegel's *Notion* or *Begriff*). As Eaton explained Hegel's theory,

> The concrete universal . . . is not a self-identical, changeless form, but a form as realizing itself in a historically changing system of sub-forms and individual things. Consider on the one hand *humanity* as an abstract character, *rational animality*, repeating itself without variation in instance after instance, and on the other, the history of mankind, the laborious rise from the Stone Age to the cul-

ture of Greece, the struggle of races—varieties of men—with one another, the priests, prophets, poets, sages, barbarians, building and destroying civilizations, procreating their kind and dying. In the one case you have the abstract universal *man;* in the other, *man* in the concrete. [270]

> Extension and intension cannot have the same meaning for the concrete as for the abstract universal. . . . The concrete universal is a principle of generation. It is the universal viewed as *generating* sub-forms and individuals, all different from itself, and yet within which it is realized. Latent in the universal *man* (as concrete) are all the forms of men, and all individual men, together with their actual historical relations, i.e., all human history. These latent possibilities of generation, all the sub-forms and even the individuals themselves, which proceed out of the universal, constitute its meaning or intensional import. Hegel insists that we can fully know the meaning of the universal only through history; its expression in time is essential to its being. Obviously, then, the most general concrete universals, those covering the widest extension, have the fullest intension in this peculiar sense of the term; namely, they have the fullest latent possibility of generating different expressions of themselves. The law of the inverse variation of extension and intension becomes nonsense as applied to the concrete universal. In fact, extension and intension seem to coalesce; for the different forms and individuals through which the universal develops into concreteness are at once its extension and intension. [271–272]

What is interesting in Hegel's view is not only the idea of evolution (albeit progressive), but that universals such as species are thoroughly concrète and historical. Of course there is apparently nothing in Hegel's conception of species to differentiate them from other (nonbiological) universals, and one can debate on many points the differences between Hegel's concrete species and the modern view of species as individuals. And yet the historical influence, I suggest, should not be ignored, given the immense influence of Hegel.

Also informative are some of the *philosophical* objections to Hegel's theory of concrete universals, as they may also apply to the modern view of species as individuals. Bertrand Russell (1914), for example, claims that Hegel's theory of concrete universals

> depends throughout upon confusing the 'is' of predication, as in 'Socrates is mortal,' with the 'is' of identity, as in 'Socrates is the philosopher who drank the hemlock.' Owing to this confusion, he thinks that 'Socrates' and 'mortal' must be identical. . . . Again, Socrates is particular, 'mortal' is universal. Therefore, he says, since Socrates is mortal, it follows that the particular is the universal—taking the 'is' to be throughout expressive of identity. . . . Again Hegel does not suspect a mistake but proceeds to synthesize particular and universal in the individual, or concrete universal. This is an example of how, for want of care at the start, vast and imposing systems of philosophy are built upon stupid and trivial confusions. [48–49n1]

Later, in §4, we shall see how proponents of the modern view of species as individuals answer logical objections of this sort.

Probably the clearest precursor of the modern individuality thesis, and certainly one of the most significant, is Julian Huxley. In a relatively unknown little book, published when he was at the ripe old age of 25, Huxley mentions "the species-individuality of which we are the parts" (24). Here, it seems to me, for the very first time (unless I am mistaken), we have biological species clearly thought of as mereological wholes. And like Ghiselin and Hull, Huxley apparently holds the view that if species are to evolve, then they must be individuals. "If Evolution has taken place," he says, "then species are no more constant or permanent than individuals" (27). Of further interest is that Huxley makes a distinction between two kinds of biological individuality, one for species and one for organisms: "wherever a recurring cycle exists (and that is in every form of life) there must be a kind of individuality consisting of diverse but mutually helpful parts succeeding each other in time, as opposed to the kind of individuality whose parts are all co-existent: the first constitutes what I shall call species-individuality" (25). Interestingly, in claiming that for "every form of life" where there is a "recurring cycle" that form has species-individuality, Huxley, unlike Ghiselin and Hull, is necessarily including *asexual* organisms as genuine *parts* of species. But beyond mereology and evolution and this implied inclusion of asexual species, Huxley really has very little to say about species individuality. He certainly does not draw out the many implications of the individuality thesis as do Ghiselin and Hull. However, and very importantly, Huxley does provide us with three minimal criteria for individuality in a biologically meaningfully sense: "the individual must have heterogeneous parts, whose function only gains full significance when considered in relation to the whole; it must have some independence of the forces of inorganic nature; and it must work, and work after such a fashion that it, or a new individual formed from part of its substance, continues able to work in a similar way" (28). But in this virtually all of Huxley's book is devoted to ascertaining the various degrees and manifestations of *organismic* individuality throughout nature. He nowhere attempts to examine how these three minimal criteria might apply to *species*. What is interesting is that, when we therefore attempt to see how these three criteria might apply to species, what we find, I suggest, is that they really do not apply at all.

As for the first criterion, Huxley tells us that "in reality it can be easily shown that nothing homogeneous can be an individual" (10). His reasoning for this is based on the necessary condition of functional unity for individuality. Take for example a mountain. On Huxley's view, any mere aggregate such as a mountain is not an individual: "Cause half a mountain to be removed and cast into the sea: what remains is still a mountain, though a different one" (9). Its definiteness, moreover, he says, "is imposed . . . from without" (11). Thus, "The inorganic system is a Particular, but not an Individual" (9). An individual organism, on the other hand, has "an inner principle of unity" (11), such that an elimination or division of some of its parts either reduces or destroys the functional unity of that organism. Thus, for Huxley, "from that very unity of the whole we can postulate diversity of its parts" (10). One can, of course, debate with Huxley whether there are or can be inorganic individuals, but I think it is clear that all *organismic* individuals (and I would define life as being minimally cellular) have a heterogeneity of parts. But now what of *species*? Huxley, in my third quotation from his book, states that species consist of "diverse . . . parts." But do they? Aside from the ontogeny of individual organisms, with sexual species there are of course males and females. But what of asexual or uniparental species? As we will see later in this chapter, there are serious difficulties involved with their exclusion. But even with sexual species, many consist of isolated subpopulations between which there is no interaction. Although, again, we will return to this problem later in this chapter, it arises in Huxley's further discussion on his first criterion.

In conjunction with his first criterion, Huxley tells us that "an individual as defined in this book cannot be cut in two without its individuality being either lost or impaired" (83–84). In the first case, as when Protozoa divide into two, "Not a jot of substance has been lost: but one individuality has disappeared and two new ones are there in its place" (17). In the second case, "Even in animals with the most astounding powers of regeneration, the working of the whole is always impaired, if only for a short time, by the removal of a part: some regulation, or remodelling, is necessary before the mutilated mass is ready to function as a whole once more" (11). A species, of course, in the first case, can indeed be cut in two by a natural barrier such as a river, but it does not thereby immediately lose its singularity as a species and become two different species. Similarly, in the second case, a species can lose a relatively small portion of itself as in a founder population, but again, it is not thereby diminished as a single species.[1]

---

[1]   Interestingly, both cases, namely speciation by splitting and by budding, are issues in the modern debate. According to Hennig (1966) and many of his followers, any branching point, whether splitting or budding, results automatically in the extinction of the parental species and in the origin of two (at least) new species. I will reserve a discussion of the cladistic species concept and its relation to the species-as-individuals view for §5. For the present, in the case of speciation either by splitting or by budding we can ask, following Sober (1984b), "When did this new species come into existence?" (339). Sober, who is a defender of the species-as-individuals view, suggests as an answer that in the case of speciation by budding (and also by implication speciation by splitting) the speciation "began at the time of the isolation event, even though the isolated organisms may have been no different from the organisms in the main population. The founders were founders of a new species precisely because of what happened later, and not in virtue of anything special about them" (339). Kitcher (1984b), a critic of the species-as-individuals view, disagrees. On Kitcher's view, "If both the isolate and the parent population remained reproductively compatible for thousands of years after the geographical isolation occurred, then Sober's proposal will be rejected—speciation did not begin until long after the event in which the geographical isolation occurred" (627). Similarly, Mayr (1987a) claims that Sober's view "is not correct biologically, since the development of isolating mechanisms, the true sign of species status, requires many generations" (162). Splitter (1988), aside from confusing geographically isolated subpopulations with subspecies (340), attempts to defend Sober's retrospective approach by claiming that the initial founder population "may not be part of any species; the speciation event itself begins at isolation and lasts until the isolate develops species status" (341). But until he tells us how a recently isolated subpopulation of a species "may not be part of any species," his defense of Sober's position cannot be taken

Huxley's second criterion of biological individuality, that there must be "some independence of the forces of inorganic nature," is equally problematic for species. According to Huxley, "When a glance is thrown over the various forms of animal life to which the name of Individual is naturally conceded, it is seen that in spite of many side-ventures, they can be arranged in a single main series in which certain characters are manifested more clearly and more thoroughly at the top than at the bottom. One of these characters is independence of the outer world and all its influences—in other words, immunity from accidents" (3–4). Even the simplest Protozoan, he says, which exists at the low end of the scale in terms of immunity from accidents, "has some power of independent movement, and is not helpless like the inorganic grain of dust" (5). According to Huxley (5–6) an individual's place in this scale of independence is dependent upon and varies with its size, its complexity, and its adaptability. Of course, to say that *species* have size, complexity, *and* adaptability is to beg the issue. What we must really ask is in what sense species are immune from accidents. In the case of geographical speciation, whether the classic dumbell model of allopatric speciation or Mayr's founder population model which he calls peripatric speciation (cf. Mayr 1988a:

---

seriously. O'Hara (1993) also tries to defend Sober's retrospective approach (241), with the claim that not only vertical species concepts but even the biological species concept "depend upon the future" (242). O'Hara tells us that on the biological species concept two populations with an interbreeding gap between them are conspecific or specific depending on whether the gap is "temporary" or "permanent": "In other words, it will depend to some extent upon prospective narration" (242). And yet O'Hara makes no mention whatsoever of the role of isolating mechanisms stressed by Dobzhansky and Mayr. Similar to O'Hara in this is Kornet's (1993) view that "The insistence that splits should be permanent in order to count as speciation events has the consequence that species of a genealogical network are defined in retrospect, i.e. looking back in history. Thus we can never, before the extinction of a network, exclude the possibility that parts of the network may still reunite into one species" (409). Once again this ignores genetic isolating mechanisms, which themselves virtually exclude the possibility that genetic isolation between two populations might somehow in the future be reversed. At any rate, on this whole issue I find myself siding with Kitcher and Mayr, though for somewhat different reasons, accepting neither the former's species pluralism nor the latter's flirtation with essentialism. First, it will be remembered from the end of chapter 2 that one of the main reasons for believing that species are real, at least at the horizontal level, is their meaningfulness for communication between biologists. But none of this meaningfulness is impaired in the period shortly after geographical splitting or budding. The splits or the buds may indeed go on to become different species, or they may not (the river, for example, may dry up). But if and until they do, they are still meaningfully conspecific (think of, for example, biological control) for quite some time after geographical separation. Second, it will be noticed that the retrospective approach has the consequence that it unashamedly admits into biology teleological thinking (specifically backwards causation, that the future determines the past). This kind of thinking is unDarwinian and contrary to the Modern Synthesis, a reproach justified alone by the fact that the retrospective approach fails to appreciate the fundamentally contingent nature of evolution. Moreover since the retrospective approach virtually ignores any of the biological properties of organisms, this shows that it is methodological only, a mere convention, so that accordingly it cannot count as a natural law on speciation.

366–367), it would appear that species have no immunity whatsoever to geographical forces. But from the viewpoint of real extinction (as opposed to pseudo-extinction), there is a sense in which species may have a degree of immunity from extinction. As noted in chapter 3.4n51, David Raup (1991), in his thesis that most species extinction is due to bad luck rather than bad genes, employs the concept of the *minimum viable population* (MVP). Quoting Dan Simberloff he says "Populations above this point [are] virtually immune to extinction, while those below this point [are] likely to go extinct very quickly" (124). Following Simberloff, Raup lists the four most common causes of extinction for species with populations below their MVP.[2] He further tells us that "Despite variability in the MVP, all studies come to the same striking conclusion: MVP size is very low, commonly in the range of a few tens or hundreds of individuals" (126). Thus species, in virtue of having an MVP, would seem to conform to Huxley's second criterion of biological individuality. But I suggest that this conclusion should not be too heartily embraced, for as we shall discover in §4, the fact that species have MVPs entails a further characteristic, namely that the concept of aging does not apply to them. What seems to confer *biological* individuality to species, then, would also seem to take it away.

Huxley's third criterion of biological individuality, that "it must work, and work after such a fashion that it, or a new individual formed from part of its substance, continues able to work in a similar way," is also problematic for species. We can see this once we look at Huxley's argument for why a solar system is not an individual in any biologically meaningful sense. "The solar system," he says, "is a whole most definitely 'isolated by Nature,' heterogeneous, and composed of parts closely inter-related in their working; what . . . prevents our calling it an individual? This, that its working is not directed to continuing either itself or other systems like itself" (21). As for species individuality, this third criterion fails, it seems to me, as a result of Huxley's explicitly (9–10) Bergsonian, and implicitly Hegelian, concept of evolution, which is the concept of progressive, or goal-directed evolution. Speaking of "disembodied spirits," he says "If such actually exist, they crown Life's progress; she has started as mere substance without individuality, has next gained an individuality co-extensive with her substance, then an individuality still tied to substance but transcending it in all directions [i.e., man], and finally become an individuality without substance, free and untrammelled" (30–31). This concept of goal-directed evolution, of course, is no longer accepted in biology. But this defect in Huxley's conception also filters back to his third criterion of individuality as applied to species. Contra Huxley, a species *is*

---

[2]   The first is *demographic stochasticity*: "if populations are already very small, a modest run of bad luck can do them in" (124). The second is *genetic deterioration*: "the species may not have the genetic variability to adapt to changing conditions" (124–125). The third is *social dysfunction*: "The deterioration that occurs in certain behavioral traits when populations become too small" (125). The fourth is *extrinsic forces*: "typified by fire, disease, . . . extrinsic forces hit populations of any size but inflict serious harm only on smaller populations" (125).

just like a solar system in that "its working is not directed to continuing either itself or other systems like itself." Each *individual organism* is clearly so directed, in virtue of its genetic program, but in addition to its genetic program there is not a further genetic program *for the species* to which it belongs. If there were, then a species, much like an organism, would have ontogeny. But a species does not have ontogeny, only phylogeny. To think otherwise is to ascribe to species what is known as *orthogenesis*, the view that the evolution of a species has a programmed direction, in spite of natural selection, a view that is, to quote Abercrombie *et al.* (1990), "Now discredited" (411).

Where Huxley *is* correct about species, however, is that a species involves "a recurring cycle" (25). The individual organism, he says, "only persists for a limited time. In spite of this, something does indefinitely continue, though it is but the kind, the species, and not the single individual itself. There is only one kind of working in the species, and this repeats itself in a recurrent cycle; but for each cycle as it recurs a new individual is required as the instrument of the working" (20). But if that is the *only* kind of working in the species, as Huxley here states, then there is no longer any reason to consider the species as an individual in any biologically meaningfully sense. Huxley does this, but it would seem to be because he is misled in his conception of evolution. Ghiselin and Hull, as we shall see, do not share this misconception, and yet interestingly many of the difficulties that I have raised here against species as being biologically meaningful individuals also applies to their thesis. But rather than repeat myself all over again in §4, I will there provide further evidence and argument against the species-as-individuals view. What has been examined here, however, adds much that is new to the debate. And for this we must largely thank Huxley for his interesting and in many ways excellent little book.

Returning to our list of precursors, an important precursor in philosophy, it seems to me, is Karl Popper. As early as 1934, Popper (1934) tells us that "words like 'mammal,' 'dog,' etc., are in their ordinary use not free from ambiguity. Whether these words are to be regarded as individual class names[3] or as universal class names depends upon our intentions: it depends upon whether we wish to speak of a race of animals living on our planet (an individual concept), or of a kind of physical bodies with properties which can be described in universal terms" (65). In claiming that it depends on our "intentions" whether species names should be used as proper names or as class names, the implication, it seems to me, is that species *themselves* may be thought of as either individuals

---

3   I suggest, in accordance with the context of the above passage, that Popper here means more properly individual *proper* names rather than individual *class* names. Cf. his moa example in his New Appendix X, where purely for the sake of preparing an argument designed to show that true, strictly universal statements need not necessarily be statements of laws of nature, Popper says "Consider some extinct animal, say the moa, a huge bird whose bones abound in some New Zealand swamps. . . . We decide to use the name 'moa' as a universal name (rather than as a proper name; *cf.* section 14) of a certain biological structure" (427).

(concrete) or classes (abstract), so that they have an ambiguous ontology. If this further claim is indeed Popper's,[4] then quite interestingly it would make Popper a precursor (possibly the earliest precursor) of *species pluralism* (cf. chapter 2.4). Interestingly, one of the principal defenders of species pluralism, Philip Kitcher (1984a), clearly echoes Popper (without mentioning him) when he says "the answer must be relative to a *prior decision* on whether or not to employ a historical species concept" (327n20). The basic problem with this approach, of course, is that if species are to be objective in the most meaningful sense of that term, and to evolve, then the nature of their ontology cannot depend on anyone's *decisions* (cf. Ghiselin 1987a: 130).

Popper is an important precursor of the species-as-individuals view in quite another way. Popper (1936) tells us that "the evolutionary hypothesis . . . has . . . the character of a particular (singular or specific) historical statement. (It is of the same status as the historical statement: 'Charles Darwin and Francis Galton had a common grandfather')" (106–107). A few pages further he tells us that "the most careful observation of *one* developing caterpillar will not help us to predict its transformation into a butterfly" (109). This is part of Popper's long-held thesis that since the statements of evolutionists, in particular for our purposes historical statements about species, are not genuinely lawlike and therefore do not allow for predictions, they therefore are not genuinely scientific.[5] To think otherwise, according to Popper, is akin to the error of historicism in sociology, "an approach to the social sciences which assumes that *historical prediction* is their principal aim, and which assumes that this aim is attainable by discovering the 'rhythms' or the 'patterns,' the 'laws' or the 'trends' that underlie the evolution of history" (3). Interestingly, it was the view that species are not subjects in laws of nature, though not from Popper but from a later philosopher, that served to convert Hull to the species-as-individuals view. As Ghiselin (1989) tells us,

> Hull changed his mind upon encountering the claim of J.J.C. Smart that there are no laws of nature in biology, and that biology therefore isn't really a science, but something more like engineering. Smart was quite correct in saying that there are no laws for *Homo sapiens*, but it was Hull who drew the proper conclusions. Species are individuals, and there are no laws about individuals in any science

---

[4]   I do not know, and it is probably too late to know. What is relevant to this question, however, is that in a later discussion Popper (1972) tells us that "I reject all *what-is questions*," his reason being that such questions are about "what is its essence, or its true nature" (195). Although Popper's antiessentialism is epistemological rather than ontological (he thinks ultimate explanations are forever beyond our grasp), and aside from whatever the worth of that view, his rejection of what-is questions seems to me greatly overdone and ultimately invalid, for the simple reason that few what-is questions are questions about essence; instead it seems to me that what-is questions are more often questions about *category*, which is an entirely different and perfectly legitimate matter. Indeed Popper himself did not refrain from categorizing with his worlds 1 to 3 (cf. §4n47).

[5]   For an extended critical analysis of Popper's characterization of evolutionary biology as a not genuinely scientific enterprise, cf. Stamos (1996b).

whatsoever. In the biological and physical sciences alike, laws have to be generalizations about classes of individuals. This was perhaps Hull's most important
contribution, but it has largely remained unexplored. [53][6]

Whether any given species taxon can be the exclusive subject of a law of nature is a topic I will explore in §2. I want now to examine the most important precursor for the individuality thesis, namely Ernst Mayr. One could also focus on
Theodosius Dobzhansky here, whose contributions to the modern biological species concept are at least equal to that of Mayr. But Mayr's formulations, auxiliary
hypotheses, and much greater attention to the species problem *per se* have made
him the clear favorite among individuality theorists.[7] Moreover, and of further
importance, Mayr has by far the most to say about the relation between the modern biological species concept and the individuality thesis which developed from
it. It is therefore to an examination of Mayr's species concept that we now turn.

Mayr (1982) quotes his 1942 definition of his species concept as:

---

[6]   Marjorie Grene (1989b) notes that "One of the seminal insights in Mayr's *Growth of
biological thought* (Mayr 1982) is the recognition that in general biologists are not interested in laws" (71). Ghiselin (1989), on the other hand, challenges Mayr (and by implication Grene and Popper) on this very point. According to Ghiselin, if, unlike Mayr, one
includes statistical predictions as laws of nature, as well as universal generalizations which
assert that something or other is impossible, then, in league with many of the putative laws
in the physical sciences, there are indeed laws of nature in biology. Among good candidates, Ghiselin cites Mayr's founder principle, which he says should be called "Mayr's
Law" (58), Bergmann's Rule, "which states that in colder climates animals get larger"
(61), and Dollo's Law, which "asserts that evolution in general is irreversible" (64). Cf.
Ghiselin (1997: 127) for his admission that this commits him (unlike his interpretation of
Buffon and Darwin) to species category realism.

[7]   Dobzhansky (1937) defines "species" as "that stage of evolutionary process, at which
the once actually or potentially interbreeding array of forms becomes segregated in two or
more separate arrays which are physiologically incapable of interbreeding" (312), to which
he immediately adds that "Species is a stage in a process, not a static unit." Eldredge
(1985b), in his excellent discussion on the earlier development of Mayr's species concept,
points out that "With his [Mayr's] definition of species, Mayr is, of course, confronting the
ontological status of these kinds of taxa. He remarks that Dobzhansky's definition is 'an
excellent description of the process of speciation, but not a species definition. A species is
not a stage of a process, but the result of a process' (Mayr 1942, 119). And Mayr's own
analogy between stages in the life of a man and stages in the development of the products
of the evolutionary process suggests that he does see species partly as a mere transient
stage in the evolutionary stream. Yet when he comes to define species, he does so in such
a way as to suggest that they are in some important sense real entities after all, so that he
can treat them as actual objects and proceed to discuss their origins" (50). Eldredge further
points out that "If it is true that Mayr did not invent the notion that species are real entities,
it is nonetheless indubitable that the underlying basis for all the various recent approaches
to looking at species as entities, or individuals, rather than as classes of individuals stems
from Mayr's species concept, first published in 1940 but established and recognized primarily upon publication of his 1942 book" (51; cf. Sober 1993: 153).

Species are groups of actually or potentially interbreeding natural[8] populations which are reproductively isolated. [273; cf. Mayr 1942: 120 for this and his longer version]

Mayr (1982) tells us that "The 'actual' vs. 'potential' distinction is unnecessary[9] since 'reproductively isolated' refers to the possession of isolating mechanisms,[10] and it is irrelevant for species status whether or not they are challenged at a given moment" (273). He then tells us that "Isolating mechanisms are biological properties of individuals which prevent the interbreeding of populations that are actually or potentially sympatric" (274).[11] Mayr's revised species definition,

---

[8] The word "natural" is necessary to distinguish interbreeding under natural conditions from interbreeding induced under laboratory or artificial conditions. The example of orchids in chapter 2.4 provides a good case in point. As Cronquist (1978) points out, without the word "natural" the definition "would, for example, sweep great numbers of different species and even genera of orchids into a single species" (6). Instead, "Here the emphasis is placed on what actually happens in nature, rather than on what can be induced experimentally" (7). One might also add the example of lions and tigers, induced in zoos to produce what are called ligers and tiglons (so named for when the male parent is either a lion or tiger, respectively). According to Wilson (1992), "during historical times the two big cats overlapped across a large part of the Middle East and India" (39). And yet, "To the best of our knowledge, no tiglons or ligers were recorded from the zone of overlap" (39). Wilson surmises that this is because lions like open savannas while tigers like forests, and because lions are social cats while tigers are solitary. Thus, says Wilson, "there appears to have been little opportunity for adults of the two species to meet and bond long enough to produce offspring" (39).

[9] Interestingly, there is a long and varied history of criticisms against potential interbreeding as a ranking criterion. Cain (1954: 87) notes that potential interbreeding is inferred solely on morphological grounds. Hull (1965: 209–210) says potential interbreeding is too vague for a great many situations, and that "In evolutionary taxonomy *unrealized* potentialities don't count." Sokal and Crovello (1970: 35) claim that potential interbreeding has never really been defined, let alone operationally, and that even inferences based on morphology are not very reliable. Mishler and Donoghue (1982: 129–130) claim that if adopted, potential interbreeding would reduce the roughly 20,000 species of orchids to only a few species. Ruse (1987) asks the question "Is the sterile worker ant even a potential interbreeder, . . . ?" (347). Clearly some of these criticisms hit very wide of their mark.

[10] Cf. Mayr (1970: 57) for a table of isolating mechanisms. This table divides into two categories: premating and postmating mechanisms. The mechanisms preventing orchid hybrids in nature, then, as well as hybrids in nature between lions and tigers, are essentially premating, while the mechanisms preventing fertile hybrids between horses and donkeys are postmating. Geographical barriers are not included as isolating mechanisms; rather the mechanisms must be intrinsic to organisms (otherwise the species concept would not be a genetic one).

[11] It should be noted that the problem of inference, which has bothered many as in n9 above, is not eliminated or diminished by the emphasis on isolating mechanisms, since according to Mayr (1996) isolating mechanisms are "invisible" and only "spring into action" in sympatric situations (273). Thus the problem of inference still remains not only for synchronic nonsympatric situations but also for nonsynchronic situations, both of which Mayr calls multidimensional situations (cf. chapter 2.3n15 and accompanying text). Mayr's

then, first given in 1963, is (Mayr 1970):

> *Species are groups of interbreeding natural populations that are reproductively*
> *isolated from other such groups.* [12]

As we have seen in chapter 3.4, Mayr later (Mayr 1982) added a niche specification to his species definition but was hesitant in doing so. It is interesting to
note that in a more recent definition (Mayr 1991) he has it removed, and defines
species as:

> A reproductively isolated aggregate of populations which can interbreed with
> one another because they share the same isolating mechanisms. [186]

Aside from some of the problems which have been raised against the biological species concept itself,[12] which in proportion with their validity would under-

---

(1987b) answer to the inference problem, which I accept, is not only that it only applies to
multidimensional situations and not to nondimensional situations, but, as he says, "And
what is so bad about inference? There are few insights in science that are based exclusively on direct observation. Nearly always inferences are also involved" (219).

[12]   Among the usual complaints are that it does not apply to asexual organisms or to fossil
species, but the more direct complaints are based on operationality and gene flow. Sokal
and Crovello (1970), for example, point out that "As direct evidence on interbreeding diminishes, the methods become increasingly phenetic" (33). Since a sufficient degree of
interbreeding is rarely if ever observed, they argue, and since it cannot work for dead samples, the biological species concept must therefore be "nonoperational" (42), meaning that
it cannot offer a useful species concept in practice. They therefore prescribe a phenetic
species concept. Sober (1993: 157–158) replies to this line of attack by pointing out all the
problems with a phenetic species concept. Mayr's reply is more to the point. According to
Mayr (1988a), "That there are operational difficulties connected with the making of such
inferences [interbreeding] does not refute the concept as such, as Hull, Wiley, and others
have pointed out" (321–322; cf. Hull 1970: 318–324, Wiley 1978: 80–82). Ehrlich and
Raven (1969), on the other hand, argue that many conspecific populations have little or no
gene flow between them, particularly in plants, moreover that "selection can override the
effects of gene flow" (64), so that the cohesive force binding species together is for the
most part "similar selective regimes" (63). Ridley (1993) provides a helpful discussion on
the empirical evidence bearing on whether gene flow or selection is more important than
the other, pointing out that the two are not necessarily mutually exclusive, that, depending
on the species situation, one might be more important than the other, and concluding that
"the evidence . . . hardly warrants any firm conclusion" (399) but that "Selection and gene
flow are probably not usually opposed forces in nature" (397). Sober's (1993: 156) reply
to Ehrlich and Raven is to point out that, although they might be right about little or no
gene flow between conspecific subpopulations, their argument relies on an inappropriate
time scale. Mayr himself, however, accepts their time scale, as well as their claims about
the absence of gene flow. Instead, he claims (1988a) that "It is highly probable that such
stasis is maintained by an internal cohesion of the genotype reinforced by stabilizing selection" (329). I will have more on what Mayr elsewhere (1975) calls "the unity of the
genotype" when I discuss the theory of punctuated equilibria in §3.

mine the individuality thesis based upon that concept, it is interesting to examine Mayr's own reasons for not fully accepting the individuality thesis. Although Mayr (1987a) tells us that "My own species concept (1942–1982) has always been that of a concrete entity" (151), he is unhappy about calling species "individuals" for a number of reasons. First, it is "quite counterintuitive to apply it to an assemblage of individuals" (158), because the Latin root of the word "individual" means "that which is indivisible" (158). Second, "It is only in the lower invertebrates and in many kinds of plants that a seriously mutilated individual can be restored as quickly as a decimated species" (159). But Mayr's most important reason appears to be that "The nature of organization (cohesiveness) is also very different. . . . the interaction of parts of a species is for most of its members quite loose and indirect, consisting only of the propensity for gene exchange. But such gene exchange is virtually nonexistent in those many species which consist of highly isolated colonies or subspecies" (159). Consequently Mayr is a little happier with the term "multiple individual" or more properly "biopopulation" (161), but even here he expresses reservations. I have briefly explored Mayr's reservations and the overall attempt to reduce species to populations in chapter 2.4. Suffice it to note that Mayr's suggestion has not gone over well.[13]

Of greater significance than Mayr's own reservations for categorizing species as individuals, and what seems to have been overlooked right across the board, is that when we examine Mayr's elaboration of his basic (minus niche) species definition we encounter serious conceptual problems for both the biological species concept and (insofar as it is based upon that concept) the individuality thesis, problems that need to be addressed. First, Mayr (1970) tells us that "The species . . . is . . . a *genetic unit* consisting of a large intercommunicating gene pool, whereas an individual is merely a temporary vessel holding a small portion of the contents of the gene pool for a short period of time" (12). A page later he tells us that "A species is a protected gene pool. It is a Mendelian population that has its own devices (called isolating mechanisms) to protect it from harmful gene flow from other gene pools" (13). Again he says "Organizing genetic diversity into gene pools, that is, species" (20). Mayr's claim here, of course, must surely be overstated. If a species literally *is* a protected gene pool, then it also cannot literally *be* "groups of interbreeding natural populations." If we are to accept the latter definition as primary, as its context indicates, then a protected gene pool can at most be only a *constitutive property* of a species; it cannot also be what a species *is* on pain of contradiction.

There is a further problem. If species are to evolve, as both Mayr and the principals of the individuality thesis want them to, then a gene pool (when thought to

---

[13] Hull (1987), for example, thinks that "this maneuver is not likely to succeed. From its inception, 'population' has been applied as readily to stars and molecules as to plants and humans, just the sort of indiscriminate usage which Mayr deplores" (172). Rosenberg (1987) believes that Mayr's proposal is "irrelevant" to the kind/individual dispute and "will only obscure the ontological issue" (197). Panchen (1992) thinks that Mayr's proposal is "ontologically too vague" (339).

be in some way constitutive of a species) must be a sort of concrete entity; it cannot be an abstract entity (as per my argument in chapter 3.5). In his Glossary, Mayr (1970) defines the gene pool as "The totality of the genes of a given population existing at a given time" (417). This certainly *sounds* like a physical conception, namely as a set extensionally defined. And indeed it is not difficult to find biologists who write of gene pools as if they have a concrete existence. For example, Jeff Doyle (1995), in a paper on allele trees and their irrelevance for species delimitation, says "Gene flow (through transmission of gametes or migration of individuals or propagules) unites the gene pool; . . . The gene pool is a historical entity, in that it not only exists at a given time, but can also show transformation by mutation, recombination, and fixation or extinction of allelic lineages; furthermore it can fragment to form new gene pools" (576). But in contrast there are other biologists who clearly think of gene pools as abstract entities. Richard Dawkins (1976), for example, says "The gene pool is a worthwhile abstraction because sex mixes genes up, albeit in a carefully organized way" (26). Similarly, Mark Ridley (1993) says "In theory, the gene pool is an abstract conception of a set of reproducing genetic units, within which gene frequencies can change" (387). What makes the concept of gene pool a concept of abstract entities, however, at least when applied to species, becomes most evident in the case of conspecific populations which are allopatric and between which there is no further gene flow. We have already seen that Mayr himself acknowledges the reality of such cases. What he seems to miss, however, is that although such populations may share the exact same isolating mechanisms for quite some time, they cannot share a common gene pool, in the concrete sense, since there is no mixing of genes between them. In such cases, then, the conspecific gene pool must be an abstract entity. The upshot is that if this analysis is correct, it follows that species, when thought to be constituted by gene pools, cannot evolve, contrary to the wishes of both Mayr and the individuality theorists.

A further curiosity is where Mayr (1988a) says "Evolution is not a smooth, continuous process but consists, in sexually reproducing organisms, of the formation of a brand-new gene pool in every generation" (99). If we take this statement and combine it with Mayr's (1970) statement that "A species is a protected gene pool," we must then conclude that every new generation is a new species! Clearly odd things happen when gene pools are thought to be constitutive of species.

To avoid these problems, we may, I suggest, in much the same way that a group of giraffes entails an average length of neck, say that each species *entails* a gene pool but is not in any way constituted by one.

Allied with his concept of gene pools, Mayr repeatedly claims that "the word 'species' in biology is a relational term" (Mayr 1970: 13; cf. 1982: 67). He tells us (Mayr 1970) that just like the word "brother," "The word 'species' likewise designates such a relational property. A population is a species with respect to all other populations with which it exhibits the relationship of reproductive isolation—noninterbreeding. If only a single population existed in the entire world, it would be meaningless to call it a species" (14). Indeed much earlier Mayr (1949) went so far as to call this relational property "the essence of the species concept"

(371). The problem here is that if species are truly *individuals*, then it is extremely odd that their ontology is relational, that the *logic* of their existence requires other individuals of the same category. Logically there is no reason why at some time in the future I may not be the only living organism left in the universe. If species are individuals, then this logic ought to extend to them as well.

At any rate, Mayr's biological species concept is not the only species concept based on sexual reproduction. Indeed, this is perhaps the best place to examine what Hugh Paterson (who first hinted about his emerging species concept in 1973) has called the *recognition concept of species*. To understand Paterson's species concept one must be clear on what he means by the *specific-mate recognition system* (SMRS). Itself a "subclass of the broader class of fertilization mechanisms," on Paterson's (1982a) view "Each component of the SMRS involves signalling between potential mating partners (or their cells) and leads to the bringing together of appropriate mating partners (or their cells) and, ultimately, syngamy. The characters of the SMRS are not isolating mechanisms, nor are they mate-selecting mechanisms . . . . With Williams I believe that 'mechanism' should be reserved for characters functioning in a role for which they were selected" (80). Nor is the SMRS to be confused with *species recognition*, which Paterson (1985) calls "a strictly philosophical process"; instead it only refers to "the recognition of an appropriate mating partner" (144–146). It should also not be confused with implying consciousness: "I strongly emphasize," he says, "that I imply no act of judgment and no act of choice on the part of the responding partner" (146). For Paterson the recognition can be purely chemical, and "might be compared to the recognition of a specific antigen by its specific antibody" (146).

What is interesting is that not only initially but through to 1982 Paterson (1982a) continued to define both species and speciation in terms of the SMRS. As for species,

> The members of a species share a common specific-mate recognition system.
> [80]

As for speciation, "a new species will have arisen when all members of a small, isolated subpopulation of a parental species have acquired a new SMRS, which . . . makes effective signaling impossible between members of the daughter and parental populations" (80).

Within a few years, however, Paterson no longer found this definition of species acceptable. Paterson (1985) tells us that "Among angiosperms the SMRS is limited to interactions such as between the pollen and the stigma . . . . Among orchids, several species, even of different genera, may share a common SMRS. . . . In these species the 'fields for gene recombination' are determined by the fertilization mechanisms other than those of the SMRS" (146). This remarkable admission, of course, implicitly concedes a distinct advantage to Mayr's biological species concept, since that concept would not make out of several species of orchids a single species. Given that more than one species may have the same SMRS, Paterson adjusted his species concept accordingly. Acknowledging once again that the SMRS is "a subset of characters of the fertilization system" (144), species are now defined as follows:

> We can, therefore, regard as a species *that most inclusive population of individ-ual biparental organisms which share a common fertilization system.* [147; cf. Paterson and Macnamara 1984: 105, 120][14]

Accordingly, speciation is "an incidental *effect* resulting from the adaptation of the characters of the fertilization system, among others, to a new habitat, or way of life" (148).[15]

According to Mayr (1988a), "In the process of the rejection of nonconspecific

---

[14]   One wonders now if the result can still be called a *recognition* concept of species. If two species can share the same SMRS and if what distinguishes them is some other part of their fertilization systems, then what we have is no longer a recognition but a *fertilization* species concept. Paterson (1988) defines the fertilization system of a species as comprising "all characters that contribute to the achievement of fertilization. These characters are diverse and include such characters in the mating partners as the design features of the gametes, those determining synchrony in the achievement of reproductive condition, the coadapted signals and receivers of mating partners, and their coadapted organs of gamete delivery and reception" (175). The problem now is that it might be difficult to stop at the fertilization system. As Templeton (1989) has pointed out, "Paterson (1985) has burdened the recognition concept with several restrictions that do not necessarily follow from his primary definition. The most serious of these is his exclusive use of fertilization mechan-isms to define a species. Obviously, a field of genetic recombination requires more than fertilization; it requires a complete life cycle in which the products of fertilization are vi-able and fertile" (8).

[15]   Paterson's change from an SMRS-based species concept to a fertilization-based spe-cies concept has failed to be tracked in much of the secondary literature on the species problem, even by some of his followers. Eldredge, for example, persists in characterizing Paterson's species concept in terms of the SMRS. "If mate recognition provides the spatial boundedness [of species]," Eldredge (1985b) asks, "what are the *temporal* bounds?" (155), which is perhaps excusable given the year of publication. But there is no excuse when Eldredge (1989) tells us that Paterson "defines species as the 'largest collectivity of or-ganisms with a shared fertilization system' (i.e., shared SMRS)" (105) and then goes on to tell us that he sees "Paterson's (1985) characterization of the SMRS as the essential ingre-dient of what a species *is*" (113). The confusion is only furthered when Eldredge (1995) tells us that "Paterson proclaims that all those factors that tie into successful mating—meeting and recognizing a prospective mate, actual mating, successful fertilization, and pro-duction of viable offspring that will eventually mature and reproduce—constitute a single integrated system: the Specific-Mate Recognition System (SMRS)" (114–115). On this interpretation, recognition of a conspecific potential mate fails to count as a function of the SMRS if the potential mate should happen to prove sterile! At any rate, the change in Paterson's species concept from an SMRS-based to a fertilization-based concept is signifi-cant for Paterson's concept of selection *for*. Paterson (1982b) tells us that "Since the char-acters of the specific-mate recognition system are adapted to the normal habitat of a spe-cies, they, too, will be shaped to be appropriate to the new conditions by natural selection" (101). According to Paterson (1985), however, it is "the characters of the fertilization sys-tem" that are subject to both "stabilizing selection" and "directional selection" (147).

mates and the acceptance of conspecific mates, one can focus attention on either one or the other of these two processes. Rejection means maintaining isolation, and acceptance means recognition. These are simply two sides of the same coin" (320). Why Mayr prefers the term "isolation" over "recognition" is because the latter implies "a degree of conscious cognitive activity that is not to be expected in 'lower' animals, and secondly because isolation among species is effected in numerous species or organisms by isolating mechanisms other than behavioral ones" (320). Others have also thought the recognition species concept merely the other side of the coin (cf. Templeton 1989: 7–8; Panchen 1992: 335; Ridley 1993: 392–393). This association has persisted in spite of Paterson's protests to the contrary. Paterson's main reason for his protest, in spite of the fact that his concept too is primarily horizontal and excludes asexual organisms, is that while Mayr's concept is *relational*, his is not. On Paterson's (1985) view, "species are not defined relationally but independently, since, of necessity, all sexual organisms must possess an effective fertilization system" (149). On Paterson's view, then, it is indeed logically possible to have a planet with only one species, the logic of which alone, I should think, warrants Paterson's species concept the status of a new coin altogether.

This distinct status should also, it seems to me, warrant Paterson's species concept the status of being a more suitable foundation for the individuality thesis. Surprisingly, however, Paterson nowhere in his writings either explicitly endorses or rejects the individuality thesis of Ghiselin and Hull. Indeed the topic does not even come up, although he does argue against the view of Dobzhansky and Mayr that "Species are individual entities with genetic 'integrity'" (Paterson 1981: 39; cf. Paterson 1986: 159, 1991: 201), since he thinks it entails group rather than individual selection (Paterson is a staunch organism selectionist), and also because he thinks a view of species integrity is a carryover from the Biblical view and subsequent creationist tradition (cf. Paterson 1982b: 93–101, 1985: 143). Among Paterson's followers, however, reception of the species-as-individuals view is paradoxically mixed. Although Lambert *et al.* (1987), for example, tell us that "species are integrated biological wholes and . . . this integration results from the operation of a SMRS" (198), they nevertheless also tell us that "this new viewpoint [that 'the SMRS is the central defining property of species' (198)] lies . . . beyond the debate concerning their [species] class or individual status" (204). On the other hand, Eldredge (1989), for example, claims not only that "the major significance of Paterson's formulation of the SMRS is that species are conceived as *self-defining (thereby entirely natural) reproductive communities*" (104) but that "A view of species based on cohesion agrees very nicely with Ghiselin's notion that particular species are individuals. Such cohesion is supplied by the reproductive plexus pointed to in the biological species concept and made even more explicit in the work of Paterson" (105).

The logical agreement between Paterson's species concept and the individuality thesis, however, is not nearly so nice as Eldredge would have us believe. This comes out rather strikingly on the issue of speciation by polyploidy.

Polyploidy (quite rare in animal species except where parthenogenetic or hermaphroditic) is a chromosomal accident which results in an offspring with three or more times the haploid (N) number of chromosomes than the parental species (N equals half the full complement of chromosomes, as in gametes). So common is polyploidy as a mode of speciation in plants that it is estimated (based on high chromosome numbers) that 70–80% of all species of angiosperms (flowering plants) were produced in the wild by polyploidy (cf. Barrett 1989: 259). Polyploidy can also be induced artificially in the laboratory. The first artificially produced allopolyploid species is the primrose *Primula kewensis*. (Allopolyploidy is polyploidy that results from interspecific hybridization; autopolyploidy involves only one parental species.) The two parental species of primrose from which it was produced, both of which have a diploid ($2N$) number of eighteen chromosomes, produce sterile hybrids ($2N$), and thus are commonly regarded as good species, reproductively isolated from each other by postzygotic mechanisms. With the application of a plant chemical called colchicine, however, the fertility of the hybrids can be restored by a doubling of the chromosomes in their gametes ($2N$), which when intercrossed result in offspring ($4N$) that are fully fertile. They are fully fertile, however, only among themselves. When backcrossed with either one of their parental species the offspring are sterile, since their triploid ($3N$) cells cannot properly go through meiosis. Thus, as Ridley (1993) put it, "it is possible to make new species, and by a method that has been highly important in the origin of new natural species" (43). Indeed as Coyne *et al.* (1988) point out, such instantaneous taxa "are considered good species by nearly everyone" (192).

Because of the nature of his species concept, however, Paterson (1981: 49–52, 1985: 149–150, 1988: 172–173) is forced to conclude that polyploids (whether autopolyploids or allopolyploids) are not examples of instantaneous speciation. Although reproductively isolated from their parental species, Paterson is forced to insist that polyploid populations are conspecific with their parental species since they still "share a common fertilization system" (Paterson 1985: 149). Consequently Paterson's species concept has absolutely nothing to do with chromosome numbers. Generally, however, others find chromosome numbers of some significance. As Ridley (1993) points out, "It is quite common for a genus, or a family, of flowering plants to be made up of species with simple multiples of a basic number of chromosomes (N): the different species might have N, $2N$, $4N$, etc. chromosomes. The obvious interpretation is that many, or all, of the species with higher numbers of chromosomes have originated by polyploidization from species with fewer chromosomes. Chromosome counts have been made for many plant groups, and it has been estimated that as many as 70–80% angiosperm species are polyploids" (43). Paterson (1988), however, claims that "Such arrays of congeneric species do not establish that polyploidy led directly to their origination as species" (173). Instead of sympatric speciation by polyploidy, Paterson theorizes that a subspecific polyploid population (subspecific in spite of being intersterile with its parental species) does not become specific in its own right until (in conformity with his view on speciation) it goes through a period of gradual speciation by allopatry and develops its own system of fertilization mechanisms.

This view, of course, is inconsistent with a view of species as individuals, as cohesive wholes, since a subpopulation genetically isolated from the rest of its species can hardly be thought of as an integrated part of that species.

Paterson's view is also inconsistent with some of his own principles. Specifically, Paterson tells us that "it is these fertilization mechanisms which determine the limit of the species' gene pool, i.e., its 'field for gene recombination' as [Hampton] Carson has aptly called it" (Paterson and Macnamara 1984: 105; cf. Paterson 1985: 144). The problem is that a new polyploid population (if sexual) constitutes not the same but instead a *new* field for gene recombination. In other words it gives us not one bigger field but two. Never mind, contrary to what Paterson (1985) points out, that between the parental and the new polyploid populations "'mating' is likely to occur at random" (149). Since Paterson no longer employs the SMRS for delimiting species, but nevertheless retains the concept of the field for gene recombination, the fact remains that what polyploidy almost instantly creates is more than one field.[16] As Coyne *et al.* (1988) put it, "If species are taken to be separate 'fields for gene recombination,' then anything preventing such recombination must be included in the species definition, whether or not such factors are 'perceived' by individual organisms" (194).[17]

There is yet a further problem with internal consistency. Both Paterson (Paterson and Macnamara 1984: 117, Paterson 1991: 206–208) and some of his followers (e.g., Lambert *et al.* 1987: 200) emphasize the importance of correct delimitation of species for biological control, especially in the case of sibling (cryptic) species. Indeed in chapter 2.4 I employed biological control, referring to the *Anopheles* example cited in chapter 1.1, as an important reason for taking species to be objectively real. Now suppose that a population arose via one of the various forms of instantaneous speciation commonly conceived (including genetic engineering), that it is far too soon for it to evolve via allopatry, but that it is nonetheless distinctive in terms of biological control.[18] From the viewpoint of bio-

---

[16] Given the widely accepted importance of polyploidy as a mode of speciation and Paterson's heterodox view, it is remarkable how silent his followers have been on this subject. Characteristic is a recent volume devoted entirely to Paterson's species concept and its implications for speciation theory. In that volume the only mention of speciation by polyploidy is by Elisabeth Vrba (1995), who says "A category that is still cited by many as valid speciation via postfertilization isolation is speciation by polyploidy, particularly in plants. But, in my view, Paterson (especially 1981: 117–118) has convincingly countered these arguments as well" (24). Clearly this assessment will not do.

[17] In reply to this passage, Masters and Spencer (1989) surmise that it is possible that Coyne *et al.* (1988) "also include geographical isolation in their species concept," which they reject as "an 'anything goes' approach" (274). In fairness to Coyne, Orr, and Futuyma, however, I do not think they meant to include geographical isolation as a mode of instantaneous speciation, and thus as a prezygotic isolating mechanism (cf. Futuyma 1986: 112–114, Coyne and Orr 1989: 80); instead their isolated sloppiness should be overlooked in accordance with the principle of charity.

[18] In a detailed review of a considerable amount of literature on polyploidy, Levin (1983) concluded that "The biological, physiological, and developmental changes which incidentally accompany chromosome doubling may immediately adapt polyploids to conditions

logical control such a population would be a "good" species, though not from the viewpoint of fertilization mechanisms.

Coyne *et al*. (1988) provide a further conceptual difficulty: "If a newly arisen allotetraploid drives its ancestors extinct, it attains species status only at the moment of extinction, leading to the absurd result that the final steps in speciation involve no additional changes in populations" (193). Masters and Spencer (1989) reply that "we are puzzled by the claim made by Coyne *et al*. that, under the RC [the recognition concept of Paterson], a newly arisen allotetraploid attains specific status if it drives its ancestors extinct. It seems to us that this is the equivalent of fixation of a new karyotype *within a species*, and no more results in speciation than the fixation of any other new mutation" (276). But this is to miss the point. The point of Coyne *et al*. is a *reductio ad absurdum* directed against the recognition concept. The point is that at time t_1, at which time tetraploid population C exists along with its parental populations A and B, C is merely a subspecies or chromosome race of species A-B (or possibly A and B?). However, by time t_2, at which time C has driven A and B into extinction, C attains its own species status without any intrinsic change whatsoever. It is the extreme oddness of this conclusion, forced upon us by the recognition concept, that constitutes a serious difficulty for that concept. We should not expect a group of organisms to go from being a subspecies to a species without some sort of intrinsic change. But that is what the recognition concept forces us to conclude. Indeed, as we shall see later in this chapter, somewhat of the same sort of *reductio* applies to the strict cladistic species concept as well.

In spite of the above consequence, however, Paterson's view on speciation is inextricably linked to natural selection. Indeed, part of the reason that Paterson and his followers are unwilling to accept instantaneous speciation by polyploidy is because such reputed speciation has nothing to do with natural selection, let alone selection at the individual level, and so would be unDarwinian (cf. Paterson 1982b: 97). As Paterson (Paterson and Macnamara 1984) put it,

> Although I believe that the pleiotropy which Mayr invokes may play an important role in speciation, just as stochastic events may, I also believe that the most common event in the process involves the "shaping," by natural selection, of characters of the fertilization system to the conditions of a new environment. Even characters that are affected by pleiotropy or drift will ultimately have to

---

other than those to which their diploid progenitors are adapted. In a sense, chromosome doubling may propel a population into a different resource or habitat space. . . . Polyploidy not only alters the adaptive gestalt of populations, it promotes a series of genetic and chromosomal changes which compound the differences between the polyploid population and its progenitors. . . . Polyploidy may evoke large, discontinuous effects, which in turn may lead to abrupt, transgressive and manifest shifts in the adaptive gestalt of populations which selection might accomplish slowly or not at all" (15–16). These conclusions and their implications for biological control are irrespective, of course, of Levin's use of the words "adapt" and "adaptive," which for most modern biologists apply in the biological world only to products of natural selection.

survive testing by natural selection. [116]

Implicit in this passage is Paterson's opposition to what is called *speciation by reinforcement*, the view that isolation mechanisms are evolved directly by natural selection. Indeed it is one of the *leitmotifs* in Paterson's writings that "the post-mating 'mechanisms' (sterility, etc.) cannot have been selected to serve the function of isolating" (Paterson and Macnamara 1984: 110), that Wallace and Dobzhansky's belief to the contrary is "teleological" (Paterson 1982b: 100), and that "As with Darwin's views, the recognition concept of species is entirely free of any commitment to teleology" (101).

While I fully agree that speciation should not be viewed teleologically, for this is to be "free of the supernatural" (Paterson 1982b: 98), and that species following Darwin should be viewed as effects (cf. chapter 5.2), there seems to me, nevertheless, serious difficulties with the critical view of Paterson. For a start, as Mayr (1988c) put it, the teleology accusation "is based on a failure to make a distinction between the selection forces that resulted in the origin of isolating mechanisms and the selection forces that favor the persistence of isolation, these being two entirely different kinds of selection forces" (432). Paterson (e.g., 1982a), however, insists that the term "'mechanism' should be reserved for characters functioning in a role for which they were selected" (80). In this he explicitly acknowledges his debt to G.C. Williams (1966), who reserves the term "mechanism" for "Whenever I believe that an effect is produced as the function of an adaptation perfected by natural selection to serve that function" (9). Indeed, much of Williams' book is permeated with the concept of selection *for* (cf., e.g., 6 and 160). At any rate, I fail to see why any subset whatsoever of causal processes should be taken to the exclusion of others for determining the status of something as a biological mechanism. Rather, I think a putative mechanism should be judged solely according to its *effects*, by what it does; its cause should be irrelevant. To think otherwise is much like insisting that to be a certain kind of organ, say, eyes, is dependent on having a certain monophyletic origin, that eyes by any other origin are not really eyes.

The same logic extends to the insistence that speciation without selection is unDarwinian. Indeed if this were true, then Darwin himself was unDarwinian, for Darwin (though for the wrong reasons), unlike for example Weismann (who we know was also wrong), never put his entire stock into selection as the one and only *vera causa* of evolution (cf. Darwin 1859: 6, 29, 132, 134–143, 167–168, 196, 209, 472), which logically leaves no reason for why every species must necessarily be the result of natural selection (though of course for Darwin it must still have natural causes). Moreover, on the issue of speciation by saltation, in spite of his insistence in the *Origin* that evolution must be gradual (cf. chapter 2.3), Darwin never persisted in thinking that all speciation must be gradual. This is evident in his replies to the botanist William Harvey. After the *Origin* came out, Harvey argued in an article that in the Royal Botanic Gardens at Kew a new species of *Begonia* had arisen via saltation. In his correspondence (1860) Darwin replied to Hooker (February 20) that "Harvey's is a good hit" but that "It would take a good

deal more evidence to make me admit that forms have often changed by saltum."
To Lyell (February 18–19) he replied that "Harvey does not see that if only a few
(as he supposes) of the seedlings inherited his monstrosity natural selection
would be necessary to select & preserve them" and that (February 23) "On the
whole I still feel excessively doubtful whether such abrupt changes have more
than very rarely taken place." And to Harvey himself (September 20–24) he
wrote: "About sudden jumps; I have no objection to them: they would aid me in
some cases: all I can say is, that when I went into the subject, & found no evidence to make me believe in jumps, & a good deal pointing in the other direction" (Burkhardt *et al.* 1993: 97, 93, 102, 373; cf. Darwin 1859: 460).

But there is a much deeper problem with Paterson's view on speciation and
selection. Eldredge (1985b), for example, finds Paterson's view attractive because "Paterson's perspective makes us see the isolation between species as an
accidental byproduct; selection is *for* mate recognition, not reproductive isolation" (156, italics mine; cf. 92). As stated earlier, I fully agree with Paterson,
following Darwin, that speciation is an "incidental *effect*," that neither species nor
the sterility between species is selected *for*, and that species (though I would
qualify this as *many* species) arise as "incidental consequences of adaptive evolution" (Paterson 1985: 149). The problem, however, as I see it, is that Paterson's
acceptance of selection for fertilization mechanisms, what he often refers to as
"directional selection" in the case of allopatric isolation (e.g., Paterson 1982b:
101, 1985: 147), is equally teleological and unacceptable.

The problem is part of a larger one, namely the concept of "selection for" in
general, even as it is found in Sober's (1984a) celebrated analysis (cf. Mayr
1988a: 116–117, 123–124). Sober distinguishes between "selection of," which
describes "the *effects* of a selection process," and "selection for," which "describes its *causes*. To say that there is selection for a given property means that
having that property *causes* success in survival and reproduction." "'Selection
for,'" he continues, "is the causal concept *par excellence*" (100). What troubles
me is that "selection for" is usually not the causal concept *par excellence*. For example, in the classic case of industrial melanism, the case of the peppered moth
Biston betularia, the real causal activity is by the predators and is selection
against the lighter phenotypic trait in the moths, which following Richard Dawkins may be called "grim reaper selection"—according to Dawkins (1986), "In
nature, the usual selecting agent is direct, stark, and simple. It is the grim reaper"
(62). The resulting darker form of moths, then, is a consequence *of* grim reaper
selection, it is not what is selected *for*. On Sober's view, however, he would say
that the darker phenotypic trait is what is selected *for*, and that any other traits
that (as in pleiotropy) go with it (dark blue eyes, postmating isolating mechanisms, etc.) would be a consequence *of* this. The darker phenotypic trait, of course,
plays an obvious role in the overall process of natural selection leading to the
darker form of moths, and one may call it a causal role if one likes, but it is no
more than mere *sine qua non* causality, since the darker phenotypic trait of the
moths is no more a cause of their improved success in survival and reproduction
than the color of the trees on which they rest or the imperfect eyesight of the birds

that eat them. To suggest otherwise is to give a thoroughly unwarranted teleological connotation to natural selection. (As per the title of Dawkins' book, the Watchmaker is, after all, *blind.*)

Moreover, the phrase "selection for" implies that natural selection is some kind of force, admittedly a view that many biologists and philosophers subscribe to today (e.g., Sober and Lewontin 1982: 174, 178–179). But as Robert Haynes (1987) aptly remarked, "selection is not a force of nature analogous to gravity or electromagnetism. Rather it arises as a consequence of the pre-existence of hereditary variation and reproductive excess within populations" (5). And indeed when pressed on this point, Darwin himself took the same view. In a letter to Asa Gray (November 29, 1857), Darwin wrote "I had not thought of your objection of my using the term 'natural Selection' as an agent; I use it much as a geologist does the word Denudation, for an agent, expressing the result of several combined actions" (Burkhardt and Smith 1990: 492; cf. Darwin 1859: 489–490).

In short, natural selection is nothing but a process of elimination, a process of blind nonrandom death, a differential reproductive "sieve" (cf. De Vries 1912: 6; Dawkins 1986: 44–45, 288)—the *effect* concept *par excellence.* Nevertheless, there may still be a meaningful sense of selection *for.* The only real selection *for,* it seems to me, is in artificial selection (breeding) and in sexual selection (female preference), though even both of these are usually blind (cf. Darwin 1859: 34–38, 87–90). Paterson, however, cannot appeal to either of these (even if not blind), since he explicitly distinguishes the recognition concept of species from mate recognition following pairing (Paterson 1985: 144) and disassociates the SMRS from sexual selection (Paterson 1982a: 79). At any rate, since almost all selection in the strict sense of cause is really selection *against,* so that most of what is left is selection *of,* there remains no reason to insist on the dependence of speciation on selection. As effects, species can arise as a result of a number of different causal processes, as per Darwin's view.

Although Paterson's species concept cannot properly figure as a precursor of the species-as-individuals view, it was important to examine it in detail at the end of this section, not only because of its close relation yet rising challenge to the biological species concept of Dobzhansky and Mayr, but also because it is closer, at least *prima facie,* to the view espoused by Ghiselin. It is thus to an examination of Ghiselin's view that we now turn.

4.2 Ghiselin, Hull, et al.

According to Endler (1989) "the *recognition species concept* was derived independently by Ghiselin (1974) and Paterson" (629). I do not think, however, that the two authors would quite agree. For a start, since, as we saw earlier, Paterson's species concept is not about mate selecting mechanisms, his species concept is therefore not about reproductive competition, so I doubt that he would think of species (contra Ghiselin, as we shall see below) in economic terms. Indeed on the related concept of empty niches, Paterson (1989) thinks it stems from what he calls "inappropriate analogies with human economic systems" (189). Moreover, Paterson (e.g., 1986) thinks of a species as having "species-specific characters such as the fertilization system" (161), a view (species-specific characters) that we shall see is totally antithetical to the species-as-individuals view of Ghiselin. Interestingly, although Ghiselin (1988), in reply to Lambert *et al.* (1987), says that he "can only applaud the effort of Paterson and his followers to shift emphasis toward the study of what holds species together rather than what keeps them apart" (66), he nevertheless also thinks that "if all that is meant is that the species is a class of individuals each of which is integrated by SMRS, then we get a version of the biological species concept which picks out the same fundamental units in nature that all the other versions do" (67; cf. Ghiselin 1997: 96–97, 108–110).

Indeed, as for reproductive isolation defining species as individuals, Ghiselin (1974) believes that such a conception is "adequate" (281). Nevertheless he believes it can be improved on. Accordingly, he provides what he would later (Ghiselin 1997: 110) call the *bioeconomic species concept:*

> Species ... are *the most extensive units in the natural economy such that reproductive competition occurs among their parts.* [281]

By "reproductive competition" Ghiselin means "competition with respect to genetical resources as such, not just any competition involving reproduction" (281), so that his view is not confined to sexual selection alone. Indeed, the emphasis on competition for genetical resources suggests a strong analogy with competition for resources in economics, which it turns out is precisely how Ghiselin prefers to think of species: "Species are to evolutionary theory as firms are to economic theory" (281). Indeed Ghiselin believes that "a host of biological problems are more readily soluble if one learns to think like an economist" (289).

At any rate, one important consequence of Ghiselin's definition is that species are not to be thought of as *relational.* "'Species' is not a relational concept like 'brother,' but something comparable to 'man.' . . . An only child cannot be a brother, but an economy might contain a single firm" (Ghiselin 1974: 282).[19] Interestingly, this approach also avoids, as Ghiselin points out, the problem of "potential interbreeding." "If a species is an individual," he says, "it hardly matters whether it is interbreeding at any given instant. All the 'members' of a species are competing reproductively with all the others, irrespective of the distance between them or the existence of temporary spatial discontinuities between their component populations" (282). Thus, conspecifics "can compete at a distance" (282).

And yet if species are to be cohesive entities, reproductive competition between their parts seems an odd form of cohesion. As pointed out earlier (chapter

---

[19]   In arguing thus, Ghiselin (1987b) explicitly rejects what Mayr calls the nondimensional meaning of species (cf. Mayr 1987a: 166n1 and chapter 2.3n15 with accompanying text): "Emphasis upon what holds species together helps us to avoid fallacious talk of the sort that Mayr keeps generating about 'nondimensional species'" (210). Interestingly, Hull (1987), on the other hand, sides with Mayr: "The nondimensional species concept is important not just because it provides our epistemological entrée into the living world, but more importantly because species interact with their environments and other species only in the specious present" (180).

3.4), according to Andersson (1994: 46–47, 248–249) competition in the form of sexual selection can not only lead to speciation but it can also lead to extinction. Ghiselin (1989), nevertheless, emphasizes that "sex integrates species. Were they not integrated they could not do anything" (57). Elsewhere (Ghiselin 1987b) he dismisses Mayr's relational or nondimensional species concept as insufficiently stressing cohesion (cf. n19 above). And even earlier (Ghiselin 1974) he makes the claim that "by shifting the emphasis from integration and gene flow to the mode of competition," criticisms such as those of Ehrlich and Raven (1969; cf. chapter 3.4 and chapter 4.1n12) are thereby "circumvented" (283). But there is a serious difficulty with this, namely that *reproductive competition in itself provides no cohesion whatsoever unless it leads to gene flow.* So now we are back to Mayr and all the problems with the biological species concept (cf. chapters 2.4, 3.4, and 4.2/3). Perhaps this is why Ghiselin (1987a) actually defends the biological species concept against detractors (138) and tells us that "one might equally well formulate" (137) the foundation for species individuality in terms of reproductive competition *à la* Ghiselin or isolation *à la* Mayr. Either way Ghiselin has a serious problem. Like Mayr, he wants to hold that species "have to be absolutely concrete" (Ghiselin 1987a: 130). But neither the concepts of reproductive isolation nor reproductive competition, as I think I have shown, provide a solid ground for this desire.

By "absolutely concrete," of course, Ghiselin means no more than that species are in some sense physical entities rather than abstract entities. As such, Ghiselin (1987b) is willing to concede that species are divisible. "Of course," he says, "I don't believe that individuals are indivisible, any more than I believe that atoms are indivisible" (208), all the while affirming that his use of the word "individual" for species is literal and not metaphorical (209). Hull's terminology, then, seems to me preferable to Ghiselin's. On Hull's (1978) terminology, species as individuals means "spatiotemporally localized cohesive and continuous entities (historical entities)" (294). Indeed for Hull (1981a) "Reference to gene transmission results in species which are spatiotemporally restricted. . . . The relevant 'identity' is identity through descent" (144)—which places him closer to Mayr than to Ghiselin, along with the attendant problems. Indeed Hull (1980) goes so far as to say "As long as the constituent populations exchange an occasional organism, such species can be considered a single, integrated individual" (324)—which takes him much further than Mayr, for Mayr (and most others) would not want to consider two species with a narrow hybrid zone between them a single individual species! At any rate, the point that both Ghiselin and Hull fail to fully appreciate is that (as we saw in examining Huxley's view in the previous section) when with any organism the whole is divided in half the whole loses its individuality (irrespective of whether the halves live on as new wholes or die off), whereas the same is not also true for species (differences in time scale notwithstanding).

The issue of cohesion, of course, as we saw in the previous section, was one of the reasons, apparently the main reason, why Mayr himself fell short of ascribing individuality *in a biological sense* to species. Hull's reply here is interesting and important. Earlier in the debate, Hull (1981a), in defending the view that species are mereological wholes, tells us that "I would think that they ['wholes' and 'highly organized entities'] were one and the same thing. At least that is how I have treated them in my writings" (143). Later in the debate, however, Hull (1987) tells us that "Spatiotemporally organized entities can be arrayed along a continuum from the most highly organized to the most diffuse. Organisms tend to cluster near the well-organized end of the continuum, while species tend to cluster near the less well-organized end." Thus, explicitly for Hull, the difference between species and organisms is "one of degree, not kind" (172).

With Ghiselin, however, the difference in their individuality is indeed one of kind rather than degree. For one thing, Ghiselin (1987b) tells us that "probably organisms become adapted but species do not, except in so far as they consist of adapted organisms" (288).[20]

This raises the question of what it is that species are supposed to do. Both Ghiselin and Hull tell us that their view on the ontology of species is not based willy nilly on evidence but that it is based on a theoretical point of view, namely evolutionary theory. As Hull (1977) put it, "If species are to be viewed as individuals, it will be because current evolutionary theory necessitates such a conceptualization" (80).[21] Elsewhere (Hull 1978) he strengthens the claim and says

---

[20]   Cf. G.C. Williams (1966): "The species is . . . a key taxonomic and evolutionary concept but has no special significance for the study of adaptation. It is not an adapted unit and there are no mechanisms that function for the survival of the species" (252).

[21]   Mary Williams argues for species as individuals along the same line. Assuming that it is gene flow that makes a group of organisms cohesive with respect to natural selection, Williams (1985) argues that "species are individuals with respect to ET (evolutionary theory) in the sense that the laws of the theory deal with species as irreducible wholes rather than as sets of organisms" (589). In reply to Williams (1985), Ereshefsky (1988a) employs Mayr's own misgivings about gene flow (as well as the criticisms of others such as Ehrlich and Raven 1969), further employs Mayr's concept of the unity of the genotype (to be discussed in the following section), and asks "Do we want to say that a species made up of reproductively isolated subpopulations kept fairly similar by having similar homeostatic genotypes is cohesive and thus an individual? I am skeptical. ... the uniformity caused by such genotypes is merely the additive result of those genotypes working independently and not the result of any interactive process between the isolated subpopulations of the species" (430). He thus concludes that "the general question of what it takes to be an individual is not settled by any particular scientific theory—it is a more general metaphysical issue" (432). Of equal interest, Mayr (1987b), in reply to Williams (1987), appeals to the priority of the nondimensional situation for species reality and makes the claim that "I am sure it is possible to demonstrate the reality of species as aspects of nature, without invoking the theory of evolution" (220). I am strongly inclined to agree with both Ereshefsky and Mayr on this issue. On the other hand, Rosenberg (1987) argues for generally the same conclusion but in an entirely different vein. According to Rosenberg, "The theory of evolution does not require that species be kinds or that they be individuals. It can provide no direct and independent motivation for the question, nor by itself an answer to it" (194). On Rosenberg's view, "There is nothing in this description of species, as kinds into which lineages enter and out of which they exit, that makes the slightest difficulty for the theory of evolution. At most we will have to say that the evolution of species is shorthand for the

"The point I wish to argue is that genes, organisms, *and* species, as they function in the evolutionary process, are necessarily spatiotemporally localized individuals. They could not perform the functions which they perform if they were not" (294). Similarly Ghiselin (1987a) tells us that "My position here is akin to the virtual truism that 'species' is a theoretical term" (134), that scientists "redefine our terms as knowledge advances" (135), and that "If species were not individuals, they could not evolve. Indeed, they could not do anything whatsoever" (129). But Ghiselin wants to say more. Basing the physical individuality of species on evolutionary theory alone may seem to invite species pluralism, since evolutionary biology is surely neither the be-all nor end-all of modern biology. Thus, Ghiselin (1987b) tells us that "I hope nobody wants to have us believe that species are individuals from the point of view of evolutionary theory, but not otherwise. Species function as individuals in evolutionary theory, but the real consideration is that they function as individuals in the processes that go on in nature" (208). Thus science, for Ghiselin, although it is a theoretical activity, it is also fundamentally an objective entity, and those who think otherwise are basically "introverts" (Ghiselin 1987a: 131).

Given such a conception, what, then (to repeat my question), are species supposed to do? According to Ghiselin (1987a), "species do very few things, . . . They speciate, they evolve, they provide their component organisms with genetical resources, and they become extinct. They compete, but probably competition between organisms of the same and different species is more important than competition between one species and another species. Otherwise, they do very little" (141). Similarly Hull (1981a) tells us that "Species are the sorts of things which evolve, split, bud off new species, go extinct, etc." (146).

But do species *compete*? In drawing an explicit analogy between economics

---

evolution of lineages through species" (194). On this view, then, when biologists speak of species as evolving they are speaking loosely, as the vulgar would and not the learned. Although Rosenberg does not himself (as we shall see) subscribe to this view, he nevertheless thinks that it is consistent with evolutionary theory. In the least I think most biologists would disagree with Rosenberg and tell us that when they speak of species as evolving they are speaking strictly and not loosely. At any rate, Rosenberg's theory is one that I reject for much the same reason that I rejected Caplan's thesis (cf. chapter 3.3). If the background assumption for species is that of extra-mental realism, then it is to make out of species independent, Platonic entities, whereas on my view abstractions cannot exist independently but must depend on the physical world. But this may not satisfy. To further the argument, then, it seems to me that there cannot be a periodic table of species analogous to the periodic table of elements simply because the latter are discrete, whereas this can make no sense for species. Species as defining kinds would have to be defined in terms of both genotypes and environments, in order to allow for the actual existence of members of each kind. (Kinds that cannot possibly have members are not worth arguing about.) But the totality of possibles here, based on possible genotypes with possible environments, entails a smeary periodic table of species, and thus not a periodic table at all. There is nothing to be gained, then, in supposing or employing such entities, and Rosenberg's claim that evolutionary theory is consistent with such entities must therefore be dismissed.

and corporations on the one hand and biology and species on the other, Ghiselin not only wants to allow for competition within a species; he also wants to allow for competition *between* species, with *an individual species itself* acting as a competitor. As Ghiselin (1974) put it, "species as well as organisms can compete," all the while allowing for "a profound difference" (282) between the two types of competition. I suppose one profound difference is that a species does not have anything like *teeth* with which to *bite* another species. Ghiselin does not get into this, but he does tell us that one profound difference between the *kind* of individual that organisms are and the *kind* of individual that species are is that "probably organisms become adapted but species do not, except in so far as they consist of adapted organisms" (288). How odd it is, however, that an entity could compete *as an individual entity* without being adapted. Competitors would seem to necessarily require adaptations in order to have *competitive advantages*. Ghiselin's view would therefore seem to be incoherent.

Interestingly, Eldredge (1985b) tells us that, unlike Ghiselin, Hull's (1980) view is that "species simply are not interactors" (158). And certainly when we turn to Hull's famous essay (1980), we find him claiming that

> In sum, entities function as interactors at higher levels of organization than those at which replication occurs, at least at the level of colonies, possibly at the level of populations, but probably at no higher levels. . . . Replicators and interactors are the entities that function *in* the evolutionary process. Other entities evolve *as a result* of this process, entities commonly termed species. [327]

Thus species, for Hull, are neither replicators nor interactors, and are individuals only in the *pattern* of evolution, not in the *process* of evolution. Oddly, however, Hull (1987) later either slips or gives evidence of a major change of view when he states that "species interact with their environments and other species only in the specious present" (180; cf. the title of that paper). But now we have the problem of incoherence again. How can species compete, or more widely interact, if Ghiselin is right and they do not have adaptations? And if Hull (1980) is correct, how can pattern entities interact? To interact must they not instead be process entities? Suffice it to say that there is much incoherence here that remains to be addressed by the principals in question. Saying that "Species are segments of the phylogenetic tree" (Hull 1978: 305; cf. Hull 1981a: 145, 1988: 79) or "chunks with[in] the genealogical nexus of life" (Beatty 1985: 278)[22] only adds to the confusion, since such temporal chunks or segments are pattern entities, not process entities, and therefore cannot *do* anything in the meaningful sense of doing.

All of this confusion results, I suggest, from a failure to perceive the logical

---

[22] Similarly Sober (1984a): "Species, from this perspective, are chunks of the genealogical nexus, not natural kinds" (165). Again: "species are individuals—i.e., spatiotemporally localized chunks of the phylogenetic nexus" (367n49). Similarly, Eldredge (1985b): "Species are profoundly real in a genealogical sense, arising as they do as a straightforward effect of sexual reproduction. Yet they play no direct, special role in the economy of nature" (160). As for Ghiselin, cf. n23 below.

and therefore ontological priority of horizontal species concepts over vertical ones (cf. chapter 2.4).[23] Species, like languages, can be real horizontally without any vertical reality. Wanting to have species equally both ways, as I think we see above, generates an enormous amount of confusion and incoherence for those especially who want to conceive of species as individuals. At the horizontal level they want them to be as real as individual process entities, but when they add the vertical dimension they find themselves forced to concede them no more reality than pattern entities, and thus the confusion and incoherence. They want species to be equally real in two very different ways, but then they find they don't get what they want and the problem is left ignored.

The whole issue becomes only more confused when it is connected with the controversy over species selection, the view that an individual species taxon can be a unit of natural selection. If species are to be individuals in a biologically meaningful sense, one would naturally assume that this view entails species selection. But oddly (Hull 1988: 219 calls it an irony), the defenders of the former view normally decouple that view from the latter view. Hull (1980) certainly does (similarly Ghiselin 1974: 288; Gould 1982: 100; Sober 1984a: 367n49), for he defines natural selection in terms of replicators and interactors (318), and as we saw from the indented quotation above, species on his view are neither. Indeed elsewhere (Hull 1978) he tells us that "Even if entire species are not sufficiently well integrated to function as units of selection, they are the entities which evolve as a result of selection at lower levels" (299).

Michael Ruse (1987), on the other hand, agreeing that "A species is not adapted" (351), argues that the species-as-individuals view logically entails species selection and that this counts against it. "First and foremost," he says, "thinking of a species as an integrated individual goes flatly in the face of the way in which the major evolutionary mechanism of natural selection is generally regarded today" (350), adding that "majority opinion today is that . . . selection works chiefly if not exclusively at least at the level of the individual organism. . . . Any species effects are just epiphenomena on individual effects, or at most, on population effects" (350–351).

---

[23] It seems to me that Ghiselin's species concept, unlike other advocates of the species-as-individuals view, entails a primarily *horizontal* dimension for species. According to Ghiselin (1987a), "Individual species are populations. Individual higher taxa are branches of the phylogenetic tree—chunks of a genealogical nexus" (129). Cf. Ghiselin (1974): "a phylogenetic nexus is the inevitable consequence of the laws governing reproductive communities" (285). Indeed, although Ghiselin (1987a) tells us that "Individuals each have a definite location in space and time" (128; cf. Ghiselin 1997: 302), his 1974 bioeconomic definition of species would seem to logically preclude a conception of species as necessarily vertical entities. This is because even if we should grant that members of a species can "compete at a distance" (282), it makes no sense to say they can *compete over time* (at least not any appreciable amount of time). I, for example, do not and cannot compete for genetical resources with people, say, a hundred years ago. And yet I and they are obviously conspecific. But now the obvious consequence is that Ghiselin's 1974 species definition entails a much too horizontal conception of species.

I am inclined, however, to disagree with Ruse. Many biologists do not accept George Williams and Richard Dawkins' view that genes are units of selection. Nevertheless, they accept that genes, defined molecularly, are physical individuals in a biologically meaningful sense, such that they can produce effects, have their structure passed on in reproduction, etc. So if we are to appeal to biologists, then, not being a unit of selection does not necessarily mean that an entity should not be thought of as a physical individual (mereological whole) in a biologically meaningful sense.

In fact, it seems to me that in some ways species may be said to have properties that are not to be found at the level of their member organisms. An MVP seems to me a good example, since an individual organism does not have anything like it except only analogously. And yet I would not want to say that such a property is an adaptation, since all species have pretty much the same MVP (cf. §1n2 and accompanying text). Adaptations must be something more, namely what the process of selection works on, so that for a property to count as an adaptation it must vary from competitor to competitor, or some competitors must have it and some not. However, there would seem to be other species-level properties which meet this criterion. Some species, for example, are broad-niched, more so than any of their member organisms, while others are narrow-niched, the former species-level property conferring an advantage in terms of the vicissitudes of environmental change.[24] Similarly, sexually reproducing species generally have an advantage over asexually reproducing species, in virtue of their greater genetic variability (the cost of meiosis notwithstanding), which would explain why the vast majority of species are in fact sexual. But the possibility certainly remains that these advantageous species-level properties can be fully explained by selection at much lower levels (cf. Dawkins 1986: 265–269). Certainly there is nothing like the peppered moth example to support species selection. As Antoni Hoffman (1992) put it, "A thorough examination of the alleged instances of species selection . . . shows that not even a single good example of species selection has thus far been described" (133).[25]

At any rate, it would appear that the huge and difficult controversy over species selection is a red herring when it comes to the controversy over the physical individuality of species. The latter view does not entail the former, while the former is rarely if ever used to support the latter.

---

[24]   Interestingly, Eldredge (1985a: 165–169) points out that narrow-niched species outnumber broad-niched species by more than three to one, while on the other hand the broad-niched species tend to avoid extinction longer. The former ratio is no doubt due to competition.

[25]   I would add a conceptual difficulty almost universally overlooked, namely that if species as individuals are defined as they often are, as we have seen above as vertical segments of the phylogenetic tree, then they cannot possibly be units of selection, since selection would have to wait until their segmenting is complete, by which time it would be too late for them to be selected. If there is to be species selection, then, the species selected must be horizontal, not vertical, with species entailing vertical consequences, though not being defined by them.

What, then, at bottom, are the main arguments used to support the species-as-individuals view? There would seem to be basically two, both interconnected. The first is based on what is perceived as the traditional distinction in philosophy between classes and individuals, or universals and particulars. "*Individuals* are single things," says Ghiselin (1987a), "including compound objects made up of parts . . . Individuals each have a definite location in space and time. In general they are designated by proper names—such as 'Ernst Mayr' or 'Canada.' *Classes*, on the other hand, are spatially and temporally unrestricted" (128). "The distinction between a class and an individual," he further tells us (Ghiselin 1987b), "is one of the most fundamental in all metaphysics—on a par with the distinction between true and false" (207). As Hull (1987) put it, "Although considerable disagreement exists over details, something like the difference between primary substances and secondary substances, particulars and universals, individuals and classes, etc. can be found in the writings of nearly all major Western philosophers. The terminology is often bewildering, but the general aim is not" (168).

Combined with the premise that species evolve, then, their syllogism is rather simple: Everything is either a class or an individual, classes don't evolve, species evolve, therefore species cannot be classes but must instead be individuals. There are other ways, of course, of putting this, perhaps a little more detailed and formal, but the basic argument would be the same. As Ghiselin (1987a) put it, "If species were not individuals, they could not evolve. Indeed, they could not do anything whatsoever. Classes are immutable, only their constituent individuals can change" (129). Ghiselin (1974) adds that "It is characteristic of individuals that there cannot be instances of them" and that "Equally significant is that individuals do not have intensions" (280). "The contrast," as Hull (1978) put it, "is between Mars and planets" (294).

Ghiselin (1966: 208–209) makes the interesting point that historically the objective reality of classes has been much more problematic than the reality of concrete particulars, such that many have thought of classes as only conceptually real or not even real at that. Since species have historically been conceived of as paradigm examples of classes, kinds, universals, their reality has always been much more problematic than that of organisms. Thinking of species as concrete individuals, however, Ghiselin (1974) points out, makes their reality much less problematic, since "nominalists believe that individuals are real" (287). "Thus nominalism," says Ghiselin (1966), whether one is an Occam or a Locke, "is fully consistent with the real existence of individual species" (208).

Much less problematic as well becomes the role of type specimens in the naming of species. "The realization that species are individuals," says Ghiselin (1974), "helps us to understand some of the important daily activities of taxonomists" (284). "It is not only difficult, but logically impossible," he says, "to list the attributes necessary and sufficient to define their [species] names. None such exist, and the only way to define these names is by ostensive definition" (284). As Hull (1978) points out, the established practice of naming new species by the discovery of a specimen has always been an enigma on the traditional view of species: "The puzzling aspect of the type method on the class interpretation is that the type need not be typical. In fact, it can be a monster" (307). As such, Hull points out, similarity is irrelevant. But now on the species-as-individuals view the actual practice of biologists makes sense:

> Just as a heart, kidneys, and lungs are included in the same organism because they are part of the same ontogenetic whole, parents and their progeny are included in the same species because they are part of the same genealogical nexus, no matter how much they might differ phenotypically. The part/whole relation does not require similarity. . . . A taxon has the name it has *in virtue of* the naming ceremony, not *in virtue of* any trait or traits it might have. . . . It is the same way in which people are baptized. They are named in the same way because they are the same sort of thing—historical entities. [307–308][26]

Although what Ghiselin and Hull say here on nominalism and type specimens makes much sense, the mere consistency in this with their view on the ontology of species is not a sufficient argument in itself, but granted they do not intend it so. The problem, as I see it, is with the class/individual distinction on which so much of their argument rests. It is because of it that Mayr (1987b) can agree that "species are individuals within the technical philosophical meaning" (219). Granted (as most will agree) that the species category is a class defined intensionally, the ontology of the members of that class, namely species taxa, remains an open question. Indeed, it is the many problems with ascribing individuality to species *in a biological sense* (partially illustrated thus far) that made Mayr refrain from ascribing individuality to them and that motivated him to find a third category for them besides classes and individuals (cf. Mayr 1987a: 159). Traditionally the members of classes can be many different things, ontologically speaking. They can be classes themselves, they can be concrete particulars, they can be sets or collections, they can be relations, they can be combinations of these, and they can be entities which problematically fit into none of these categories. Indeed part of the problem, of course, as I have pointed out earlier and shall point out again, is that the words "class" and "set" have meant many different things to different philosophers, and there is no common usage upon which one can draw. Moreover, as Marjorie Grene (1989b) pointed out, "Nor does 'individual' in logic, let alone in 'philosophy' in general, signify historical entity, and only historical entity. The logical muddle consequent on this misunderstanding has been horrendous" (71). But even if there were a clear consensus in all of this, arguments from logic (or metaphysics or language) to ontology have, historically, proved notoriously untrustworthy. Ghiselin himself (1987b: 207–208) basically makes the same point in his opposition to Kitcher's use of set theory for species ontology (to be discussed in §5), but in his own use of traditional metaphysics he is just as guilty as Kitcher. The problem is that the distinction between classes and individuals, universals and particulars, secondary and primary substances, rests on the more fundamental distinction between abstract and concrete objects, a dis-

---

[26] As Hull (1978: 314n5) correctly points out, this theory of naming should in no way be taken to associate Ghiselin and Hull with the Kripke-Putnam view on natural kinds.

tinction in the modern tradition of philosophy, if not the ancient, that is by no means clean. G.J. Warnock (1956), for example, writing before any such argument as that of Ghiselin and Hull, and without any reference to or apparent knowledge of the species problem, raises a protest "against the alleged dichotomy between concrete objects and abstract entities." In reference to "(1) 'gravitational field'; (2) 'the North Pole'; (3) 'the Heaviside layer'; (4) 'the Common Law'; (5) 'shadows'; (6) 'rainbows'; (7) 'the Third Republic,'" Warnock says

> None of these things could be called a Universal; none has 'instances'; some require the definite article; yet none would naturally be called 'concrete'; and it is at least uncertain which, if any, should be labelled 'particular'. What is referred to by (2) or (3) has a definite position; shadows and rainbows have dimensions; and the Third Republic had a definite duration. But shadows and rainbows, though visible, cannot be touched, heard, or smelt; the Common Law cannot be seen, and also has no position, shape, or size; the Heaviside layer can move, but cannot be seen or heard or felt to be moving. And so on. [83]

It is significant to add here that no less a philosopher than Quine (1960), himself the class realist *par excellence*, implicitly concedes Warnock's point:

> General definition of the term 'abstract,' or 'universal,' and its opposite 'concrete,' or 'particular,' need not detain us. No matter if there are things whose status under the dichotomy remains enigmatic—"abstract particulars" such as the Equator and the North Pole, for instance; for no capital will be made of the dichotomy as such. It will suffice for now to cite classes, attributes, propositions, numbers, relations, and functions as abstract objects, and physical objects as concrete objects *par excellence*, and to consider the ontological issue as it touches on such typical cases. [233]

To Warnock's list above one might add the case of any natural language, which only makes the application to the modern species problem more interesting since, as I shall briefly discuss in §4, it was the evolution of languages that served as an informative and influential analogue in the development of Darwin's views on the evolution of species. At any rate, what should be clear from all of the foregoing is that Ghiselin and Hull's argument, based as it is on the class/individual distinction, raises as many problems as it solves, and by itself leaves the ontology of species taxa an open question.

Closely related to Ghiselin and Hull's argument above is their argument from laws of nature. Ghiselin (1989) tells us that "Species are individuals, and there are no laws about individuals in any science whatsoever." He then adds that "This was perhaps Hull's most important contribution, but it has largely remained unexplored" (53). Hull (1978) tells us that "spatiotemporally unrestricted classes . . . [are] the sorts of things which function in traditionally defined laws of nature" (294). He tells us moreover that "Species may be classes, but they are not very important classes because their names function in no scientific laws. . . . as they should not if species are spatiotemporally localized individuals." "In point of fact," he continues, "no purported evolutionary laws refer to particular species. . . . Evolutionary theory refers explicitly to organisms and species, not to Hitler and *Homo sapiens*" (309–310). Instead Hull (1981a) tells us that "statements about particular species are descriptive statements, not laws of nature" (150). Similarly, Rosenberg (1987) tells us that "at the lowest level, there are no laws about *Didus ineptus* or *Cygnus olor*," and that "The simplest explanation of this fact is that species are not kinds, they are individuals" (196).

Hull (1978) draws a number of consequences which follow from the acceptance of this view. First, it explains why "evolutionists can make no specific predictions about the future of humankind *qua* humankind" (309), a view we found in Popper. Second, species membership is not a matter of degree, so that "retarded people are just as much instances of *Homo sapiens* as are their brighter congeners" (313). Third, "No species has an essence. Hence there is no such thing as human nature" (313). Fourth, Hull (1989) tells us that "Any ethical system that depends on all people being essentially the same is mistaken" (2).

There are a number of ways in which one can reply to the argument from laws of nature. To begin, I suggest a logical difficulty. The class of all laws of nature, as a prime example, cannot itself be a subject in a law of nature, on pain of infinite regress. But clearly it is not therefore a spatiotemporally restricted individual. What this shows is that even if we were to accept the class/individual distinction as clean, which I do not, not being the subject of a law of nature cannot be a sufficient condition for being a spatiotemporally restricted individual.

Another possible line of reply is to produce a counterexample in the sense of a putative law of nature that has a particular species taxon as its subject. What is surely significant is that though some of Ghiselin and Hull's detractors claim that this should be possible, not one (as far as I know) has produced a real example from a biology journal or textbook. Kitcher (1984b) resorts to a hypothetical example. Interestingly, Kitcher understands laws of nature as "prohibitions" (622). No less interesting is that Ghiselin (1987b) also thinks of laws in this way. He tells us that "Laws of nature cannot tell us more than what they prohibit" (211). This is, of course, Popper's concept of laws of nature as well as his defining characteristic for genuine science (cf. Stamos 1996b). As Popper (1934) put it,

> The negation of a strictly universal statement is always equivalent to a strictly existential statement and *vice versa*.
>
> The theories of natural science, and especially what we call natural laws, have the logical form of strictly universal statements; thus they can be expressed in the form of negations of strictly existential statements or, as we may say, in the form of *non-existence statements* (or 'there-is-not' statements).
>
> In this formulation we see that natural laws might be compared to 'proscriptions' or 'prohibitions.' [68–69]

Using modern logic notation, we can represent Popper's view as follows:

$$(x)(\phi x \supset \psi x) \equiv \sim(\exists x)(\phi x \cdot \sim \psi x)$$

It will be noticed that the right side of the biconditional more evidently expresses the form of a law of nature on Popper's view (or rather its epistemology). At any rate, the point is that at least Ghiselin (I'm not so sure about Hull) and Kitcher share a common ground on which to disagree. Now what of Kitcher's hypothetical example? It is supposed only to open up a logical possibility, for example where a law of the form "'All S are P' corresponds to situations in which absence of P would generate a new species." Employing polyploidy as a mechanism of instantaneous speciation, Kitcher says "Suppose S_2 with twenty-four chromosomes evolves from S_1 with twelve chromosomes. . . . Then I take it that it might be a law that all members of S_1 have less than twenty-four whole chromosomes, and the lawlike status of this generalization would be founded on the evolutionary fact that if members of S_1 give rise to an offspring with twenty-four chromosomes then that organism belongs to S_2" (623).

Aside from the fact that Kitcher confuses speciation by polyploidy with evolution,[27] one obvious problem with Kitcher's example is that it is not necessarily species-specific. S_1 could well stand for many different species with twelve chromosomes, so that the "law" would more properly be a law of speciation by polyploidy, or of chromosomes and their relation to reproduction, etc., rather than a species-specific law. Another problem is that some species have what are called "chromosome races," races with different chromosome numbers but which are nevertheless interfertile (cf. Mayr 1970: 224; Futuyma 1986: 236–237). But perhaps the most serious problem is that laws of nature are traditionally supposed to be ahistorical, which is to say they are not supposed to change over time. Hull (1977: 82–83) notes the odd consequence that if a particular species taxon is thought to be a proper subject of a law of nature, and if it is agreed that species evolve, then it must be concluded that laws of nature (or at least some of them) evolve as well. Thinking of species as individuals, says Hull, saves us from this highly problematic consequence.[28]

Another line of approach is represented by Mayr. As Mayr (1987b) says,

> I still insist that the question of laws has nothing to do with choosing class or individual (population) as the status of the species taxa. If I were to deny the existence of regularities in nature, I would not be a scientist. What I question is whether the term law, as defined and used in physics, is the appropriate term for the regularities found in the living world. To me it seems that the statistically calculated regularities found in living nature are not the same things as the laws of physics and therefore should not be designated by the same term: laws. [214–215]

---

[27] Cf. Mayr (1991): "Sudden new origins, however, are not evolution. The diagnostic criterion of evolutionary transformation is gradualness" (18).

[28] It would, of course, create a problem for Hull's argument if evidence should ever arise suggesting that at least some of the laws of nature (e.g., the speed of light in a vacuum) are slowly changing over time (cf. Nagel 1961: 378–380 for discussion). I do not take this possibility seriously for I am inclined to believe that abstractions do not change (cf. chapter 3.5).

Although Mayr (e.g., 1982) is well aware that many processes in the inanimate world are stochastic in nature, his concept of laws in physics is nevertheless that of "strictly deterministic" laws which "permit precise predictions" (37). This view, however, suffers from being old-fashioned and no longer tenable. As Hans Reichenbach (1951) put it, with the advent and acceptance of quantum physics "the *if-then* of classical physics was replaced by an *if-then in a certain percentage*" (175). Even the prediction that an ice cube will make the water in the glass colder, which may be expressed in universal form, is only a statistical prediction, a law of probability, since it is physically possible (albeit highly improbable) that the heat in the ice cube could temporarily warm the water (159–160). Similarly, with radioactive decay, radium^{226}, for example, has a half-life of 1,599 years (cf. Lide *et al.* 1995: 11:131), and yet no one radium atom need decay at that time. The problem now concerns the role of explanation in laws. As John Hospers (1956) pointed out, "we would still want an explanation of why 25 per cent of A's are not B's. . . . and in order to answer *this* question, we would need a non-statistical law" (99–100). But of course if the popular interpretation of quantum physics is correct and there is genuine chance at the subatomic level, then no such law or laws will ever be forthcoming. Perhaps then we should not expect such laws for life. As Robert Haynes (1987) put it, "Quantum mechanics, thermodynamics, and molecular biology collectively reveal that the laws governing the fundamental transformations of matter and life are, at bottom, statistical in character. They are the laws of aggregates and averages, based upon chance events, statistical fluctuations, and molecular accidents" (3).

And yet at the macro level determinative laws seem clearly to *emerge*. Unlike the decay of a particular atom of radium, one can always count on water to boil at 100°C at sea level or magnetic iron to lose its magnetism at 770°C (the Curie point for iron). Nevertheless, in biology no such determinative laws seem to emerge; instead the putative laws are like the statistical micro laws in physics. Ghiselin (1989) tells us that "The peculiarly biological laws are indeed approximately coextensive with those of economics" (57). But at least some of his putative good examples of genuine laws of biology are really laws about what goes on at the molecular level. Dollo's Law, for example, which "asserts that evolution in general is irreversible" (Ghiselin 1989: 64), is really a statistical law about the improbability of an incomprehensibly long series of mutations reversing in order. Even Mayr's founder principle—which according to Ghiselin (1989) states that "under ordinary conditions speciation will not occur unless the ancestral species gets broken up into populations that are separated by some extrinsic (usually geographical) barrier" (58), and which Ghiselin calls "Mayr's Law" (cf. Hull 1987: 169)—involves lower-level laws since it invokes Mayr's concept of the cohesion of the genotype for the parent population (to be discussed in the following section) and genetically stochastic processes for the founder population (cf. Mayr 1988a: 445–449).

Moreover, prohibitive generalizations, statistical propensities, and dispositions can also be ascribed to specific individuals. David Johnson (1990: 73), for example, assures us that he would never throw himself from the top of the Eiffel Tower. For myself, on the other hand, I can give no such assurance, except to say that the chance is very low though not impossible given the right circumstances. Are either of these laws of nature? In reply, Ghiselin (1989) tells us that

> Ontologically speaking, the laws of nature are regularities among what goes on in the material universe. Accordingly they are formulated as general statements —assertions about more than one thing. But 'general' is insufficient, because what we really mean is that they are about classes of individuals. . . . Laws of nature are also necessary truths, not contingent ones. By this we mean physical necessity, not logical or metaphysical necessity.
>
> So if we are to find laws of nature in biology, we must identify some regularities in the behavior of biological objects that are necessarily true of everything to which they apply, irrespective of time and place. [56–57]

Of course many of us want a criterion or criteria to discern accidents of history, such as the genetic code, from genuine laws of nature (however statistical), such as the boiling point of water or the Curie point of iron. According to Hull (1987), "Any philosophy of science that does not distinguish between statements about the coins in my pocket and Newton's Laws is sorely deficient" (174). Although he admits that "No analysis has proved totally satisfactory," he remains firm that "I do not see how anyone who wants to understand science can dismiss these distinctions as being of no consequence" (169). Of course Popper, in his 1959 Appendix to Popper (1934), thought we could only ever discover genuine laws of nature "in a *negative* way" (433), only when a putative law had been falsified. Thus we can never know when we've got one but only when we haven't got one. If both Kitcher and Ghiselin (and Hull?) truly believe that laws are fundamentally prohibitive, then they must also agree. But now that leaves open the possibility that there are species-specific laws of nature that have yet to be discovered (unless one begs the issue by presuming species to be individuals).

But there is a more serious problem. Appealing to our instincts, which laws conceived as prohibitions entails, simply will not do (and what else do counterfactuals and subjunctive conditionals do?). As Joseph Agassi (1975) put it, "we may have an intuitive idea of a natural necessity, yet without knowing what is a true instance of it. For, and this I contend is a historical fact, what looks a natural necessity one generation may look an accidental universality at the next" (239).

Trends, of course, are not laws (as Popper was wont to point out). But if we follow the trends, who can say for certain where they will go? One strong trend, instrumentalism (cf. Popper 1963: 97–100 for a short history), which holds that scientific theories are neither true nor false in correspondence but instead better or worse as tools for prediction and research, has in recent years received added life (cf. Rosenberg 1994: 129) from the so-called semantic conception of scientific theories, according to which scientific theories are really sets of abstract models which are neither true nor false in correspondence but instead more or less isomorphic with the phenomena to which they are intended to apply. As Paul Thompson (1989) describes the role of laws of nature on this conception, "laws do not describe the behavior of objects in the world; they specify the nature and behavior of an abstract [semantic] system" (71).

Such a trend, of course, would bode ill for any argument from laws of nature to the thesis that species are individuals. At any rate, from all of the foregoing I suggest along with Mayr (though for quite different reasons) that any appeal to laws of nature in favor of the ontology of species as individuals is misconceived in the extreme.

4.3 Punctuated Equilibria

According to Darwin (1859), evolution entails "an infinite number of those fine transitional forms, which on my theory assuredly have connected all the past and present species of the same group into one long and branching chain of life" (301). One of the major criticisms of Darwin's view, however, a criticism that he faced and which continues to this day, is that the fossil record fails to support this contention. Darwin called this "probably the gravest and most obvious objection which can be urged against my theory" (280). Instead of providing us with a plethora of transitional forms, the fossil record is full of gaps, and the transitional forms that have been found are comparatively rare. Darwin attributed this to "the extreme imperfection of the geological record" (280). (For a summary of his reasons for believing it is extremely imperfect, cf. 463–465.)

In 1972 Niles Eldredge and Stephen Jay Gould challenged this interpretation of the fossil record, which by then had become the received view. In so doing they began a controversy that has raged ever since. On their view (Eldredge and Gould 1972), in brief,

> The theory of allopatric (or geographic) speciation suggests a different interpretation of paleontological data. If new species arise very rapidly in small, peripherally isolated local populations, then the great expectation of insensibly graded fossil sequences is a chimera. A new species does not evolve in the area of its ancestors; it does not arise from the slow transformation of all its forebears. Many breaks in the fossil record are real.
>
> The history of life is more adequately represented by a picture of 'punctuated equilibria' than by the notion of phyletic gradualism. The history of evolution is not one of stately unfolding, but a story of homeostatic equilibria, disturbed only 'rarely' . . . by rapid and episodic events of speciation. [193]

Before I examine this view further, however, I want to establish its connection with the species-as-individuals view. Interestingly, there is no mention whatsoever or even the slightest indication in Eldredge and Gould (1972) that they thought their view implied or supported the view that species are individuals. Indeed Eldredge (1985b) tells us that "after reading Ghiselin's paper [1974] when it was first published, I could not see its relevance to evolutionary theory, and I thought the idea, to the extent that I grasped it, was trivial, hardly warranting so dramatic a description as radical" (138n3). He then immediately tells us with the gift of hindsight that he was "wrong," while elsewhere (1985a) he tells us that "the implications of this notion for evolutionary theory in general are profound"

(101). Its profundity, I suggest, lies mainly in its consistency with the theory of punctuated equilibria! If species are truly to be individuals, then they require discrete beginnings and endings, precisely the desiderata the theory of punctuated equilibria delivers. Moreover, the more subscribers to the individuality thesis, or the more it becomes a trend (as indeed it has), the more the theory of punctuated equilibria benefits. Indeed the one theory supports the other. At any rate, exactly where along the line both Eldredge and Gould made the connection with Ghiselin's thesis, Eldredge does not say.[29] Clearly it had nothing to do with the original formulation of their thesis. Later on, however, it would become a symbiont (or rather a parasite, depending on one's point of view). Gould (1982), for a starter, after praising the thesis of Ghiselin and Hull as containing "great utility and richness in implication" (109), tells us that "The theory of punctuated equilibrium allows us to individuate species in both time and space; this property (rather than the debate about evolutionary tempo) may emerge as its primary contribution to evolutionary theory" (109–110). Eldredge is even more enthusiastic. According to Eldredge (1985b), "Allochronically, . . . Mayr believes that species give up the ghost ontologically" (52). On the other hand, says Eldredge (1985a), "Species seem to us [Eldredge and Gould] very much as 'real' in time as Mayr himself saw them in space" (115). Indeed Eldredge (1985a) goes so far as to say that "punctuated equilibria puts the icing on the cake in the argument that species are real historical entities, comparable in a formal manner to individual organisms" (122).

Ghiselin and Hull, however, are not quite so symbiotic. Ghiselin (1987b) does tell us that "The efforts of Eldredge to clarify the ontological status of objects ranked at various categorical levels has been prosecuted in the same spirit that led him to question the received wisdom about why the fossil record is incomplete. Good philosophy in science is the sort that leads to important discoveries. It needs no other justification" (210). More recently, however, Ghiselin (1997) has made it clear that there is "no necessary connection" (128) between his own theory on species and the theory of punctuated equilibria. Moreover he goes on (268) to seriously question the empirical evidence for the latter. Hull (1978) tells us that "if Eldredge and Gould are right, the case for interpreting species as historical entities is even stronger" (300). However, Hull (1981a) also tells us that "Although I think that the Eldredge and Gould model of speciation looks highly promising and prefer the cladists' decision on the treatment of lineages . . . I see no incompatibility with gradualistic models and the view that species are individuals" (144). The reason for this is that he thinks that similarity is a red herring when it comes to individuating both organisms and species. Hull (1978) says "an organism can undergo massive phenotypic change while remaining the same organism" (301). What individuates it, of course, is not only spatiotemporal continuity, but its beginning and its end. Similarity is irrelevant, for otherwise identi-

---

[29] I suspect it was when they came to support the concept of species selection, which, judging from their publications, would be shortly before 1977 (cf. Gould and Eldredge 1977). Certainly the concept of species selection is not to be found in their 1972 paper, for there they explicitly tell us that "We postulate no 'new' type of selection" (221).

cal twins would be the same individual, which they clearly are not. Hull then tells us that "If species are historical entities, then the same sorts of considerations which apply in the individuation of organisms should also apply to them, and they do" (303). Thus on Hull's view the individuality of species does not at all require phenotypic stasis, so that any prevalence of stasis in the fossil record in no way adds icing to the individuality cake. Gould (1982), on the other hand, tells us that "If new species usually arose by the smooth transformation of an entire ancestral species, and then changed continuously toward a descendant form, they would lack the stability and coherence required for defining evolutionary individuals" (109). Similarly Eldredge (1985a) tells us that so-called pseudoextinction, extinction by transformation, "muddies the waters in our search for an analogy between organisms as individuals and species as individuals. . . . Species, in this light, can only be anatomical abstractions, by no means the discrete sort of affairs—the *individuals*—that organisms are" (107). So although both Eldredge and Gould connect the fate of the individuality thesis with the fate of the theory of punctuated equilibria, it is apparent that Ghiselin and Hull do not.

But perhaps Ghiselin and Hull are wrong and Eldredge and Gould are right. This possibility warrants a closer look at the latter's thesis.

Although Eldredge and Gould's theory of punctuated equilibria has varied in the telling,[30] such that some have thought it a moving target,[31] there is a clear enough common denominator that pervades their view and which consists of basically two theses: First, that the vast majority of the lifetime of species is characterized by stasis (stability), in which they undergo little if any evolution. As Eldredge (1985a) put it, "Gould and I claimed that stasis—nonchange—is the dominant evolutionary theme in the fossil record" (128). Raup (1991) tells us that "The average life span of species in the fossil record is about four million years" (108). Eldredge and Gould usually give estimates of between 5 and 10 million years, but the estimates are close enough not to matter. Second, Eldredge and Gould claim that speciation occurs in a geologically rapid period of time. According to Eldredge (1985a), "5,000 to 50,000 years became the comfortable yardstick" (121).[32] He tells us, moreover, that "most (if not all) anatomical change in evolution, adaptive though it may be, happens not throughout the bulk of a species' history, but rather at those rare events when a new reproductively isolated species buds off from the parental species" (121; cf. Gould 1982: 106). Thus, on this view, for the vast majority of the lifetime of a species it undergoes very little or no evolution, certainly not *speciation*.

It should be noted, however, that neither Eldredge nor Gould claim this for *all*

---

[30]   Cf. Hoffman (1992) who distinguishes three versions and Ruse (1992) who distinguishes three phases.

[31]   Cf. Somit and Peterson (1992: 7), Coyne and Charlesworth (1997).

[32]   Gould (1980) tells us that "The process may take hundreds, even thousands of years; . . . But a thousand years is a tiny fraction of one percent of the average duration for most fossil invertebrate species—5 to 10 million years" (184). Apparently Gould's lower estimate of the time required for speciation reflects his strong Marxist leanings.

species. Eldredge and Gould (1972) instead claim this for "most species" (207), "more the rule than the exception" (15) as Eldredge (1985a) put it. As Gould (1992) more recently put it, "The question cannot be resolved, either pro or contra, by single cases, however well defined or documented. . . . we never made (as often charged) a claim for the exclusivity of punctuated equilibrium. Our assertion has always been for a relative frequency sufficiently high that the predominant amount of morphological change accumulated in evolutionary trends must be generated in punctuational events of cladogenesis" (74).

At the very end of my Popper paper (Stamos 1996b) I made the claim that the theory of punctuated equilibria is in principle neither verifiable nor falsifiable, so that on my criterion of the demarcation between science and metaphysics (cf. chapter 3.4n47) it must be metaphysical rather than genuinely scientific. I based my claim mainly on what Eldredge and Gould say in their 1972 paper. In that paper they say "we recognize that there is little hard evidence to support either view" (207). They further tell us that "the data of paleontology cannot decide which picture is more adequate" (208). They then tell us that the chief merit of their theory is that "it is more in accord with the process of speciation as understood by modern evolutionists" (208). This latter statement seemed to me a red herring. If the processes of natural history do not leave behind enough evidence to decide the issue, then in principle the issue is empirically undecidable and therefore metaphysical. One may still read the rocks, but how one reads the rocks will depend upon one's prior theories, so that one will never know for sure whether one is reading the rocks correctly or not.

To this conclusion (here elaborated a little further) an anonymous reviewer for *Biology & Philosophy* (I believe it was Ghiselin) replied that my claim "will amuse, amaze, and enrage a lot of people." Indeed it seems to me that a closer inspection of the view of Eldredge and Gould is warranted, if only to amuse, amaze, and enrage even further. What should come out of this is that it must forever remain hopelessly impossible to determine whether the greater proportion of evolutionary history proceeded by punctuated equilibria.

In creating the theory of punctuated equilibria, Eldredge and Gould employed two interrelated theories put forward earlier by Mayr. Although Eldredge and Gould (1972) persistently refer to the model of *allopatric* speciation as their paradigm model, it is clear that the model of geographic speciation that they employ is not the so-called dumbbell model but rather the budding model as developed by Mayr (1954), what he later dubbed *peripatric* speciation. As Mayr (1988a) briefly describes this model, "The major novelty of my theory was its claim that the most rapid evolutionary change does not occur in widespread, populous species, as claimed by most geneticists [e.g., Fisher and Wright; cf. Mayr 1988a: 480], but in small founder populations. . . . Living in an entirely different physical as well as biotic environment, such a population would have unique opportunities to enter new niches and to select novel adaptive pathways. My conclusion was that a drastic reorganization of the gene pool is far more easily accomplished in a small founder population than in any other kind of population" (461). Interestingly Darwin, in a letter to Lyell (February 18, 1860), expressed his belief that evolution would generally be slower in smaller populations, since "where there are few individuals, variation almost must be slower" (Burkhardt *et al.* 1993: 94; cf. Darwin 1859: 105). At any rate, the important point is that Mayr's claim is only that evolution generally occurs faster in small founder populations rather than large populations; it is not that evolution occurs only or mostly in the former. Indeed for Mayr peripatric speciation is not the only form of speciation (cf. Mayr 1988a: 444), since he also allows for speciation by splitting (Mayr 1988a: 366) as well as speciation by anagenesis (phyletic transformation). In a classic passage on the latter, Mayr (1970) says "An isolated population on an island, for instance, might change in the course of time from species *a* through *b* and *c* into species *d* without ever splitting into several species. In the end there will be only one species on the island, just as at the beginning" (248). Mayr also allows for instantaneous speciation by polyploidy (Mayr 1988a: 262, 375). Thus Mayr's position on speciation is largely pluralistic.

At any rate, according to Eldredge and Gould (1972) one consequence of Mayr's model of peripatric speciation is that "new fossil species do not originate in the place where their ancestors lived" (203). Another is that "Many breaks in the fossil record are real; they express the way in which evolution occurs, not the fragments of an imperfect record" (205). A third important consequence is that if a peripheral isolate does not meet the fate of most peripheral isolates (i.e., extinction) but goes on to speciate and flourish, then it should suddenly appear in the fossil record, perhaps even sympatric with its ancestor species. As Gould (1980) put it, "In any local area inhabited by ancestors, a descendant species should appear suddenly by migration from the peripheral region in which it evolved. In the peripheral region itself, we might find direct evidence of speciation, but such good fortune would be rare indeed because the event occurs so rapidly in such a small population. Thus, the fossil record is a faithful rendering of what evolutionary theory predicts, not a pitiful vestige of a once beautiful tale" (184).[33]

Undoubtedly the most novel and controversial part of Eldredge and Gould's theory is their claim about stasis.[34] The controversy, of course, is not over wheth-

---

[33] I agree with Mayr (1988a) in that "The incompleteness of the fossil record is thus as much part of the argument of the punctuationists as it is of the gradualists" (474).

[34] Or at least as far as I am concerned in the present section. Eldredge and Gould have made other equally controversial claims, namely in characterizing Darwin as a strict phyletic gradualist, in claiming that the theory of punctuated equilibria constitutes not an extension of but rather a replacement of the Modern Synthesis, and in employing species selection to account for at least some of the patterns of punctuated equilibria. The second and third of these do not concern me here. But allow me to comment on Darwin. Although Mayr thinks Darwin was a pluralist as regards rates of evolution (cf. Mayr 1988a: 205), an assessment with which I am in complete agreement, my *agape* for Mayr does not prevent me from stating flatly that throughout his writings Mayr propounds a particular claim about Darwin which I believe should be regarded as a myth. It is the influence of this myth, I believe, that largely contributed to Eldredge and Gould's characterization of Darwin as a strict phyletic gradualist. According to Mayr, although in his early evolutionary writings Darwin adhered to both the biological species concept and geographical specia-

er the fossil record gives the *appearance* of stasis. As Futuyma (1986) put it, "Except in unusually complete fossil sequences, the impression is one of apparent stasis for long periods, 'punctuated' by periods of very rapid shift to a new stable morphology" (401). The question, rather, is whether the appearances are for the most part misleading or not. If it is agreed they are not misleading, a further question arises concerning their cause. Eldredge and Gould (1972) explicitly reject gene flow, accepting the claim of critics that "in most cases, gene flow is simply too restricted to exert a homogenizing influence and prevent differentiation" (221). But neither do they accept any appeal to niches or common selective regimes or stabilizing selection.[35] Instead they appeal to the concept of species as

---

tion as the main mode of speciation, his later association with botanists caused him to employ a morphological/nominalist species concept and to use the word "variety" ambiguously to denote either individual variants or geographical varieties. According to Mayr (1982), "instead of using the term 'variety' consistently for geographic races, he frequently employed it, particularly in his later writings, as a designation for a variant or aberrant individual. By this extension of the meaning of the term 'variety,' Darwin confounded two rather different modes of speciation, geographic and sympatric speciation" (268; cf. 288, 415, and Mayr 1991: 32–33). When I read the *Origin*, however, I get a very different impression. When writing on what it is that natural (and by analogy artificial) selection is supposed to work on, Darwin repeatedly says that it is "variations," never that it is "varieties" (cf. 1859: 30, 61, 80, 84, 102, 108, 169, 454–455, 467, 469). Moreover, Darwin always says that it is "varieties" that are incipient species, never that it is "variations" or "variants" (cf. 1859: 52, 54, 59, 111, 133, 169, 176, 325, 404, 469–470). And as for the related role of local varieties and geographical speciation, Darwin in his recapitulation states clearly that "Widely ranging species vary most, and varieties are often at first local,—both causes rendering the discovery of intermediate links less likely. Local varieties will not spread into other and distant regions until they are considerably modified and improved; and when they do spread, if discovered in a geological formation, they will appear as if suddenly created there, and will be simply classed as new species" (464–465; cf. 301). Mayr (1988a) tells us that "isolation (at least during the process of speciation) had become unimportant for him [Darwin] owing to his adoption of sympatric speciation. I have been unable to discover in Darwin's writings any connection between allopatric speciation and change of evolutionary rate" (460). Surely the passage from Darwin here quoted, as well as the one I quoted in the text above, though perhaps not sufficient to outright refute Mayr's claims, should give one pause in a simple acceptance of them (cf. also the passages examined in chapter 2.3).

[35] In reply to the punctuationists, Ayala (1983) states that "it seems that stabilizing selection may be the process most often responsible for the morphological stasis of lineages" (122). Futuyma (1986), however, asks "how could selection favor the same morphology in the face of environmental change?" (405). Ridley (1993), on the other hand, argues that "Stabilizing selection does not have to mean that the environment is constant. . . . we saw how beetle populations migrate south and north as Ice Ages and interglacials come and go. The beetles remain constant in form, but move about as the conditions change: the beetle population as a geographic whole could then be experiencing stabilizing selection even though the environment at any one place was not constant" (520). Eldredge (1985a), however, in consideration of the same beetles, thinks the most probable explanation is that "Whole biotic associations—not simply a 'beetle community,' but entire ecosystems, . . .

having homeostatic systems, a concept defended by Mayr as early as 1954, what he would later call the unity or the cohesion of the genotype. As we saw in §1, the main reason why Mayr refrained from categorizing species as individuals in a biologically meaningful sense was because the stability of species at the horizontal level often could not be adequately ascribed to gene flow. Arguing against the reductionistic or atomistic view he calls "beanbag genetics," according to which genes operate independently and can be selected for or against accordingly, Mayr argues for a "holistic" view, which he (1975) briefly summarizes as follows: "Most genes are tied together into balanced complexes that resist change" (71). This cohesion not only makes the individual organism the unit of selection, according to Mayr, but it also accounts for the stability of large populations and species in spite of lack of gene flow and variations in selection pressures, what Mayr (1982) calls "a species-specific unity to the genetic program (DNA) of nearly every species" (297).[36] Moreover, it accounts, according to Mayr, for the rapid rate of evolution in founder populations. As Mayr (1988a) more recently summarized his 1954 theory, "I postulated . . . that certain events in the founder population might help loosen this cohesion and liberate the founder population

---

are simply relocated. . . . Species track *the same environment as it moves around in space.* And if a species cannot track its environment, its usual fate is extinction" (139–140; cf. Eldredge 1985b: 186). Paterson (1985) claims that species tracking and species equilibrium are "well explained by the stabilizing selection due to the coadaptation of the SMRS between the mating partners in their normal habitat" (151), in other words "the coadaptation that exists between the mating partners with respect to signals and receivers constitutes an effective buffer to change" (149; cf. Paterson 1978: 27; Lambert *et al.* 1987: 199–203). G.C. Williams (1992), however, thinks that the thesis of species tracking is a "fable": "The shifting climatic zones of recent millennia were not followed by shifting distributions of intact communities. Different species responded to changes in different ways. . . . It seems unlikely that the community composition of a particular time and place could be closely matched anywhere after a major climatic shift. Any population of a continental habitat today must be dealing with a rather different array of resources, competitors, predators, and parasites from those of the most recent Pleistocene temperature minimum" (130). Moreover, Coyne *et al.* (1988), contrary to the claim about SMRS stability, point out that "a large proportion of studied species show geographic variation for mate discrimination" (198; cf. 194–195). In reply, Masters and Spencer (1989) claim that "because of the hegemony of the IC [the isolationist concept of Dobzhansky and Mayr], there have been very few studies of intraspecific competition from the point of view of the RC [the recognition concept of Paterson]. No adherent of the RC expects the entire system of communication to be stable—only those particular aspects which are relevant to specific-mate recognition. For example, dialects [e.g., in bird songs] are characters peripheral to the SMRS that are free to vary locally, while significant underlying characters may be strongly stabilized" (275). At any rate, I suggest that what the problem of stasis illustrates is the often striking resemblance between debates in biology and debates in metaphysics.

[36] Kitcher (1984a: 319) calls this "a flirtation with essentialism," a claim that does not seem to me justified. Mayr's claim really amounts to no more than the claim that an established gene pool has internal relations which make it resistant to qualitative change. In this there is not the slightest flirtation with essentialism.

from the straitjacket imposed on it by the epistatic balances of its genotype. I designated such an event a genetic revolution" (473).

The evidence for the cohesion of the genotype does not concern me here, although I will say it is somewhat impressive (cf. Mayr 1975). Similarly, I am not concerned here with the evidence for rapid evolution in founder populations, which of course is very impressive (cf. Mayr 1970: 303; Giddings *et al*. 1989). What concerns me here is the use of both of these concepts in Eldredge and Gould's theory of punctuated equilibria. As they (1972) put it,

> the importance of peripheral isolates lies in their small size and the alien environment beyond the species border that they inhabit—for only here are selective pressures strong enough and the inertia of large numbers sufficiently reduced to produce the "genetic revolution" (Mayr, 1963, 533) that overcomes homeostasis. The coherence of a species, therefore, is not maintained by interaction among its members (gene flow). It emerges, rather, as a historical consequence of the species' origin as a peripherally isolated population that acquired its own powerful homeostatic system.
>
> Thus, the challenge to gene flow that seemed to question the stability of species in time ends by reinforcing that stability even more strongly. . . . That local populations do not differentiate into species, even though no external bar prevents it, stands as a strong testimony to the inherent stability of species in time.
>
> The norm for a species or, by extension, a community is stability. Speciation is a rare and difficult event that punctuates a system in homeostatic equilibrium. [223][37]

It is precisely in applying the concept of the unity of the genotype (or for that matter any other concept to explain stasis) to the vertical dimension for species rather than just the horizontal dimension that constitutes the untestability of the stasis portion of Eldredge and Gould's theory. Certainly any appeal to so-called living fossils (a term interestingly coined by Darwin 1859: 107) won't do, for as Raup (1991) tells us, "in none of the examples is the living species the same as the fossil species" (41).

Dawkins (1986) tells us that "It is the emphasis on stasis that is the punctuationists' real contribution, not their claimed opposition to gradualism, for they are truly as gradualist as anybody else" (243). He then goes on to claim that not only is the thesis of stasis—that "species have genetic mechanisms that actively resist change, even if there are forces of natural selection urging change" (247)—falsifiable, but that it is in fact falsified. Dawkins, being a true Darwinian, takes his evidence from breeders. According to the stasis thesis, "if we try to breed for some quality, the species should dig in its heels, so to speak, and refuse to budge, at least for a while." However, "the fact that whenever we try selective breeding,

---

[37] As Gould (1980) stated their philosophy (or rather his own interpretation of it) more generally, "change occurs in large leaps following a slow accumulation of stresses that a system resists until it reaches the breaking point. Heat water and it eventually boils. Oppress the workers more and more and bring on the revolution" (184–185).

we encounter no initial resistance to it, suggests to me that, if lineages go for many generations in the wild without changing, this is not because they resist change but because there is no natural selection pressure in favour of changing" (247–248). Dawkins, of course, may in fact be right. Interestingly, Darwin (1859), in taking his evidence for evolution by natural selection from breeders, tells us that "Breeders habitually speak of an animal's organization as something quite plastic, which they can model almost as they please" (31; cf. 12 and 80). However, it seems to me that there is a fundamental flaw in Dawkins' argument, which resides in the fact that breeders do not work on whole species. Instead, they work on small populations, and usually with stronger selection than normally found in the wild, all of which may in turn serve to support rather than refute the stasis thesis.

The evidence from ring species would seem to be more to the point. As Ridley (1993) points out, "the existence of ring species shows that geographic variation within a species can be large enough to produce speciation" (520). However, it seems to me that this argument will only work if many of the original populations from which ring species developed were large. If most or all of the original populations were the size of (or were in fact) peripheral isolates, then the stasis thesis remains untouched.

The problem with testability, it seems to me, comes from the impossibility of inferring real or false stasis from the apparent stasis in the fossil record. As Hoffman (1992) argues, "The claim that long-term stasis is the norm in the evolution of species cannot be either confirmed or refuted. It is untestable" (129). His reason, in brief, is that "the evolutionary change may have occurred chiefly in components of the phenotype that are not preserved in the fossil record" (130). To this line of argument Ruse (1992) replies that "One cannot seriously argue that all change occurs in the soft, nonfossilizable parts of organisms. Evolutionists today see no reason to draw distinctions between hard and soft. Why should they do so for the past?" (147). But Hoffman's line of argument certainly undermines Eldredge and Gould's claim for relative frequency, "more the rule than the exception" as Eldredge put it. Mayr (1988a) is quite correct in stating that "The question 'What percentage of new species adopts one or the other of these two options?' cannot be resolved either by genetic theory or through the study of living species. It can be decided only through an analysis of the paleontological evidence. And this poses great methodological difficulties" (469).

Indeed, evidence for the untestability of the stasis thesis may be garnered from opposing interpretations of fossil data initially thought favorable to the theory of punctuated equilibria. As Ridley (1993) points out, one of the recent classic examples is Peter Williamson's 1981 data on fossil snails in the Lake Turkana region of Kenya. According to Williamson the relatively complete fossil record strongly supports Eldredge and Gould's thesis. However, as Ridley further points out, controversy surrounds this case because the pattern of punctuation may not have been caused by geologically rapid speciation but instead by distinct ecophenotypic variants of a single species, which he describes as follows: "A snail's adult phenotype depends on the environment it grows up in, as can be shown by experimentally rearing snails under different environmental treatments. Some treatments can switch the snail phenotype between strikingly different forms. . . . To change one 'species' into the other is just a matter of rearing the eggs under appropriate conditions" (516). To add to this, Hoffman (1992) points out that Williamson later came to believe that the period of speciation for the snails was more around 50,000 generations, which Hoffman says "definitely is not rapid by biological standards" (127).

Many other controversial examples are to be found in the literature. Ridley (1993) discusses opposing interpretations of Eldredge's favorite example, trilobites (516–517). Ruse (1992) discusses the same for the *Homo* line (148).[38] That such fundamental controversies could attend the very same data suggests in itself that, like the much older (and still alive) problem of universals, the issue is basically metaphysical rather than empirical.

This is not to say that there have not been some relatively fine-grained analyses from particularly good strata that seem to give strong support to the punctuated equilibria model, such as Cheetham's (1986) study of marine bryozoans from the genus *Metrarabdotos* (but cf. Hoffman 1989: 108–109). On the other hand, there have been some equally fine-grained analyses from equally good strata which strongly support the traditional gradualist model, such as Hunter *et al.*'s (1988) study of foraminifera from the genera *Subbotina*, *Planorotalites*, and *Globorotalia*. More recently, Erwin and Anstey (1995) have provided a review of fifty-eight studies that were designed to test the theory of punctuated equilibria. The results were equivocal, as with the two studies above, with about a quarter of the studies reporting a third pattern that combines both gradualism and stasis.

In spite of the above, the competing models are ultimately untestable, for as stated earlier it is a matter of frequency, and the fossil record, all in all, is just too incomplete to provide the data that would ultimately settle the matter. But what clinches my claim, it seems to me, is the problem of inferring reproductive compatibility over time. Ayala (1983), for example, refers to the fact that sibling species are undetectable in the fossil record so that in that record such speciation would appear as stasis. And sibling species, he points out, "are common in many groups of insects, in rodents, and in other well studied organisms" (121; cf. Mayr 1988a: 468). Gould (1992), however, is not impressed by this, calls it "perhaps the most common point advanced against us," and thinks it irrelevant: "If many periods of stasis include multiple species, then our argument is only strengthened—for the constraints upon change then affect several related forms, there-

---

[38] It is surprising to see how some theorists have used the theory of punctuated equilibria to further their metaphysical agendas. This is especially the case with Sir John Eccles (1989: 12–38, 237–243), a most distinguished brain scientist, who applies the theory of punctuated equilibria to human evolution in order to support his belief that human evolution was (and still is?) guided by the hand of God. Like so many post-*Origin* scientists in Darwin's day, who could not find it in themselves to free the new science from the shackles of their old religion, and who thereby tried to pour the new wine into the old skins, such a move on Eccles' part is only too familiar to modern historians of biology.

by providing affirmation in replication, whereas single incidents are only episodes" (73). But this misses Ayala's point, which is that sibling species provide strong evidence that "it is no way apparent how the fossil record could provide evidence of the development of reproductive isolation" (121). *And that is the real issue!* In their 1972 paper Eldredge and Gould (201) stress that their theory cannot be decided by the paleontological data, but that that data is more in accord with the prevailing neontological species concept, namely Mayr's biological species concept, and much that it involves. As they put it, "The biospecies abounds with implications for the operation of evolutionary processes. Instead of attempting vainly to name successional taxa objectively in its light, we should be applying its concepts. . . . [namely] the theory of allopatric speciation" (202; cf. Gould 1992: 73–74). The problem of testability, however, is that, as Mayr has repeatedly emphasized, "The biological species concept, expressing a relation among populations, is meaningful and truly applicable only in the nondimensional situation. It can be extended to multidimensional situations only by inference" (Mayr 1982: 273). Applied to fossil species over time, such inferences can at best be little more than guesses. What ecophenotypic switches and sibling species illustrate are the extremes of the fact that morphological divergence and intrafertility divergence do not necessarily covary (cf. Stebbins 1989: 114). *For the claims of Eldredge and Gould to be more than mere words, then, they would have to claim that the stasis that constitutes the history of most species is such that a population at the beginning of a long period of stasis is reproductively compatible (in spite of the separation over time) with its descendant population at the end of that period of stasis.* If they are not prepared to state their claim in this way (and I have yet to see it), then their claims about stasis and evolution become trivial, being of a piece with the more traditional claims of phyletic gradualism. For most species throughout history, of course, my restatement of what their claim should be may in fact be true. For most it may in fact be false. But either way, *the claim is metaphysical since there is in principle no possible way to test it.*[39] In every case it must forever remain an open question.

The lesson to be relearned in all of this is the logical and therefore ontological priority of horizontal over vertical species concepts. Eldredge and Gould, in trying to fit a basically horizontal concept into a vertical mold, have failed to appreciate this point. But it is a point that underlies many of the attempts to think of species as analogous to individual organisms. Unlike organisms, species need not have a reality over time, but that does not make them unreal. In trying to make the reality of species depend on the vertical dimension of punctuated equilibria, Eldredge and Gould have attempted, in effect, to place the reality of species on an utterly precarious foundation, a foundation, fortunately, that lacks both empirical and logical support.

At any rate, I believe I have demonstrated that the species-as-individuals view

---

[39] I take it as axiomatic that time travel is precluded on logical grounds. If I am wrong, then in principle there may be a way of conducting the requisite tests, though I can think of better things to do with a time machine.

has nothing to be gained from the controversy generated by the theory of punctuated equilibria. Nevertheless, this is not to say that the latter theory is without value. As Hoffman (1992) put it, "it stimulated exciting empirical research that has resulted in some fascinating data, and it also enforced much higher standards of paleontological work on the tempo and mode of evolution in species-level lineages" (131). In this way alone, perhaps more than any other, the theory of punctuated equilibria stands as a prime example of the value of metaphysics for actual practice in biology.

4.4 Problems with Species as Individuals

In this section on the individuality thesis I want to examine a number of problems not dealt with in previous sections. I will begin with problems often raised but which seem easily dismissed and will graduate to problems that are increasingly troublesome.

The first to be discussed is the problem of hybrid zones, which might seem in itself a good reason for dismissing the individuality thesis. On Ghiselin's (1974) view, "That limited exchange of genes may occur between species creates no more difficulty for the species concept than does the fact that someone might work for two firms creates difficulties for the notion of a firm" (283). But as mereological wholes, the issue is over the sharing of parts. Can individual organisms share parts while retaining their individuality? If not, then there might be a problem with species being individuals since, as we have seen earlier, most biologists agree that two species may remain distinct even though there is a hybrid zone between them. Sober (1993), in his defense of the species-as-individuals view, provides a positive reply. According to Sober, "The idea that distinct individuals may share parts with each other is not altogether alien. We count 'Siamese twins' as two organisms, not one" (156). The reason why we count Siamese twins as two organisms instead of one, however, is because there are two brains. Should a human give birth to an organism with only only brain connected to two backbones each with four limbs, I doubt we would consider it two individual organisms rather than one. In the case of Siamese twins, then, the analogy with species is a bit faulty, since species do not have anything analogous to privileged (in the sense of individuating) parts such as brains. Nevertheless, it would seem inappropriate to dismiss the analogy altogether.

Another objection might be made based on the "logical geography" of species words. It will be remembered that Aristotle argued, based apparently on no more than a simple observation of language, that the names of primary substances (individuals, the subjects of proper names) never function as predicates, while the names of secondary substances (species and genera) function as both subjects and predicates. As Aristotle in the *Categories* put it, "Things that are individual and numerically one are, without exception, not said of any subject" (1^b6–7), "only they [species and genera], of things predicated, reveal the primary substance" (2^b30–31), while "the species is a subject for the genus" (2^b20).[40] The same basic

---

[40] It will be remembered from chapter 3.1 that on the analyses of Balme and Pellegrin

idea is repeatedly found in modern philosophy. Although the distinction is sometimes muddied by pointing out that proper names sometimes function in predicate
expressions (cf. Ramsey 1925, cited in chapter 1.1n2), proper names never function as predicates *simpliciter* (cf. Strawson 1959: 173–179). Indeed we never say
that this is a Socrates or that that is a Socrates. Nor in dealing with mereological
wholes do we use the name of the whole (the proper name) as a predicate for any
of its parts. Of Socrates' left foot, no one would ever say that it is a Socrates. But
clearly with species names the situation is quite different. One might therefore
object that species are not individuals because species names commonly function
as predicates *simpliciter* whereas proper names do not. Perhaps, then, one could
reject the species-as-individuals view on the basis of linguistic confusion. In §1
we saw that Russell rejected Hegel's notion of concrete universals on the basis of
linguistic confusion, charging that he (Hegel) confused the "is" of predication
with the "is" of identity. Indeed there is a strong tradition in modern philosophy
which holds that syntactic functions in natural languages can actually tell us
much about the real world. Russell (1940), for example, strongly believed that
"partly by means of the study of syntax, we can arrive at considerable knowledge
concerning the structure of the world" (347). Indeed it was largely by means of
syntax that Russell argued for the nonreductive, extra-mental reality of relations
(as we shall see in the next chapter). Following in this tradition Gilbert Ryle
called any error in logical geography (the syntactic function of a word or phrase)
a "category mistake," and he used this methodology in *The Concept of Mind*
(Ryle 1949) to argue against the concept of mind which he called the "myth" or
"dogma of the Ghost in the Machine" (17). Perhaps, then, like that myth, the species-as-individuals view is one big category mistake.

Hull (1981a) argues that any such line of argument is irrelevant: "I fail to see
the relevance of ordinary language to the points at issue. . . . Scientists depart
frequently from ordinary usage—as they must. . . . Evolutionary biologists are no
more bound by ordinary usage in their pursuits than are physicists" (142). In a
fascinating related discussion, Jagdish Hattiangadi (1987), employing the biological point of view, makes a strong case for what he calls *grammatical evolution* (173). He reveals that not only words but "even the rules governing word
usage which we call 'grammatical rules', . . . are not exempt from possessing a
theoretical background. Even they mask points of view, which we recognize, if at
all, only after they are challenged by the growth of knowledge" (13). Thus, theoretical advances in science need not be restricted by but may actually affect
changes in the meaning of not only words but also grammatical rules. Hattiangadi
(7–9) conjectures that in the future such may actually be the case in ordinary language as a result of Einstein's special theory of relativity and the theory of wave-
particle duality in quantum physics. Someday in the future it may well be gram-
matically correct to speak of speeds as having mass and chairs as being events. At
any rate, the point to notice is that what is a category mistake for one generation

---

genos and *eidos* in Aristotle's works are not taxonomically fixed, so that what is called a
genos may in turn function as an *eidos* and *vice versa*.

need not necessarily be a category mistake for another.[41] Not only, then, need the species problem not be restricted let alone decided by ordinary language, but if the species-as-individuals view should ever become entrenched, ordinary language may well follow suit. Interestingly, Ghiselin (1974), in concert with his view that many biologists have implicitly treated species as individuals for many years, thinks that the conventional geography of species words among biologists is already undergoing such a revolution. He points out that "some biologists will never say 'John Smith is a *Homo sapiens*,' but insist that correct usage demands 'John Smith is a specimen of *Homo sapiens*'" (280). Indeed, perhaps some day in the distant future when we are both biologically and politically correct, a future Neil Armstrong will declare as he first steps onto Mars: "That's one small step for a specimen of *Homo sapiens*, one giant leap for *Homo sapiens*." At any rate, it ought to be clear that neither ordinary nor technical language should ever be used to decide, pro or contra, whether species are in fact individuals.

Closely related to the issue of species names as proper names is the issue of whether species themselves can literally be seen with the eye. Ghiselin (1966) tells us that "As species are individuals, there is but one rigorous way to define their names: ostensively, in a manner analogous to a christening" (209; cf. 1974: 284, 1987a: 134, 1997: 66–67). Thus if species are physical individuals and species names are proper names, one would naturally think that, just like Socrates 2,400 years ago, one ought to be able to point to individual living species, so that it ought to be possible to actually see them.

This is in fact the explicit claim of many biologists and philosophers who subscribe to the species-as-individuals view. Their claim is that, given the right perspective, it ought to be possible for some type of being to actually see species. Hull (1981b), for example, points out that "Given our relative size and duration, we can see the distance between organisms, their diversity and gradual replacement. If we were the size of atoms, organisms would look like clouds of atoms, comprised mainly of empty space. If we were the size of planets, species would take on the appearance of giant amoebae, expanding and contracting over the face of the earth" (144; cf. Hull 1992: 183). The main error committed by his detractors, then, claims Hull (1981a), is in "assuming a single frame of reference which is appropriate for organisms and applying it inappropriately to species" (149). Indeed it is this very error that might alone compel one to conceive of species either as sets or classes or as unreal. Similarly, Eldredge (1985a) refers us to the

---

[41] Thus Ryle's now popular phrase "logical geography" is itself a category mistake. According to Ryle (1949), "To determine the logical geography of concepts is to reveal the logic of the propositions in which they are wielded, that is to say, to show with what other propositions they are consistent and inconsistent, what propositions follow from them and from what propositions they follow. The logical type or category to which a concept belongs is the set of ways in which it is logically legitimate to operate with it" (10). The use of "logical" in such a conception, however, entails a static concept of language, since logic itself is timeless. Given language evolution, then, I suggest that "conventional geography" would be a much more appropriate phrase for what Ryle had in mind.

movie *Fantastic Voyage* (in which Raquel Welch is shrunken in size along with a research team and is sent into a human's bloodstream) and claims that "One reason, I think, why biologists persistently and consistently refuse to see species as entities is that we are all in the position of Ms. Welch in the bloodstream" (100). In fact he says that "Failing to 'see' species is precisely like missing the forest for the trees" (100). But like Hull, though explicitly because species lack anything like skin, Eldredge suggests that "perhaps the best analogy after all is with an individual amoeba" (104). In a similar vein, Mary Williams (1987) urges that "we must cut our intuition loose from its moorings in human experience; let us do this by imagining the world of a scientist about the size of an atom and with a correspondingly small lifespan. . . . If Micro could travel very fast through space, he would perceive organisms as regions of space with a higher density of molecules than the surrounding space. . . . One can imagine a Micro Ghiselin arguing that these denser aggregates of molecules are individuals. In answer to this a Micro Mayr might argue that these aggregations, although not classes, are not sufficiently like atom-sized micropersons to be called individuals; in fact, he might argue, since they are deeply important natural phenomena they represent a new ontological category" (204–205). In a different vein, Wiley (1989) uses ostension to distinguish between species and higher taxa. "We can point to *Homo sapiens*," he says, "but we cannot point to Vertebrata since "once the first speciation event split that ancestor into two species, Vertebrata ceased to be a cohesive whole" (293). In a mitigated vein, Ghiselin (1981) tells us that "If I wanted to observe a species I could. I would only need to find a very small, localized one, perhaps consisting of but a single organism" (305). Similarly Edward Wilson (1992), who also believes that each species is "a unique individual" (45), suggests that the only time one can actually see a species is when it is endangered, such as "the Devil's Hole pupfish, barely hanging on in a tiny desert spring at Ash Meadows, Nevada. You can stand at the entrance to Devil's Hole, look down 15 meters to where the water laps over a sunlit ledge, and see the species swimming around like goldfish in a bowl" (64; cf. Cracraft 1989a: 51n8).

With all of this I am *extremely* skeptical. It seems to me that with a highly populous species with a relatively extended range, there is absolutely *no* perspective from which *any* entity could possibly *see* a species. Certainly if we were the size of planets we would not see species as amoebae. We would not see what we call species at all! Can anybody pick out a widespread species from a plane? Or from a spacecraft orbiting the earth? Even if each member of a species were to project from its being a particular frequency of electromagnetic wave, much like beacons in the dark, with each species having a different frequency, I doubt that there could be any kind of entity that could *see* species. What only adds to the problem is that there are literally millions of species on earth, crisscrossing and overlapping each other in numerous and diverse ways. Individual organisms, on the other hand, with very few exceptions, are relatively discrete, and not just as a matter of perspective: they are *ontologically* discrete, just like planets. I am in agreement with Occam, then, when he says "no one sees a species intuitively [i.e., with the senses]" (cf. chapter 2.2). Even with Wilson's example, the time dimension for the *seeing* is much too horizontal, much too thin. As generational entities species require for their horizontal reality a span of time much greater than that which can be accomplished by any *seeing*.

What is especially curious is that Ghiselin would allow a species to consist of only one organism. As such, a species would no longer be a mereological whole, since a whole with only one part is not really a whole. (I will have more on the problem with a species of one in chapter 5.2.) What is surely significant is that when it comes to ostensive definitions for species Ghiselin relies mostly on type specimens: "by 'pointing' to the entity which bears the name. . . . Attention is drawn to only a part or aspect of the individual the name of which is being defined" (1974: 284).[42] The problem now is that if as in most cases (as per Ghiselin's view) one cannot see a species as a whole, how can one know when one points to two or more type specimens whether they are parts of the same or different wholes? If similarity is irrelevant for species ontology, as it is on the view of Ghiselin and Hull, and if interbreeding and reproductive competition are highly problematic (as we've seen earlier), then with their view the road seems paved toward species nominalism.

Of course if it is true that species cannot possibly be seen, this does not necessarily refute the species-as-individuals view, but it does give one a further reason to be skeptical. One would think that there ought to be a perspective from which one could possibly see a physical mereological whole as a whole, rather than inferring it simply from its parts. That species do not fit this desideratum suggests that they have some other sort of ontology. But does it follow, then, that they are not real, as Occam thought, or that they are abstract classes, as Aristotle thought? I do not think so, and I believe that the ontology I put forward for them in chapter 5 will demonstrate why. Species, it seems to me, are neither intuitively obvious nor frivolously invented. Instead they are something that biologists often

---

[42]   More recently Ghiselin (1995) has stated that "When defining the name of a whole, we can do so by 'pointing at' or 'showing' only a part of that whole, provided that it be understood that this is what we do. Definition of a species by type designation involves showing a component, which is understood to be a component of the species that is named. Similarly, when defining the name of an organism, we might 'point at' just a part of it, for example a beard" (221). There is a subtle distinction underlying these statements, easily overlooked, but profound. In the first sentence, a universal proposition on ostensive definitions of wholes, it is stated that we "can," in the sense of "may," define a whole ostensively by pointing only at a part. In the sentence which follows, however, given to the example of a species, the implication is stronger, more like a "must" than a "can." But notice that for the second example, that of an organism, the "can" in the universal proposition reverts to a "might," the implication being that with an organism we might ostensively define it by referring to one of its parts but we need not necessarily do so, in other words we might also ostensively define it by pointing to the whole. The fact of the matter, of course, is that we often do. But do we also with species? I think most biologists will agree that typically, if not always, they do not. Perhaps this is simply because of the human perspective. A more likely reason, however, as we will have repeatedly seen, is because species lack the requisite cohesion.

carefully infer, and the nature of this inference suggests to me that they are real but that they are neither abstractions nor concrete entities. Just what I think it suggests, of course, will have to wait until the next chapter.

Another problem for the species-as-individuals view is that species have nothing comparable to aging. The view that species age and have an age limit in much the same sense as organisms is a view that finds its earliest adherent, as far as I can tell, in an early nineteenth-century geologist by the name of Giovanni Brocchi. In 1814 Brocchi argued that every species is constituted in such a way as to have a predetermined lifespan, so that a species must eventually go extinct even if its environment remains favorable. Lyell (1832: 128–130) provides an interesting discussion on Brocchi's view, and even though he does not accept it he speaks highly of Brocchi as a geologist. The aging thesis, however, reached its zenith beginning in the late 1860s through to the 1890s with the theory of "racial senility" put forward by the American paleontologist/evolutionist Alpheus Hyatt (cf. Bowler 1989: 260–264). On Hyatt's view a species approaching near the end of its predetermined lifespan degenerates to a state much like its original, analogous to "second childhood." Interestingly Hyatt's theory is rooted firmly and explicitly in nineteenth-century Continental embryology, specifically in the so-called Meckel-Serres law, or the law of parallelism, according to which the ontogenetic stages in the development of the embryo of a higher organism sequentially recapitulate the stages below its species in the progressive scale of being. Thus a human embryo recapitulates all the main stages in the animal *scala naturae.* There is some controversy over whether these early embryologists combined this view with a view of species evolution. Mayr (1982: 471–472) says no. More recently, however, Robert Richards (1992) argues that the recapitulation theories of Kielmeyer (19), Tiedemann (43–45), Treviranus (45–46), and Meckel (54)—but excluding Serres (167)—were wedded to a view of species evolution. (The passages he cites, however, do not seem to me all that convincing.)

Interestingly, Lyell (1832: 62–64) discussed the recapitulation theories of Tiedemann and Serres, but rejected them as providing any support for species evolution. Darwin, of course, as pointed out in chapter 2.3, read volume II of Lyell's *Principles of Geology* (1832) during the latter part of his *Beagle* voyage, and so was well acquainted with both the species aging thesis of Brocchi and the recapitulation theory of the Continental embryologists. What is so interesting is that subsequent to his *Beagle* voyage and his rapid conversion to evolutionary thinking Darwin quickly combined the species aging thesis with recapitulation theory, as evidenced in his early notebooks on evolution (Barrett *et al.* 1987). The former seems to have come first. In his Red Notebook (1837), while contemplating on what he believed to be an extinct species of llama, Darwin states that he is "Tempted to believe animals created for a definite time:—not extinguished by change of circumstances" (129), and also that "There is no more wonder in extinction of species than of individual" (133; cf. Notebook B: 22–23). Recapitulation seems to have been accepted very shortly afterward, the earliest evidence being page 1 in his Notebook B (1837), with the clearest passage being "An originality is given (& power of adaptation) is given by *true* generation, throughe means of every step of every progressive increase of organization being imitated in the womb, which has been passed through to form that species" (78).

Darwin quickly gave up the aging thesis, however, within a few days or weeks, coming to the conclusion that "death of species is a consequence . . . of non adaptation of circumstances" (Notebook B: 38–39). For how long and to what degree and form Darwin retained the recapitulation theory is today a topic of considerable debate (cf. Bowler 1989: 180, Richards 1992, Ruse 1993). To what degree the theory itself is still valid is, of course, no longer an issue (cf. Mayr 1982: 475–476). The thesis of species aging has fared only worse. Peter Ward (1992) tells us that "This concept of 'racial senescence,' of gene pools that grow old and begin to introduce bizarre and ultimately lethal morphotypes, has been thoroughly discredited" (6). Indeed it is not even necessary that species must always be evolving. Mayr (1988a) tells us that "owing to the genetic turnover in populations, all species are evolving all the time" (321). But even if evolution is defined, as it often is, as a change in gene frequencies within a population, then, as Sober (1984a) points out, "This genetic criterion for the occurrence of evolution implies that populations can change in composition without evolving" (29). A greater average height, for example, may be due to nothing more than improved nutrition, while a doubling in population may conceivably result in unchanged gene frequencies. But even if it were the case that all species are evolving all the time, it does not follow that it would be in any way comparable to aging. According to Raup (1991), "there is absolutely no basis for equating the life spans of species with those of individual humans. There is no evidence of aging in species or any known reason why a species could not live forever" (6). Presumably this has something to do with MVPs. As we've seen earlier, Raup tells us that a species above its MVP is "virtually immune to extinction" (124). Certainly it is logically possible, and for all I know maybe even physically possible, that individual organisms could someday be genetically engineered so as not to age, making their mortal coil potentially immortal. If that possibility is real, then in a sense we are all potentially immortal. But species are *naturally* potentially immortal, while we are not. Surely this is a crucial difference that makes suspect the individuality of species in a *biologically meaningful* sense. It suggests that their ontology is something other than that of biological individuals.

More important is the issue of multiple origins. Kitcher (1984a) raises an interesting possibility in reference to *Cnemidophorus tesselatus*, a unisexual species of lizard known to have arisen through hybridization and which, in spite of its unisexual character, is categorized by biologists as a species. What if, supposes Kitcher, the speciation event which produced this species occurred independently more than once (although he knows that biologists think it probably hasn't), occurring on different occasions when the parental species came into contact?[43] Or even more striking, says Kitcher, what if the original population of *C. tesselatus* went extinct, and then the parental populations happened to meet again, repeating

---

[43] Mayr claims that this has actually happened with the parthenogenetic grasshopper species *Warramaba virgo*; cf. chapter 3.3n37.

the original result? With individual organisms, of course, each has a singular origin (which Kripke raised to the status of an essence), and multiple origins produce siblings, not the same individual. But with the hypothetical multiple origin scenario for *C. tesselatus*, says Kitcher, "what biological purpose would be served by distinguishing two species? To hypothesize 'sibling species' in this case (and like cases) seems to me not only to multiply species beyond necessity but also to obfuscate all the biological similarities that matter" (315).

Sober (1984b) replies that "The problem is that this interpretation is not forced on the view that species are individuals" (340). To Kitcher's first scenario, he says "An individual may have parts that had their separate origins; a fleet of ships may have its component boats constructed in different ship yards." To Kitcher's second scenario, he says "the view that species are individuals need not maintain that every small budding on the tree of life counts as a distinct species."

Both of Sober's replies seem to me to have missed the point. To strengthen Kitcher's first scenario, we need, first of all, to consider sexual species. (As we shall see, on the individuality thesis asexual organisms do not form parts of species.) Accordingly allopolyploidy (cf. §1 above) provides us with a better scenario for Kitcher's counterexample. Suppose that two independent allopolyploid populations were produced in two different locations by the same two parental species.[44] Suppose further that the two allopolyploid populations produced lineages which never exchanged genes but which nevertheless slowly evolved for the next million years in a parallel fashion due to common selection pressures. Both populations would certainly not be buds, so we can dispense with Sober's second reply. But I think we can also dispense with Sober's first reply. Since there is no gene exchange between the two lineages, their member organisms taken together cannot be considered as conspecific parts. Moreover, at every horizontal level I think most biologists would treat the two populations as conspecific (cf. Wiley 1978: 86–87, quoted later in this section), especially if they are thought to be re-

---

[44]   I chose allopolyploidy rather than autopolyploidy because the latter is much less likely to produce the two geographically distinct yet interfertile populations that my scenario requires. This is not because autopolyploids, unlike allopolyploids, need not be sexual, but because, as A.J. Richards (1986) points out, "autopolyploidy is a rare condition in natural populations. Most polyploids, perhaps half of all Angiosperm species, are alloploids that have obtained genomes, for each of which they are usually diploid, from two to several different species" (210). However, there is a difficulty which reduces the likelihood of my scenario. For reasons that need not be discussed here, Barrett (1989) points out that "diploids are often primarily outcrossing, whereas polyploids tend to be selfers" (260). The problem is that my scenario requires two populations of outcrossers, not selfers, since I want them to easily qualify as *parts* from the perspective of the individuality thesis. But from what Barrett says, allopolyploids are normally not outcrossers. Although all that Barrett says reduces the likelihood of my example actually occurring in nature, it is nevertheless not science fiction. Indeed given the prevalence of angiosperms, of allopolyploidy among angiosperms, and the roughly 120 million-year history of angiosperms, I dare say that something like my scenario has probably happened many times (and not just among angiosperms; cf. Werth *et al.* 1985 for a strong case among ferns).

productively compatible (cf. §5n58). It will also be remembered from chapter 2.4 that one of the main reasons for thinking that species are real is the evidence from biological control. From that viewpoint alone there is certainly no reason to think of the two populations as separate species and every reason to think of them as one species. Indeed thinking of them as two species could actually work against the purposes of biological control.[45]

Closely related to the problem of multiple origins is the problem of extinction. According to Darwin (1859), "When a species has once disappeared from the face of the earth, we have reason to believe that the same identical form never reappears" (313; cf. 352), which suggests that for Darwin it is at least not impossible for an extinct species (however unlikely) to reappear. This conclusion is further supported by one of Darwin's letters to Lyell (June 21, 1859), in which he wrote that "it is, I think, next door to an impossibility that the same species should have been formed identically the same in any two areas" (Burkhardt and Smith 1991: 307–308). "Next door to an impossibility," of course, is nevertheless to affirm a *logical possibility*. Thus for Darwin, to borrow from Lyell (1830), it is logically possible that "The huge iguanodon might reappear in the woods, and the ichthyosaur in the sea, while the pterodactyle might flit again through umbrageous groves of tree-ferns" (123). According to Ghiselin and Hull, however, the species-as-individuals view precludes even the logical possibility that an extinct species might re-exist, since the logic of individuality entails that the death or extinction of an individual is necessarily forever. Hull has the most interesting things to say here. As Hull (1978) put it,

> Organisms are unique. When an organism ceases to exist, numerically that same organism cannot come into existence again. For example, if a baby were born today who was identical in every respect to Adolf Hitler, including genetic make-up, he still would not be Adolf Hitler. . . . But the same observation can be made with respect to species. If a species evolved which was identical to a species of pterodactyl save origin, it would still be a new, distinct species. . . . Darwin presents . . . [his] point as if it were a contingent state of affairs, when actually it is conceptual. [305]

Certainly if species are indeed individuals, then this would be true. If I were to have a monozygotic twin, it would no more be me than the man next door. Even in the science fiction examples beloved by philosophers working on the problem

---

[45] To the Kitcher-type of scenario we might also add the problem posed by genetic engineering. As Ruse (1987) pointed out, "Today, through recombinant DNA techniques and the like, biologists are rushing to make new life forms. Significantly, for commercial reasons the scientists and their sponsors are busily applying for patents protecting the new creations. Were the origins of organisms things which uniquely separate and distinguish them, such protections would hardly be necessary. . . . this suggests that origins do not have the status claimed by the s-a-i boosters" (353). Although I agree with Ruse here, it seems to me preferable to provide plausible scenarios from "real life" rather than from what scientists can do, if only because the former have a certain priority.

of personal identity, a copy of me, no matter how perfect, no matter even if it replaces me and fools everyone else into thinking it is me, would still not be me. *I* am not a type, and copies of me are not *me*. But as we've seen, the analogy between species and organisms is in many ways defective, so that any argument about extinction based on that analogy must be equally defective.

Hull has another argument for believing that extinction is necessarily forever, an argument designed for those who do not believe that species are individuals. This argument appeals to a taxonomy desideratum of the Modern Synthesis (cf. Mayr 1942: 276), namely that all taxa (species and higher) should ideally be monophyletic (i.e., have a single origin, although there is disagreement on what exactly this should mean). As Hull (1981a) put it, "If taxa must be monophyletic, then once a species is extinct, numerically that same species cannot re-evolve. . . . 'Extinct' does not mean 'temporarily does not have any members'" (147).

But many traditional taxa, of course, have proved to be either polyphyletic (i.e., they have originated from more than one group) or paraphyletic (i.e., they do not include all of the descendants from the originating group). The class Reptilia, for example, is not strictly monophyletic but instead paraphyletic, since Jurassic reptiles produced a branch that became so different from the rest of reptiles (both ancient and modern) that it was accorded the status of a class itself, namely Aves (cf. Futuyma 1986: 288–290). Moreover, some have thought that the class Mammalia is not monophyletic but polyphyletic (cf. Simpson 1961: 121). Even certain species have been thought to have multiple origins. For example, biologists agree that all our domestic breeds of dog form one unified species (*Canis familiaris*), because of their interfertility (cf. Ridley 1993: 40), in spite of also agreeing that they probably originated from more than one wild species (cf. Maynard Smith 1975: 45). Indeed as we have seen above, hybrid species are very common, particularly in plants, and may have multiple origins in more than one way.[46] At any rate, the above should not only illustrate what taxonomists know only too well, namely that the concept of monophyly is problematic, but also that it is not a strong foundation on which to base a concept of extinction.

But Hull has another approach, irrespective of monophyly. If it is agreed that species evolve, it must be agreed that species are historical entities. But the extinction of historical entities is necessarily forever. As Hull (1988) put it, "In order to count as Baroque, a building must have been built at a particular time and place. It also helps if it is built in a particular style. However, no matter how similar a building might be to a Baroque building, it cannot count as Baroque if it

---

[46] The matter is only further complicated by modern endosymbiotic theory, championed most vigorously by Lynn Margulis with now wide support, according to which all eukaryotic (plant, animal, fungal, and protist) cells trace their ancestry back to episodes of permanent bacterial symbiosis, hence polyphyly. According to Margulis (1991), such "rampant polyphyly . . . wreaks havoc with 'cladistics,' . . . [and] invalidates entire 'fields' of study" (10). In reply, of course, one could deny the legitimacy of asexual species, which we will see is the view of Ghiselin and Hull, but we will also see a further difficulty that returns from the field of bacteriology.

was built at the wrong time or in the wrong place. Later structures can be built in the Baroque style, but they cannot actually be Baroque. In short, the Baroque period is a historical entity" (77). Thus, "One of the main messages of this book is that species, if they are to play the roles assigned to them in evolutionary theory, must be treated as historical entities. *Dodo ineptus* is conceptually the same sort of thing as the Baroque period. Both are gone and can never return. Extinction *is* necessarily forever" (79).

This argument, however, is not as strong as it might appear. Certainly the Baroque period was a historical period, and a replica of a Tiffany lamp will not fetch the same price as a genuine Tiffany lamp (Hull 1988: 78). But is it true that the extinction of *any* historical entity is necessarily forever? If so, then Hull has a strong argument. But I do not think that what is true of some members of the class of historical entities is true for them all.

Consider languages.[47] Languages, of course, evolve (Ghiselin 1987b: 136;

---

[47] In the following discussion I forbear on the ontology of languages, except to say that they are objectively real. Neither fully abstract nor fully concrete, languages would seem to be a hybrid category of some sort. Indeed the topic may deserve a book in itself. At any rate, since the language analogy is not only a recurring theme throughout the present work but an important one for many of my arguments, I suppose I should address a common criticism at this point. It will be replied by many that language is a man-made entity, not a natural entity, so that it is conceptually real at best and therefore makes an inappropriate analogy to species if the latter are to be thought of as extra-mentally real. While it is true that language is a man-made entity, it certainly does not follow that it is in the same category with fictions such as centaurs and unicorns, or even constellations. As Darwin pointed out in *The Descent of Man* (1871), "no philologist now supposes that any language has been deliberately invented; each has been slowly and unconsciously developed by many steps" (55); or again, "But it is assuredly an error to speak of any language as an art in the sense of its having been elaborately and methodically formed" (61). Second, unlike fictions, human language *per se* (syntax in addition to semantics) is a *biological adaptation*, since it is genetically grounded and confers a competitive advantage (cf. Bickerton 1990: 146–147). Even different individual languages can be seen as having differences that confer competitive advantages and disadvantages. Differences in grammatical structure alone may result in different worldviews, some more accurate than others. This, of course, is the Sapir-Whorf hypothesis. As a biological adaptation, then, language is just as objective as any other biological adaptation, like the beak of a woodpecker or the shell of a tortoise. This is not to say, of course, that a *particular* human language is genetically based. A Chinese baby raised in an English family learns English just as well as his English counterpart, and *vice versa*. But nevertheless there is a certain objectivity to even natural languages taken individually. Perhaps the greatest evidence of this is the bare fact that *languages evolve*, and as per Darwin their evolution is not goal-directed or consciously produced but has a life of its own, which as such is subject to forces analogous to those that drive species evolution (i.e., chance variation and natural selection on words and concepts, semantic drift, and group variation following geographic isolation; cf. Dawkins 1986: 217–218 for an interesting discussion on language evolution). In this way languages have a certain autonomy over and above the language carriers on which they depend. Much of this conforms, interestingly, to the three worlds thesis of Karl Popper (cf. Popper 1972: 74), according to which world 1 is the physical world, world 2 is the subjective world of our con-

Hull 1977: 82). Moreover, language evolution is not only gradual (anagenesis) but also variational (cladogenesis). Just as a widespread species does not tend to stay the same but produces geographical varieties, which themselves are "incipient species," so likewise widespread languages produce geographical varieties or dialects, which themselves are "incipient languages" (cf. Lyell 1863: ch. 23). The bottom line is that languages provide a much better analogy to species than organisms. An individual organism, of course, has ontogeny, but it does not have phylogeny. In other words an individual organism does not evolve. On this virtually everyone is in agreement.[48] Languages, on the other hand, evolve. In fact it

---

scious experiences, and world 3 is the world of the abstract products of world 2, such as stories, mathematical and scientific theories, computer programs, music, paintings, etc. Although world 3 is a product of world 2 and is often codified in world 1 objects, it nevertheless has a certain degree of autonomy from the two worlds below it, not only because world 3 objects can lead to unexpected connections and discoveries, but also because they can have a downward causal effect on worlds 2 and 1 (e.g., the evidently false belief in life after death and God not only makes many people happier but it also often leads to truly magnificent works such as the music of Bach and the Sistine Chapel—not to mention also, of course, wars and famines on the one hand and greater overpopulation on the other). As Popper (1972) put it, "The world of language, of conjectures, theories, and arguments—in brief, the universe of objective knowledge—is one of the most important of these man-created, yet at the same time largely autonomous, universes. . . . The idea of *autonomy* is central to my theory of the third world: although the third world is a human product, a human creation, it creates in its turn, as do other animal products, its own *domain of autonomy*" (118; cf. Popper and Eccles 1977: 38–50). The above may also be compared with Richard Dawkins' (1976) concept of the *meme*, e.g., "tunes, ideas, catch-phrases, clothes fashions, ways of making pots or of building arches" (192). Memes are the imperfect replicators whose differential survival results in cultural evolution. Memes, of course, remain dependent on human minds, but nevertheless like other obligate parasites they easily take on a life of their own quite independent of anyone's intentions. Moreover memes are not genetically coded, and accordingly they need not confer any biological advantage at all (cf. Dawkins 1976: 190–191). But now the distinction between language on the one hand and fictions such as centaurs, unicorns, and constellations on the other hand might seem hopelessly blurred. But blurred it is not. A language is not a fiction; instead it is what Bickerton (1990) called it, namely "a system of representation" (5). It is a system, moreover, that is based on a biological adaptation, namely the human language capacity *per se*. And given the nature and agents of language evolution, such evolution makes for a very close analogue to species evolution. Fictional memes, on the other hand, are fictions (if the tautology may be excused); they have no representational role but are merely intentional products, products moreover that tend not to evolve (have the respective memes for centaurs, unicorns, and constellations evolved?) and that from a biological point of view have no adaptive value whatsoever. In the end they are conceptual entities only.

[48]   Cf. Dawkins (1986: 268), Futuyma (1986: 7), Wilson (1992: 75), Sober (1993: 1), Ridley (1993: 5). Maynard Smith (1975) put it best: "although evolutionary changes are usually described in terms of the differences between successive adults, i.e. as phylogenetic changes, the differences between those adults were the consequence of differences between the paths of development which gave rise to them, i.e. of ontogenetic changes; phylogenetic changes are the result of changes in ontogeny" (310).

was the analogy between species and languages—and not any analogy between species and organisms—that played the most important role in the development of Darwin's theory of species evolution. Indeed the evolution of languages, and their "phylogenetic" relationships, was a topic much in the air during the intellectual development of the young Darwin. Sir John Herschel, the most prestigious scientist in England during Darwin's youth, and whose opinion Darwin respected most, urged that "Words are to the Anthropologist what rolled pebbles are to the Geologist—Battered relics of past ages often containing within them indelible records capable of intelligent interpretation." Moreover, Darwin's older and scholarly cousin Hensleigh Wedgwood, whose sister Darwin would marry, "was himself a philologist, looking for the 'laws' by which alphabets slowly change. He praised the Germans for understanding the 'organic' development of language and for tracing 'every descendant' of their own Gothic tongue" (Desmond and Moore 1991: 215–216).[49] Indeed the influence on Darwin was profound. In his early notebooks, although there is no developed connection made between species and languages, there are a number of relevant and interesting passages, including an analogy between nonblending in both (B: 244) and an argument for why the evolution of languages must be gradual (OUN: 5). In the *Origin* (1859), however, the connection is explicit and developed. We are told, for example, that "a breed, like a dialect of language, can hardly be said to have had a definite origin" (40), that the classification of both species and languages must be genealogical in order to be natural (422–423), and that "Rudimentary organs may be compared with the letters in a word, still retained in the spelling, but become useless in the pronunciation, but which serve as a clue in seeking for its derivation" (455). But it is in *The Descent of Man* (1871) that we find the most striking passages. Therein Darwin writes:

> The formation of different languages and of distinct species, and the proofs that both have been developed through a gradual process, are curiously the same. ... The frequent presence of rudiments, both in languages and in species, is still more remarkable. . . . . Languages, like organic beings, can be classed in groups under groups; and they can be classed either naturally according to descent, or artificially by other characters. Dominant languages and dialects spread widely and lead to the extinction of other tongues. . . . The same language never has two birth places. Distinct languages may be crossed or blended together. We see

---

[49] Interestingly, neither Herschel nor Hensleigh came to accept Darwin's theory on species evolution. Hensleigh quickly rejected it as "absurd" (Darwin 1839, Notebook E: 144), and later stated in a letter to Darwin (January?, 1860) that "As far as I see the only positive evidence we have is of degradation, not of advance as in the loss of sight in cave animals moles &c." (Burkhardt et al. 1993: 3). Herschel, who in a letter to Lyell (February 20, 1836) called the origin of species "that mystery of mysteries" (cf. Darwin, Notebook E: 59 and n59–2), rejected natural selection as "the law of higgledy-piggledy" (cf. Darwin's letter to Lyell, December 10, 1859), which Darwin took as being "very contemptuous" (Burkhardt and Smith 1991: 423). Needless to say their rejection of Darwin's theory was surely motivated by reasons other than any disanalogy between species and languages.

variability in every tongue, and new words are continually cropping up; but as there is a limit to the powers of memory, single words, like whole languages, gradually become extinct. . . . The survival or preservation of certain favoured words in the struggle for existence is natural selection. [59–61]

Given these passages alone, and the clear horizontal reality of natural languages, one might find it almost incredible to believe that modern scholars would interpret Darwin as a species nominalist (cf. chapter 2.3)![50] At any rate, given the close analogy between language evolution and species evolution, what about language extinction? Is language extinction necessarily forever? Darwin (1871) tells

---

[50] It is the failure to investigate languages as the best analogue to species that undermines not only Rieppel's (1986) interpretation of Darwin but more significantly his critique of the species-as-individuals view. Although Rieppel is not concerned to determine whether species are classes or individuals (283), he does argue that "essences need not define timeless classes only" (288) and that "As was shown by the analysis of three paradigms of individuality—the particular organism, the firm, and *Hydra*—identity in the course of change depends on some *individual* essence if it is not to be the product of mental abstraction only" (300). Thus, "The alternative is nominalism" (312). In the case of particular organisms Rieppel maintains that "The genome remains unchanged throughout the life span of an organism and thus represents the element of 'being'" (289), in other words the organism's individual genome [genotype in my terminology] "is the individual essence or underlying trait that bestows identity upon the ever-changing aggregation of atoms" (289). As such, the individual genotype that defines the numerical identity of an organism (multi-cellular) must be an abstract entity, otherwise it would have to change with every mitosis event (cf. §4.5n61). But it is precisely here that Rieppel's view fails. First, a DNA sequence or genetic program cannot serve to individuate an organism, as monozygotic twins and clones amply illustrate. Second, it is simply not true as a matter of fact that an organism's genetic program as determined when it was a zygote remains unchanged throughout its lifetime. With every mitosis event there is the possibility of mutation, such that it is conceivably possible that at the end of its lifetime an organism's original DNA program will not be instantiated in even one of its cells. And yet no one would wish to say that it is not numerically the same organism. From his false essentialism-versus-nominalism dichotomy Rieppel then argues that if species are to be individuals they too must have an essence, such that "The essence of a species (individual essence of a species *qua* individual) is made up of homologous characters (universals) abstracted from particular organisms that are held constant in the course of time; they characterize the species as a closed system" (301). He then tells us that "Strictly speaking, it is impossible for essences to evolve," so that "Species as closed systems evoke an essentialistic species concept that is incompatible with Darwinian evolution" (307). "However," he continues, "the species as unit of an open-ended process of continuous change can no longer be objectified" (308). Thus Darwin was a species nominalist (304 and 307) and evolutionary biology requires species nominalism: "Continuity dictates a nominalistic view of species—it emphasizes process, thus rendering pattern a matter of arbitrary lines of demarcation" (313). That all of this is misconceived in the extreme becomes especially clear once languages are considered, as they were for Darwin, as the closest analogue for species. Languages are clearly real (in particular horizontally), they have (as we shall see) a criterion of identity, and they gradually evolve, and yet in all of this there is nothing essentialistic. Indeed languages are probably the clearest refutation of the essentialism-versus-nominalism dichotomy.

us that "A language, like a species, when once extinct, never, as Sir C. Lyell remarks, reappears" (60). But I suggest that the word "never" here should not be taken very strongly. In the *Origin* (1859) Darwin tells us that "When a group has once wholly disappeared, it does not reappear; for the link of generation has been broken" (344; cf. 475). But as we have seen a few pages above, Darwin makes it clear elsewhere in the *Origin* (313) and in his correspondence that he does not mean that extinction is necessarily forever, only that the reemergence of an extinct species is extremely unlikely. When Darwin, then, says above that a language that goes extinct *never* reappears, I strongly suggest that his use of the word "never" means no more than what it means on page 313 of the *Origin*, as we have seen above; in other words it means *probably never*, not *necessarily never*.[51]

And indeed, how could the extinction of a language be necessarily forever? Languages, unlike many other kinds of entity, have a crystal-clear criterion of identity: two languages (minimally different if only in the sense of having different names) are identical *to the degree* that users of each understand each other (without accompanying gestures). Thus, there is in principle no reason why a dead language may not by some means be resurrected or even (however incredibly unlikely) re-evolved. (As regards the former scenario, I've been told that this has actually occurred with both the ancient Hebrew and Egyptian languages, although the experts disagree.) Moreover, given the ontological priority of the horizontal dimension over the vertical, an extinct horizontal language made horizontal again may the second time around take a different evolutionary path. But that, of course, would in no way diminish the identity (in degree) of the two horizontal manifestations.

If languages are much better analogues to species than organisms, then it could be argued that, just like languages, species extinction is *not* necessarily forever. (Thus the movie *Jurassic Park* was not founded on a logically false premise.) Indeed genetic engineering, just as in the issue over multiple origins, may play a decisive role here as well. As Dawkins (1986) put it, "In theory, if we were skilled enough at genetic engineering, we could move from any point in animal space [genetic hypervolume] to any other point. From any starting point we could move through the maze in such a way as to recreate the dodo, the tyrannosaur and trilobites" (73). If only! At any rate, if we could ever do such things I can see no good reason, given the many problems with the species-as-individuals view, for

---

[51] Lyell's remark is to be found in *The Antiquity of Man* (1863), in which he says "a language which has once died out can never be revived, since the same assemblage of conditions can never be restored" (467). It may not be clear whether Lyell's claim here is a claim of physical necessity or merely only probability. However, it is surely significant that much earlier Lyell (1838) stated only as a simple matter of geological observation that no species "ever reappeared after once dying out" (275; cf. Darwin, Notebook E: 105). Of much greater significance is Lyell's (1830) postulation of a vast, geologically recurring "great year" such that, as quoted earlier, "The huge iguanodon might reappear in the woods, and the ichthyosaur in the sea, while the pterodactyle might flit again through umbrageous groves of tree-ferns" (123).

refusing to name the new creatures by the older specific names (so that the obstacle to their recreation is technical, not ontological). Thus a genetically engineered dodo would not be like a replica of a Tiffany lamp; it would be like a Tiffany lamp, the real thing. That it does not belong to the original period is disanalogous and irrelevant. If it looks like a dodo, if it moves like a dodo, if it has the DNA of a dodo, etc., then it's a dodo! And if we were to make a population of them, they might similarly go extinct, or they might go on to a new evolutionary path. But either way, the two horizontal populations separated in time would both be dodo populations.[52]

Finally, there is the problem of asexual species. On the individuality view, asexual organisms cannot form species because each organism is reproductively isolated from every other organism. This view, of course, is not new. Generally, those who accept the biological species concept also accept that asexual organisms do not form species. We have already seen Mayr's view on the subject. We could, of course, distinguish between different kinds of species, namely sexual and asexual. But as Dobzhansky (1937) put it, "all that is saved by this method is the word 'species'" (321).

Within the ranks of the species-as-individuals view, however, the situation is much more interesting. Ghiselin certainly finds no difficulty with rejecting the possibility that asexual organisms may be parts of species. As he put it (Ghiselin 1974), "It seems rather like creating imaginary firms for the self-employed" (283). Elsewhere (Ghiselin 1987a) he says "Not every elementary particle in the universe is part of an atom" and that asexual species should be called "pseudo-species" (138). Most recently (Ghiselin 1997) he has reaffirmed that "An asexual biological species is a contradiction in terms" and that "There is no compelling reason to insist that every organism should be a part of a species" (305). Sober (1993) states that "The asexual organisms that exist at a time do not comprise a breeding population. They may trace back to a common ancestor that resembles them all, but the organisms at a given time go their own ways. A lineage of asexual organisms constitutes a historical entity, but it isn't a biological individual" (155). Wilson (1992) states that "The vast majority of species are in fact sexual" (49), so by implication he is at least willing to use the label, but nevertheless he thinks that asexual species are arbitrary. He recognizes that "Asexual and self-fertilizing forms tend to maintain a remarkable integrity," but he finds himself forced to conclude by the lack of gene pools that "In the end, . . . the lines drawn

---

[52] Languages are not the only good analogues. Theories would do the trick as well. At one point Hull (1978) was willing to consider theories as "the most likely analog to species" (311). Theories in some analogous way appear to evolve. But theories need not be monophyletic. Mary Williams (1985: 589), for example, claims that she arrived at the conclusion that species are individuals quite independently of Ghiselin. Moreover, a theory no longer believed in might be believed in at some later date. This is because theories, like languages, have a crystal-clear criterion of identity: two theories are identical *to the degree* that they share the same predictions and prohibitions. So once again we have entities which are historical and which evolve (so that they are not abstract classes) but for which extinction is not necessarily forever (so that they are not concrete individuals either).

by the biologist around such species must be arbitrary" (48).

Not everyone who accepts the individuality thesis is so sure, however. Hull is more sensitive to the issue. According to Hull (1987), "Just as not all organisms belong to colonies or some other form of kinship group, not all organisms belong to species. But from a taxonomic perspective, this conclusion is dissatisfying. Just as every library book must be placed on some shelf somewhere in the library, there is a strong compulsion among systematists to insist that every organism must belong to some species or other" (179). Indeed he can even be found saying that "Ideally, an adequate theory of biological evolution must apply to all organisms—plants as well as animals, clonal and aclonal forms alike" (Hull 1992: 185). Eldredge and Gould (1972), more interestingly, although they did not yet subscribe to the species-as-individuals view, claim that "The arrangement of many asexual groups into good phenetic 'species,' quite inexplicable if interaction is the basis for coherence, receives a comfortable explanation under notions of homeostasis" (223). Indeed even Gould (1977) can be found saying that "All organisms must belong to a species" (233).

Most interesting, however, is Wiley's position. As we saw in chapter 3.4, Wiley explicitly subscribes to the species-as-individuals view. However, Wiley's ontology involves three categories, namely classes, individuals, and historical groups, spatiotemporal restrictedness being what distinguishes the second and third categories from the first and the absence of cohesiveness being what distinguishes the third category from the second. Thus the species category, Wiley (1981) tells us, is an example of a class since "we might expect to find evolutionary species anywhere in the universe where organic evolution has occurred" (74). And species taxa are individuals because they "are restricted to particular spatiotemporal frameworks and have both cohesion and continuity" (74). Higher taxa, however, although they are historical entities, are not individuals: "there is no active cohesion within a natural supraspecific taxon because it is comprised of individual evolutionary units which have the potential to evolve independently of each other" (75). Thus higher taxa are "historical groups derived from individuals" (75). What is so interesting is that Wiley includes asexual species as individuals, even though "In asexual species identity may be manifested only in phenotypic and genotypic similarity" (25). Wiley (1978) explains his reasoning as follows:

> asexually reproducing lineages have existed and do exist. . . . I suggest that asexual species can be accounted for under the evolutionary species concept in the same way we place allopatric demes into a single sexually reproducing species. We may ask, "why are the Siberian and North American populations of wolverines considered the same species?" . . . One might argue that there must be migration which keeps the populations from diverging morphologically. There is no evidence at hand for this, just the a priori assumption . . . . I suggest that Siberian and North American wolverines are considered the same species because we have no corroborating evidence that they have reached a point of divergence where we can deduce that they are following separate evolutionary pathways. . . . It is not the evolution of asexual reproduction which "permits" us

to consider genetically separate clones as a single species, but their lack of significant evolutionary divergence. Any species composed of allopatric demes or
asexual clones has the potential for splitting into two or more separate independent evolutionary species, but it does not follow that any significant divergence
must occur. Lack of differentiation is as valid a historical fate as differentiation
and it may have real genetic, epigenetic, ecological, or other bases. [86–87]

Wiley's species concept is objectionable on a number of grounds, as argued in
chapter 3.4. What is of interest here is Ghiselin's response. According to Ghiselin
(1987a), "What makes an entity a whole is not 'identity,' in Wiley's sense of
mere distinctness from other things, but 'integrity,' or 'integration.' Wiley (1981)
rightly asserts that such integration is a necessary condition for a group to do anything. Clearly, however, mere 'similarity,' be it genotypic, phenotypic, or whatever, does not integrate a group" (140). As Ghiselin (1989) put it, "sex integrates
species. Were they not integrated they could not do anything, such as evolve"
(57). But given the many problems with gene flow that we've examined earlier
(including my argument that Ghiselin's criterion of reproductive competition fails
to take its place), surely Wiley has a point, so that similarity should not be dismissed so easily. Add to this Templeton's (1989) point that "the asexual world is
for the most part just as well (or even better) subdivided into easily definable
biological taxa as is the sexual world" (8).

At any rate, it is interesting to note the category into which Ghiselin places so-
called asexual species. In a personal communication (1995) Ghiselin wrote to me
that "I do not deny that asexual species are individuals. Rather they are spatiotemporally restricted entities without the sort of integration that makes biological
species play the role in theory and in nature that they do. I would compare them
to such higher-level non-integrated things as phyla or orders or monophyletic
groups of lineages of languages." Thus Ghiselin's ontology involves "two kinds
of individuals," sexual species being one kind and asexual species and higher
taxa another (cf. Ghiselin 1997: 54).

There are two issues that need to be carefully examined here. The first is
whether it is really true that sex is necessary for species integration. The second is
whether integration itself, and not just the kind provided by sex, is really necessary for biological evolution.

Beginning with the first issue, Ghiselin has recently conceded (Ghiselin 1997)
that for species proper (i.e., sexual species) "For all we know something else [in
addition to sex] may hold them together too, and such possibilities are well worth
exploring," in other words, that his definition of species "can be cast in terms that
are rather noncommittal about what the cohesive forces are" (104). Nevertheless
Ghiselin refrains from extending such possibilities to so-called asexual species.
Instead he maintains his view, as we have seen above, that "asexual species" is a
contradiction in terms, the reason being that "a species has to be a population . . .
in the broad sense of a reproductive community" (93).

If it could be shown, however, that in the biological world the cohesion that
Ghiselin requires for genuine species not only can, but in fact is, accomplished by

means other than sex, and by asexual organisms at that, then Ghiselin's exclusion of asexual organisms as genuine parts of species falls by the wayside, as well as the requirement in his species concept (at least his bioeconomical version) that the defining mode of reproduction be sexual.

Although controversial, it is quite possible that the kind of cohesion I am referring to is in fact instantiated in the bacterial world (kingdom Prokaryotae). Bacteria, of course, are unnucleated, primarily unicellular organisms which were the earliest and relatively for quite awhile the only forms of life (cf. Margulis and Schwartz 1988: 25–30). Although bacteria routinely exchange genes by means such as transduction, transformation, and conjugation (means which I need not get into here), gene exchange does not play a role in their reproduction. Instead they reproduce by fission (splitting), so that from the viewpoint of reproduction they are clearly asexual. In spite of this asexuality, the traditional practice among bacteriologists, traditional still today, is to divide bacteria into species and genera with Latinized binomials based partly on their morphology and partly on what they do, along with the admission that the divisions are basically arbitrary (cf. *Bergey's Manual of Determinative Bacteriology*, the most authoritative book on the subject).

However, in an interesting twist, a new view has been emerging, headed mainly by the bacteriologist Sorin Sonea, a view which has gained amongst its following the likes of Lynn Margulis (cf. n46 above). On Sonea's view, "bacterial species" (plural) is a misnomer. (No problem here, since Ghiselin and Hull would agree.) Not only is the gene exchange between bacteria such that "we cannot reconstruct their phylogeny" (Sonea and Panisset 1983: 123), but "bacterial cells are genetically incomplete" (8). As Sonea (1991) put it, the largest and most stable piece of DNA in a bacterium, the "large replicon" or "chromosome," "usually possesses fewer genes than are necessary for a bacterial strain to exist alone in any natural environment" (99). The nature and frequency of gene exchange between bacteria is such that "Each bacterium has access by undirectional transfer (with diminishing probabilities as metabolisms differ) to most or probably any bacterial gene in nature; gene exchanges are frequent, normal phenomena in bacteria" (Sonea and Panisset 1983: 4). Moreover, "each bacterial cell benefits from a very large gene pool (all the genes of a population that exchanges genes among its members) that probably extends to all bacterial genes on Earth" (8–9). Thus the concept of closed gene pools within the bacterial world, each with a separate evolutionary fate, does not at all apply. As Margulis (1993) put it, "'Species' that change up to 15 percent of their genetic constitution on a daily basis can no longer be called species. . . . 'speciation of bacteria' is an oxymoron, since all bacteria are linked through their promiscuous genetic system" (99). Most importantly, Sonea argues that this global bacterial gene pool makes the bacterial world a single species, or, similar to the species-as-individuals view, a single integrated individual. As Sonea and Panisset (1983) put it, "The specialized bacteria are not separate biological species, just as the different cells in our blood are not. Bacterial strains are the specialized cells belonging to one superorganism—the planetary bacterial entity—just as our blood cells belong to one superorganism, the body" (122–123).[53] Moreover this entity, this "vast communications network" (9), has been increasing in integration over its roughly 3.5 billion-year history, revealing "a unifying tendency in its evolutionary pattern" (Sonea 1991: 102) such that it "seems to have reached its own type of adulthood" (100). Part of the evidence and consequence of this adulthood is the increasing problem in bacterial control, in other words the often decreasing effectiveness of antibacterial drugs, notably penicillin. Given the ability of genetic intercommunication between different strains of bacteria, genes that confer resistance to penicillin are eventually and increasingly passed to the strains which are the target of biological control (Sonea and Panisset 1983: 22, 86–87).

All of this, of course, does not entail that, say, asexual plants may likewise be parts of biological species/individuals. But it does add to the many difficulties of necessarily excluding asexual organisms as members of such species. Perhaps what so worries Ghiselin and others is that once asexual organisms are admitted as being members of genuine species, genuine species can no longer be viewed as genuine mereological wholes, with the individuality thesis consequently unraveling at the seams. But mereology, as we've amply seen, is a huge problem with sexual species as it is.

At any rate, although sex might be sufficient for species integration, we have seen from bacteriology that it is not necessary. But is integration, in turn, necessary for evolution? This is the second issue that I raised above. And once again the answer would seem to be in the negative.

One way of looking at the issue is to focus on the concept of gene pool, since evolution is often defined, predominantly in population genetics, as a change of gene frequencies within a population (cf., e.g., Futuyma 1986: 551; Sober 1993: 1). Of course much depends on how one defines "gene pool" and "population." For some biologists, interestingly, and not just for bacteriologists such as Sonea, asexual organisms can form or be part of gene pools. Simpson (1961), for example, argues that horizontal gene transfer is not the only way a gene can increase in frequency in a gene pool: "In both cases ['biparental and uniparental populations'] the spread of genes occurs by differential reproduction, genes increasing if their possessors have more offspring and decreasing or eventually disappearing if they have fewer. In other words, the role is determined and is circumscribed by natural selection equally in the two cases. Biparental reproduction produces recombinations of genes, increases variability, and makes the species more adaptable, but it has no decisive bearing on gene spread within the population" (162).[54] Simpson's concept of gene pool, of course, lacks what it takes to be what Mayr (1970) calls a *protected* gene pool, namely "its own devices (called isolating

---

[53] Although Sonea began publishing his views in 1971, there is no explicit evidence that his individuality thesis for the bacterial world was influenced by Ghiselin's individuality thesis for the sexual world. If there was no influence, the coincidence and its timing are truly remarkable.

[54] Although in a minority, Simpson is by no means alone in this view; cf., e.g., Templeton (1989: 8–9), Foottit (1997: 301).

mechanisms) to protect it from harmful gene flow from other gene pools" (13). However, since asexual populations do not (or may but need not) have horizontal gene flow, it might be argued that they don't need isolating mechanisms to protect them, so that they nevertheless meet Mayr's (1970) basic definition of gene pool: "The totality of the genes of a given population existing at a given time" (417). At any rate the point I wish to make is that if one agrees that asexual organisms generally subdivide, as Templeton and so many others have claimed, into easily recognizable taxa as well as if not better than sexual taxa, and if in virtue of this fact these taxa may be said to have (in a sense) gene pools, as Simpson has claimed, then it's difficult to see why asexual organisms cannot also belong to species which in turn evolve in the sense of evolution defined in population genetics. Ghiselin (1997) claims that the term "population" is best understood as "a group of things which interact with one another" and that "The best test of whether or not a group forms a population is to ask whether or not it is possible to affect one member of the group by acting on another" (15). But this is a prime example of a persuasive definition, and at any rate it fails to apply for many sexual species, as Ehrlich and Raven (1969), Levin (1979), Templeton (1989), and so many others have pointed out. The bottom line is that from the viewpoint of natural selection and biological evolution there is no reason why asexual organisms cannot form populations complete with gene pools (in the abstract sense argued for in §1) and evolve in the sense of evolution defined in population genetics.

Indeed strange things happen when the opposite point of view is taken. Hull (1988), for example, states that "Species did not come into existence until sexual reproduction succeeded in evolving. For the first half of life here on Earth, no species existed—just organisms" (215; cf. 429n3 and accompanying text). Indeed there is something odd in saying that for a great part of the history of life on this planet there were organisms but no species. It's like saying that until the advent of syntactic languages there were no languages at all, that purely semantic languages don't count. Indeed it doesn't say much for the bacterial world.

But what is even more interesting is that on the view of Ghiselin and Hull asexual organisms can indeed in some sense evolve. Ghiselin (1989) is willing to admit that "Individuals that are not populations can evolve and undergo other kinds of change—asexual lineages are a good example" (58; cf. Ghiselin 1997: 55). Similarly Hull (1980) holds that "strictly asexual organisms form no higher-level entities; organism-lineages are the highest-level lineages produced. They alone evolve as a result of replication and interaction. . . . these organism-lineages *are* the species in asexual organisms" (328). But if asexual organism-lineages are spatiotemporally localized and can evolve, then one has to wonder what all the fuss is about! The inescapable conclusion is that although integration is sufficient for species to do things such as evolve, it is apparently not necessary. Consequently modern *evolutionary* theory does *not* require species to be individuals, at least in the sense of cohesive wholes (cf. §2n21 and accompanying text). Indeed it arguably does not need species at all, only lineages (but that is a view fraught with difficulties itself; cf. the end of §5).

In closing this section, Ghiselin (1981) tells us that "to judge from the current literature, my views are the wave of the future" (305) and later that (1987b) "We might say that a kind of Copernican revolution has occurred. Much as the earth is no longer the center of the universe, the priority relationship between classes and individuals has been reversed. The mere fact that Aristarchus of Samos or Le Comte de Buffon entertained some analogous notions does not affect that issue" (209). Similarly Hull (1987), speaking obliquely of his own role, says "this is one of the rare instances in which professional philosophers have played a salutory role in promoting an important conceptual shift in science" (183).

It seems to me, however, upon surveying the whole scenario of the modern species problem, that it is indeed premature to characterize the revolution, though indeed begun by Ghiselin, as a *fait accompli*. It's much too early to tell. The counterrevolutionaries, though not united, are numerous and powerful. There are many Russian fronts and possible Waterloos. The revolution will probably be protracted. The old guard may have to pass, leaving the fate of the revolution to future generations. From my own vantage point, the tide appears to have already begun to turn, and the revolution, though brave and heroic, seems destined to fail, its two manifestos (Ghiselin 1966, 1974) and its two *magnum opuses* (Hull 1988, Ghiselin 1997) fated along with so many others for the dusty shelves and micro-texts of posterity. Quite simply the species-as-individuals view has far too many problems to guarantee its success. But time may prove me wrong. At any rate, its lasting value, it seems to me, is in making one think of species more as concrete entities rather than as abstract entities. Exaggeration, though, has its place. Eventually the time must come for a less radical view.

4.5 Species as Sets, Clades, and Lineages

As we've already seen, the species-as-individuals view of Ghiselin and Hull is not the only physicalist conception of species. In this section, I will examine a number of other physicalist conceptions, namely that species are sets, clades, and lineages.

The view that species are sets has been defended most prominently by Philip Kitcher. As Kitcher (1984a) put it, "Species can be considered to be sets of organisms, so that the relation between organism and species can be construed as the familiar relation of set-membership" (309). We have seen in chapter 2.4 that Kitcher combines his species-as-sets ontology with species pluralism. As Kitcher (1984a) put it, "Species are sets of organisms related to one another by compli-cated, biologically interesting relations. There are many such relations which could be used to delimit species taxa. However, there is no unique relation which is privileged in that the species taxa it generates will answer to the needs of all biologists and will be applicable to all groups of organisms" (309). Thus the rela-tions which the paleontologist studies may result in different sets of species than the sets which result from the relations studied by the anatomist, or the evolution-ary biologist, or the population geneticist, or the ecologist, or the ethologist, etc., each of these producing "alternative legitimate conceptions of species" (319). Accordingly, the set ontology serves as the common denominator underlying each of the various species concepts, each species being a set of organisms picked out in a legitimate biological pursuit.

Needless to say, Kitcher's view has not been well received. Mayr (1987a) complains that "According to his definition, any aggregate consisting of more than a single item is a set, no matter how heterogeneous. To illustrate his concept of a set without a defining property Kitcher lists 'Queen Victoria, the manuscript of *Finnegan's Wake*, and the number 7.' . . . Frankly, a biologist is utterly bewildered. For more than 100 years biologists have worked very hard to discover and describe how biological species differ from other phenomena of nature and whether there is only one kind of species or several different ones. All this is obliterated if all variable phenomena of nature are dumped into a single highly heterogeneous receptacle, the set" (149). Hull (1987) points out that "Although Kitcher thinks that restricting the term 'species' to apply just to those things that evolve is too monistic, he is willing to reject two widely-held species concepts— those of Creationists and pheneticists" (177). Ghiselin (1987b) argues that "Set theory, being a purely logical apparatus, treats it as indifferent whether things be individuals or classes. In consequence thereof, all the individuals get turned into classes" (207).

All in all, these criticisms miss their mark. Kitcher makes it clear that his set ontology is not meant to apply to everything (I've been unable to find where Mayr finds the Queen Victoria example), nor that he believes that every perspective on the biological world is scientifically legitimate (there is thus no self-contradiction, since similar to Hull his view is theory dependent), nor that he is indifferent as to whether things are concrete or abstract (as we shall soon see). My own problem when I first read Kitcher (1984a, b) was that he failed to clearly define what he meant by the word "set" and to disassociate his meaning from the different uses of that word by others.[55] But Kitcher was certainly to later come clean on this score. According to Kitcher (1987), "the conception of 'class' that is foisted off on the philosophical opposition is a bastard notion that deserves no place in anybody's ontology." (Thus Kitcher rejects the concept of abstract

---

[55] Kitcher is by no means alone on this score. Cf. the interesting discussions by Max Black (1971) and Ruth Marcus (1974) on earlier examples mostly from logic and set theory textbooks. Within the literature on the species problem, another good example along with Kitcher is Mary Williams (1985), who, in defending the species-as-individuals view, muddies her own waters when she repeatedly makes statements such as "Thus if species X (which is a collection of organisms) is an individual with respect to ET, then in any axiomatization of evolutionary theory there must be a primitive term that denotes a collection of organisms" (580). Surely Williams does not mean what she says here and elsewhere, for otherwise her view is basically the same as Kitcher's, *viz.* that species are collections of organisms from a theoretical point of view. (As we shall see, Kitcher did not use the word "collection" for species until Kitcher 1987.) What only adds to the muddle is that, as Marcus (1974) points out, the word "collection" connotes "A group of things that have been gathered together" (91). The muddle is that the respective species realisms of both Williams and Kitcher require (for quite different reasons) that species be more than collections, that they must in some strong sense be extra-subjective.

classes found in Quine, Ghiselin, and so many others.) "The respectable concept," Kitcher continues, "is that of a set. In the usage of mathematicians and philosophers sets are collections of elements, and the elements in the collection bear
the relation of membership to the set" (185).[56] Kitcher's use of the word "set,"
then, is clearly in accord with mine (cf. chapter 1.3). A set is nothing more than
its members, sets are identical when their members are identical, and any change
in a set's membership is necessarily a change in that set (186). What is especially
interesting is that Kitcher thinks that the notion of class, abstractly conceived,
should "be banned from any future discussion on the topic of species" (186). In
fact even the species category, on Kitcher's view, should not be conceived as a
class! Instead, "The species category is a set of sets (the species taxa)" (187; cf.
n60 below for my reply). This still seems to me a subjectivistic policy, since species as sets do not exist independently but rely on being picked out by biologists
(since all organisms constitute one big set). But I will not repeat my arguments
from chapter 2.4. It need only be noticed here that on Kitcher's view the biological relations which figure in those choices do not themselves constitute any part
of a species' ontology. A species *just is* its member organisms and nothing more.

What is significant is that Kitcher does not think that a set ontology has universal application. When it comes to organisms themselves, Kitcher (1984b) says
"I hold no brief for the view that organisms are just sets of cells. . . . My proposal
was *local*; I sketched a simple way to understand our talk about evolving species
by construing species as sets. Blanket Pythagoreanism is not my project" (617).
That organisms are not just sets of cells, however, points to an important issue.
Hull (1987) says "I sympathize with Mayr's . . . bewilderment in reading Kitcher's (1984a, b) discussion of species as sets. . . . he is unwilling to do the same
for organisms. Organisms are not 'just' sets. But if species as lineages can be
construed as sets defined in terms of organismal descent, I see no reason to preclude the definition of organisms in terms of cellular descent. . . . until Kitcher
explains why set theory is so impoverished when it comes to treating organisms,
his view remains a mystery to me" (173).

Kitcher (1989) provides two answers. First, "it is in principle possible for the
organization of the cells that make up a complex organism to be destroyed while
each cell persists. The set of cells remains but it is no longer an organism." On
the other hand, "Does a species continue to exist when we disrupt the relations
among the organisms that are (on the set-theoretic view) members of it? I believe
that a case can be made for an affirmative answer. If an endangered species becomes scattered so that human intervention is required if its remaining members
are to reproduce, then there remains a chance of preserving the species: that, of
course, is what motivates efforts that people sometimes make" (186). Second,
Kitcher argues that if interbreeding relations must be taken into account, then a
species can still be a set-theoretic entity by employing "the ordered pair of a set
and a relation," while on the other hand in the case of the cells of an organism

---

[56]   Indeed Kitcher (1987) claims to have developed "a method of rewriting set theory so
as to avoid commitment to abstract objects" (191n1).

"we can only guess at the complexity of the relations that would have to be adduced. . . . another disanalogy between organisms and species" (186).

There are a number of problems with these replies. To begin, Kitcher's first reply seems flawed simply for lack of a proper time scale. Given the right time scale, all of my cells could conceivably be scattered and then brought back together without harming my being. Notice also that both of Kitcher's replies imply that species are something more than just sets (collections) of organisms. Instead it seems now that relations are creeping into the ontological makeup of species. (I will have more on relations and ordered pairs in chapter 5.3.)

Why is it that an organism is more than just a set of cells? This is an important question. Throughout the history of philosophy there has been a strong tradition against the reality of relations, along with a tradition of reductionism which became especially prevalent with the rise of modern physics. Ontological reductionism, of which eliminative materialism in philosophy of mind is only a side branch, says nonphysical things are reducible to physical things and physical things are nothing more than their parts.[57] On this view, then, neither you nor I are real, nor anything else commonly thought real. Instead all that is real are the wee little things, the ultimate elementary things: formerly electrons, protons, and neutrons, then quarks followed by an assortment of other odd entities, and now extremely tiny vibrating strings if one accepts the latest theory in elementary particle physics (cf. Weinberg 1992: 212–219, whose ontology is nevertheless more inclusive). How is it, then, that we too are real, and not just our parts? Or are we all, and all that we see, no more real than phlogiston? Granted that we may agree with Quine (1981) and "see all objects as theoretical" (20), surely these tiny strings are far more theoretical than you or I, for *we* at least can be pointed to. But how is this possible, if string theory (or something like it) is true? As with Odysseus' descent into the Underworld and his return to the world above, there seems to me a way out of the world of ontological reductionism, radical or otherwise, a way that brings us back to the sunny world of rocks and trees and common sense and even, yes, abstract entities. The answer, it seems to me, is only possible if, unlike most nominalists, we admit relations into our ontology. In the case of physical things, what makes a whole a whole, whether it is an organism or a rock, is not just the sum of its parts but also the relations between those parts which contribute to making it what it is. (Holism is thus vindicated, without requiring any vitalism in the case of living things.)

---

[57] This view, of course, was given classic expression by the astrophysicist Sir Arthur Eddington (1928), who in his famous discussion on two tables (the table of the common man and the table of the modern physicist) argued that only the table of the physicist is "really there," by which he meant "mostly emptiness," "numerous electric charges rushing about with great speed," and "fields of force" (ix–xii). Equally classic is Hempel's (1966) reply to the view expressed by Eddington—the view that "the underlying structures, forces, and processes assumed by well-established theories . . . [are] the only real constituents of the world" (77)—namely that "to explain a phenomenon is not to explain it away. It is neither the aim nor the effect of theoretical explanations to show that the familiar things and events of our everyday experience are not 'really there'" (78).

What is true for cells and organisms, however, is also true for species. A species cannot be constituted just of organisms. For one thing, if a species is just its member organisms, then the only way a species can change is quantitatively, by undergoing a change in membership. There is now no possible sense in which a species could stay the same even though its members come and go. All qualitative change, including any stasis, is an illusion and biologists are wrong. This is surely unacceptable. Mere quantitative change does little justice to the nature of species change and to the intuitions and observations of biologists.

What is required, then, is something more: not an ontology just of organisms, but an ontology that also includes relations. As Holsinger (1984) put it, "a [species] taxon is not just a collection of individual organisms. It is composed of individual organisms that interact with one another, are related to one another in particular ways, and participate in biological processes in similar ways" (296). This, of course, does not necessarily mean that species are individuals (indeed problems for that view abound), nor does it justify species pluralism (as Holsinger thinks). The great mistake of Kitcher (and many others) is that he failed to appreciate the priority of the horizontal dimension over the vertical dimension. Failing that, he embraced species pluralism. But he wanted species realism as well, and the only way he could imagine having both was by employing a set-theoretic ontology as the common denominator. One mistake led to another.

Nevertheless there is value in Kitcher's view, in that for him (as with the species-as-individuals view) individual organisms play a constitutive role in the ontology of species. This seems to me important, for otherwise much of what biologists say and do makes little sense. Moreover, Kitcher is clearly sensitive to the reality and importance of objective relations in the biological world, including both interbreeding relations and similarity relations (cf. Kitcher 1984a: 319, 1984b: 625). His problem is that he has not been able to fully free himself from the overbearing philosophical tradition of relations denial, so that he can't quite bring himself to admit into the ontology of species any of the important biological relations that he otherwise recognizes.

Much more congenial to the species-as-individuals view of Ghiselin and Hull (though not necessitated by it) is the group of species concepts known as *cladistic*. These species concepts are based on cladistic taxonomy, a school of taxonomy which is founded on the work of Willi Hennig. Of the three main schools of taxonomy (the other two being numerical taxonomy and traditional evolutionary taxonomy), I think it is safe to say that (for better or for worse) cladistic taxonomy has come to be the most widely accepted.

As the word "clade" is Greek for branch, a cladistic species is a branch of some sort on the phylogenetic tree of life. Although Hennig (1966) recognized the reality of species at the horizontal level according to the biological species concept of Dobzhansky and Mayr, he did not stop with that concept but utilized it in constructing his own species concept for the vertical dimension, "the temporal duration of a species" (66), which he perceived as primary.[58] Rejecting the "pale-

---

[58] The impression given for why Hennig did not rest satisfied with the horizontal dimen-

ospecies" and "chronospecies" of the paleontologist as arbitrary, since these temporal slices are "purely morphologically defined concepts" (62) and would become "impossible" (63) with a perfect or complete fossil record, Hennig conceived of the full reality of a species as follows:

> The limits of a species in a longitudinal section through time would consequently be determined by two processes of speciation: the one through which it arose as an independent reproductive community, and the other through which the descendants of this initial population ceased to exist as a homogeneous reproductive community. [58]

Elsewhere he wrote of "two successive processes of species cleavage . . . [as] the temporal delimitation of the existence of a species" (64), admitting all the while that "Since the speciation process . . . requires a long time, there is a certain unsharpness in the delimitation of species" (66).[59]

One interesting feature of Hennig's cladistic species concept is that the cleavages which delimit the temporal ends of a species may be either the splitting kind of cleavage or the budding kind of cleavage (58–59). That Hennig made no mention of real extinction here, only branching speciation as delimiting the temporal ends of a species, is certainly the most glaring oversight in his presentation of the cladistic species concept. Of course, it is an oversight that is easily remedied. What remains peculiar about Hennig's species concept, however, is that according to it a species automatically goes extinct not only when it speciates by splitting but even when it produces a new species by budding. Mark Wilkinson (1990) has named this "Hennigian extinction" (436), a name that I think is particularly apt, not only because Hennig originated the concept but because no one outside of Hennig and his followers would ever be inclined to subscribe to it.

---

sion of Dobzhansky and Mayr is because of the many problems he found with species defined only at that dimension, such as the lack of actual interbreeding between allopatric populations of the same species (45), the phenomenon of ring species (47), and the problem of operationality (52). Hennig apparently seemed to think that these problems dissolve once the horizontal dimension is extended into the vertical dimension and the latter taken as the defining dimension of species (52–53).

[59] Interestingly, Hennig went some distance in characterizing species as individuals. The definition of individuality that he accepted is that individuals are "operational units" such that "an individual is a unit both externally (in its effects on others), and internally (in the reciprocal action of its components)" (82). He then added that "It is more difficult to determine the particular individuality character of the species. There can be no doubt that, like the higher categories of the phylogenetic system or of any other divisional hierarchy, they have 'place or duration in time.' It is questionable, however, whether the species can be regarded as operational units within their environment" (83). Where the difficulty arises is that although some species comprise "a true closed population and reproductive community," others "consist of incompletely isolated vicarying reproductive communities [that] form a transition to the categories of higher rank" (83). Thus for Hennig some species are individuals (much like organisms) and some are not (much like higher taxa), though all are historical entities (cf. Wiley and Ghiselin in §4 above).

Although Hennig is often unclear in his exposition of the cladistic species concept and goes only a little way to defend it, these shortcomings are largely made up for in the clear exposition and able defense of the cladistic species concept provided by Mark Ridley (1989). Ridley defines species on this concept as follows:

> A species is . . . *that set of organisms between two speciation events, or between one speciation event and one extinction event, or that are descended from a speciation event.* [3]

The first point to notice, a minor quibble, is that since the definition entails that a speciation event is *ipso facto* also an extinction event, the phrase "one extinction event" in the second disjunct of the definition should be preceded with a qualifier to distinguish it from Hennigian extinction. It is apparent enough that the definition means non-Hennigian extinction, commonly known in biology as *real* extinction, which for clarity we might rename *dead-end* extinction (since for cladists Hennigian extinction is just as real).

A second point of interest is that the third disjunct in the definition is an absolute requirement, for without it only dead species would be real species and accordingly species could not possibly evolve. Hennig's definition opens itself to this criticism, which alone makes Ridley's revision a clear advance.

A third point of interest is that Ridley's definition explicitly uses the concept of a *set* of organisms so reminiscent of Kitcher's set ontology for species. Even the species category, following Kitcher, is defined not as a class but as a set: "The cladistic definition of the species rank is the set of all the longest possible branches in the phylogenetic tree that do not themselves have splits in them" (6). However, unlike Kitcher's pluralism when it comes to the legitimate different ways in which sets of organisms may be picked out and collected into species, Ridley obviously recognizes only one legitimate way and is therefore a strict species monist. Nonetheless, for both Kitcher and Ridley the species category, in virtue of being a set, is constantly changing as the members of the set come and go![60]

Of further interest is Ridley's claim that cladistic species are fully physical individuals: "The individuality of species demands the cladistic species concept. . . . Species, in evolutionary theory, are individuals, not classes. The reason

---

[60]   As we shall soon see, Kornet (1993) also defines the species category in the above way, namely as "a set of sets" (423). The problem with this view is that there clearly is a sense—just as with the class of gold atoms, the *kind* gold—in which the species category does not change even though its members come and go. That sense cannot be captured if the species category is a set, but it can be captured if it is a class. Moreover, the species category cannot be a subject of laws of nature if it is a set (since laws of nature don't change); to be such a subject it must be a class. Since a strong case can be made for the view that the species category is the subject of laws of nature, as in Universal Darwinism (cf. chapters 2.3, 3.5, and 4.2), it would seem highly inappropriate to conceive of it as anything but a class.

is their cladistic nature: species are branches of the phylogenetic tree, independently of their characteristic attributes" (14).

Perhaps the most radical part of the cladistic solution is the almost complete decoupling of evolution from speciation. Speciation by transformation (anagenesis) is completely excluded in favor of speciation by branching (cladogenesis). It therefore does not matter how much change a species undergoes between its terminal points: "Cladistic species may undergo infinite evolution" (10).[61] In spite of such transformation, a species would retain its numerical identity just so long as it does not undergo any splitting or budding. Similarity, therefore, although it plays a role in the discovery of a cladistic species, plays absolutely no role in the constitution or ontology of a species. This, claims Ridley, makes the cladistic species concept thoroughly objective: "The virtue of the system is that phylogenetic relations undoubtedly do exist, uniquely, objectively, unambiguously. They are out there in nature" (2). Again: "The virtue of the cladistic definition is its perfect objectivity. Species are defined unambiguously as branches" (4). Whether branches exist unambiguously and can be defined accordingly is a matter I will take up a little later. What need only be mentioned at this point is that the complete elimination of similarity from the ontology of species entails not only that "the shared derived characters of a lineage may change without any change in the cladistic species," but that "different species may not differ phenetically" (5).

Of further interest is the use Ridley makes of other species concepts, in particular the biological species concept and the ecological species concept. The cladistic species concept, says Ridley, "defines a species as a lineage between two speciation events, but it has neither explained what a lineage is, nor what speciation is. It needs to do so, and it is here that the other concepts are needed" (7). As for the ecological species concept, Ridley tells us that "Adaptive zones do actually exist in nature: they must do, otherwise species would not exist either" (9). Moreover, an adaptive zone "sets up selection to control the degree of permissible interbreeding with other types of organisms; in this sense the ecological concept explains the biological one" (7). Thus the ecological concept helps explain "why species lineages exist at any one time" (10). The ecological concept also helps explain speciation: "If the species has ecologically split then it has formed two monophyletic species, and that is that" (11). As for the biological

---

[61] This constitutes a major difference between the species concept of punctuated equilibria and cladistic species concepts. While the former denies speciation by anagenesis (or the prevalence of it) in fact, the latter denies it (all of it) in principle. Moreover, as we saw in §3, while Eldredge and Gould think that even moderate amounts of phyletic change would muddy the waters and destroy the individuality of species, Ridley thinks it makes no difference whatsoever to the individuality of a species whether it remains unchanged between its terminal points or changes radically. Given the radical change yet retained numerical identity in organismal metamorphosis, the nod here must go to Ridley. However, apart from the issue of individuality, that for Eldredge and Gould rapid evolution is speciation, even if it does not involve branching (as in peripatric speciation in a bottleneck situation, which they must concede; cf. Mayr 1988a: 462), seems to me a major point in their favor, as we shall see.

species concept, "Interbreeding matters in a species in much the same way as cellular interaction matters in an organism; it may be needed to cause the repro-duction of lineages, but that does not mean that the interbreeding makes the spe-cies an individual" (7). Moreover, "so long as there is a continuous, unsplit line-age, defined as a biological species at each time instant, there is a single cladistic species. Whether the earlier and later forms could interbreed is no more relevant than whether they have the same ecological adaptations" (12). Thus, says Ridley, "both the biological and ecological concepts offer a sufficient account of lineages and speciation in the full cladistic theory of species. The cladistic concept is therefore not theoretically incomplete" (8). Moreover, both the biological and ecological concepts "indicate a practical criterion by which the members of a spe-cies lineage can be recognized at any particular time" (8).

Ridley's use of both the biological and ecological species concepts, of course, does not entail a form of species pluralism, since both concepts are subordinated under the cladistic species concept: "The 'pluralism' is only a multiplicity of (ex-tremely-formulated) explanations of the cladistic lineage; it is not a pluralism in the cladistic species concept itself" (8).

What is especially interesting is what Ridley says happens when either the bi-ological or ecological concepts are taken in isolation to define species. Agreeing with Ghiselin that if the ecological concept is so taken it results in a class con-cept, Ridley disagrees that it precludes evolution; his claim instead is that it strictly limits it: "Strictly speaking, the ecological definition would not make evolution impossible, but would only underestimate it. Thus, if a species of par-rots were defined by a set of frugivorous adaptations, then the species indeed could not evolve for those defining adaptations without changing its species: but it would not be prevented from evolutionary change in its other characters. It could evolve, so long as the evolution were of non-defining characters" (10). Thus, says Ridley, the ecological concept "is not a complete species concept. It is subordinate to the cladistic species concept. Cladistic lineages have to be unified, in a particular niche, *at any one time*. But, through time, the adaptations posses-sed by a species can change, without the species itself changing into another spe-cies" (10). Thus when subsumed under the cladistic concept, the ecological con-cept cannot function as a class concept, since a species cladistically defined is a monophyletic lineage: "The adaptive zone may change through time; but so long as it exists, and does not split, the species exists too. The fact that the species may change in any way [i.e., qualitatively] without the species itself changing [i.e., numerically] gives the result that Ghiselin really wants—the species is not a class" (10–11).

Similarly with the biological species concept, since a later population in a lin-eage may not in principle be able to interbreed with an earlier population in that lineage, Ridley argues that in isolation the biological species concept severely limits the amount of evolution a species may undergo, indeed that "we should say that the biological species cannot evolve" (11–12). The argument is the old fa-miliar claim that any division of a lineage over time "is necessarily arbitrary be-cause a continuously changing lineage cannot naturally be divided into biological species" (12).[62]

There is a problem, however, with Ridley's subordination of the biological and ecological species concepts, in particular his supposed congruence for them when defining branching points. The problem is whether the supposed congruence is justified. Apparently because of his confidence in the congruence, Ridley finds no difficulty with the common objection that sexually dimorphic species might "occupy different niches." His reply, that "it would be a trivial statistical exercise, or rotation, to unify the 'different' niches" (10), is an *ad hoc* move we've seen in others in chapter 3.4. In fact, when Ridley says that an adaptive zone "sets up selection to control the degree of permissible interbreeding with other types of organisms; in this sense the ecological concept explains the biological one" (7), he is in effect contradicting himself. This is because he uses interbreeding to define niche width and therefore to partly define niches, as we have just seen in his unification of the niche of a sexually dimorphic species. This practice only serves to highlight the *ad hoc* nature of such a move, not to mention that it has the undesirable consequence of precluding asexual organisms, or a lineage of such, from possibly occupying or having a niche, which clearly serves as its *reductio*. At any rate, Ridley's use of interbreeding to define niche width raises a serious problem for branching points in his cladistic theory of speciation. It is of course quite possible, as we have seen in chapter 3.4, that the two sexes of a single species might differentiate ecologically while not (of course) reproductively. This sharply indicates that the two "subordinate" species concepts are not necessarily congruent when it comes to defining branching (the case is only more severe with cyclomorphic species), so that they are not necessarily congruent when defining both cladistic speciation and Hennigian extinction. That Ridley's supposition to the contrary is thus undermined entails serious damage to the cladistic species concept upon which it is built, for that concept necessarily requires a coherent theory and definition for branching points.

Indeed, the issue of branching points discussed above raises the question of whether Ridley's cladistic species concept includes asexual organisms. On the one hand, he mentions dandelions "which have proliferated into an asexual taxonomic chaos" (8). On the other hand, however, he immediately tells us that "Maybe shared selection pressures are more important in some species, and interbreeding in others: the controversy does not affect the cladistic concept" (8). The issue is especially important, since if asexual organisms are admitted into cladistic species, then Lynn Margulis' theory (cf. §4n46) of "rampant polyphyly" in the endosymbiotic origin of eukaryotic cells undermines one of the basic premises of cladism, namely that all taxa must be monophyletic.

---

[62] Interestingly, Ridley claims that the same result obtains if one restricts oneself to Paterson's recognition concept of species, though now with the interesting further claim that by itself it is a class concept: "the SMRS is being used to define species as a class and classes can only change within limits. But if the SMRS is only an explanation of why cladistic lineages are distinct at any one time, and the cladistic concept is accepted, then the change in the SMRS is irrelevant and the lineage forms only one species" (14).

A further problem with branching points concerns Hennigian extinction, particularly in the case of speciation by budding. As Ridley tells us, "the ancestral species automatically goes extinct when it splits. Cladistic species are monophyletic in the sense of including all the descendants from a speciation event, up to the next speciation event, when the cladistic species comes to an end" (4). Many have found this consequence absurd. According to Mayr (1988a), for example, "their claim that every speciation event inevitably terminates the life of the parent species is clearly wrong. . . . For instance, when the New Guinea kingfisher *Tanysiptera galatea* produced daughter species on Koffiao, Biak, Numfor, Tagula, and the Aru islands, this did not affect the parent population on the mainland of New Guinea. It would have been quite absurd to have ranked the mainland population as a new species every time it sent off a founder population to one of the adjacent islands" (325). Similarly Wilkinson (1990) finds Hennigian extinction absurd in consequence of a thought experiment based on species budding, wherein the continuance or extinction of the parental species depends solely upon whether the founder population goes extinct or survives to speciate: "If one isolated population subsequently goes on to found a new species, or alternatively goes extinct without forming a new species, this need have no logical bearing on the nature of change in the residual population, and, to my mind, should have no bearing on our recognition or definition of species boundaries" (435; cf. §1n1).[63]

Although Ridley's view on this matter is standard in Hennig (1966) and is shared by other strict cladists such as Kornet (1993), some cladists have taken a less strict view on the matter. Wiley (1981), for example, claims that cladism need not "preclude a particular ancestral species from surviving a speciation event. In cases where the mode of speciation is ecological—sympatric, via hybridization—or involves a small isolate, nonextinction of the larger population might be expected since the removal of these localized populations or the 'wasting' of gametes might have little or no effect on the larger population" (34). Again: "Actually, Hennig's methodological principle is not essential to phylogenetic systematics regardless of the practicalities of ancestor identification. . . . the evolutionary species concept is completely compatible with phylogenetic systematics and Hennig's methodological principle must be set aside" (53). Similarly Cracraft (Eldredge and Cracraft 1980), whose *phylogenetic species concept* will be examined below, holds that in the case of peripatric speciation (speciation by budding) the parental species "continues on unchanged" and "hence [is] not diagnosable as a 'new' species itself" (125; cf. Cracraft 1989a: 36–37; Mishler and Brandon 1987: 408). Although such a move makes much sense from a biological point of view, it should nevertheless be seen for what it is, namely a serious deviation from cladism proper, such that it may not even deserve any longer to be called cladism. To allow a parental species to numerically survive the budding off of a new species is to employ a loose sense of monophyly known as

---

[63] Cf. Eldredge and Cracraft (1980): "strict application of cladistic methodology to the problem of species recognition is not entirely consistent with the concept of species as individuals and evolutionary units" (90).

paraphyly. In effect it is to allow similarity to slip in through the back door.

Closely related to the problem above is the view of cladists (both strict and loose) that a species retains its numerical identity through what Ridley above calls "infinite evolution" (the proviso being that the species must not undergo any branching). In spite of the fact that most biologists and philosophers would probably reject such a view if it were put to them, it is a remarkable fact that virtually all cladists, whatever the variety of their cladism, submit to the above view (indeed it would seem an essential feature of cladism). Obviously implicit in Hennig (1966), it is similarly entailed by the views of strict cladists such as Kornet (1993), as we shall see below, and as we have seen above it is fully explicit in Ridley (1989). It is even *de rigueur* among less strict cladists such as Wiley and Cracraft. As Wiley (1981) put it, "The evolutionary species concept is, of course, diametrically opposite to the notion that phyletic evolution . . . can produce new species" (39). Similarly Cracraft (Eldredge and Cracraft 1980) holds that speciation "can only result from a pattern of splitting, or budding off from, an ancestral species. . . . The only alternative, the wholesale, complete phyletic transformation of one entire species into another, is inadmissable. Separately *diagnosed* species resulting from such phyletic transformation are rather to be regarded as end members of an unbroken continuum" (114; cf. Cracraft 1987: 340).

Ridley, as we have seen, defends the above view on the basis of objectivity, since numerical identity through infinite evolution does not involve similarity. But the denial of similarity, as we shall see, involves even further odd consequences, especially from a practical point of view. Wiley, on the other hand, defends the above view for quite different reasons—as we have seen in §4 above, he is not uncongenial to similarity. His first two reasons (Wiley 1981) are that *"The recognition of phyletic species is an arbitrary practice"* (39–40) and that *"Arbitrary species result in arbitrary speciation mechanisms"* (40–41). However, these objections fall to the wayside once the horizontal dimension is taken as primary (instead of judging that dimension from the vertical dimension) and pattern is given priority over process (all of which I shall argue more fully in chapter 5.2/3). Wiley's third reason is that *"Phyletic speciation has never been satisfactorily demonstrated"* (41). In support of this claim Wiley appeals to the theory of punctuated equilibria, but in so doing he necessarily inherits the many problems that beset that theory (cf. §3 above).

To me, none of these defenses are strong, and the absurdity of the view that a species might retain its numerical identity through infinite evolution becomes strikingly apparent when we apply it to the closest analogue to species, namely languages. Suppose a language undergoes "infinite evolution" without branching. Although it is historically true that linguists have been interested in the "phylogeny" of languages earlier than have biologists in the phylogeny of species, surely few if any outside of cladism and its cognates would ever want to defend a *cladistic language concept*, in other words the view that a language which undergoes "infinite evolution" without branching is numerically the same language from start to finish. Quine (1986), for example, calls "later English," in contradistinction to "earlier English," "another language" (14). This is because languages have a crystal-clear criterion of identity: two languages (minimally different if only in name) are identical *to the degree* that a user of one (given the simplest of sentences, and without any gestures) understands a user of the other and *vice versa*. As Fodor (1975) put it, "a natural language is properly viewed in the good old way: viz., as a system of conventions for the expression of communicative intentions" (106). Derek Bickerton (1990) also informs us that language "is widely regarded as a means of communication" (7). Nevertheless Bickerton, himself both a Chomskyan and an evolutionary linguist, holds that language "is not even primarily a means of communication. Rather it is a system of representation" (5). In other words, "Before language can be used communicatively, it has to establish what there is to communicate about" (21). Surely a system of representation, however, which undergoes "infinite evolution" without branching, would not numerically remain the same system. And indeed when it comes to species, Bickerton tells us that "it is . . . true by definition that species in general can't undergo really radical changes—if they do, we just call them new species" (237).

All of this seriously undermines the cladistic language concept. And yet given the close analogy between species evolution and language evolution, a defender of numerical identity through infinite evolution for species might feel compelled as a matter of consistency to defend the same for languages. Ghiselin is a prime example of this. In his recent discussion on language evolution, Ghiselin (1997) maintains a cladistic concept of language. For instance, he says "English has never undergone the analogue of speciation. The language of *Beowulf*, the *Canterbury Tales, Hamlet*, and *Huckleberry Finn* has changed a great deal without ceasing to be one and the same individual language" (141).

But consider some of the implications of this. In the diachronic case of infinite evolution without splitting, it means that a person living at the temporal end of the continuum would not at all be able to understand (given the simplest of sentences) a message sent by a person living at the temporal beginning of the continuum. And yet on the cladistic language concept they would both be users of numerically the same language.

A similarly strange consequence arises if we focus on the monophyletic aspect of the cladistic language concept. In the synchronic case where qualitatively the same language is either multiply invented or multiply evolved (and the logical possibility is not precluded by its improbability), then on the cladistic language concept it would be possible for two or more people to completely understand the linguistic utterances of each other and yet speak numerically different languages.

These are not merely odd consequences that might possibly be embraced by a future generation. Instead what these absurd consequences show is that the cladistic language concept is not really a language concept but instead a *language lineage* concept and that the two are very different things altogether. And given the close analogy between languages and species, all the rules of argument by analogy lead inexorably to the same conclusion for species. This should alone become evident when we think of intercommunication between biologists (cf. chapter 1.1), especially in the case of biological control (cf. chapter 2.4). Indeed as Walter Bock (1994) put it in a different context, "Species are not phyletic lineages and these two concepts must be sharply distinguished if progress is to be made in comprehending evolutionary theory and its application to practical taxonomy" (303).

Closely related to the problem of numerical identity through infinite evolution, cladistic species concepts (both strict and loose) also have a serious problem with multiple origins, in particular instantaneous speciation by hybridization. According to Ridley (1989), "If only some of the members of *d* and *k* hybridize, then, the remaining members of *d* and *k* themselves form new cladistic species and the hybridization event has produced three new species in all" (5). What this means is that for every hybrization event two species automatically go extinct and three new species are born. In the case of repeated hybridization events, as in seasonal allopolyploidy, this would entail multiple extinction and speciation, irrespective of the fact that the great majority of the organisms in the parental species had absolutely nothing to do with the hybridization events. What this further means is that many organisms will be a member of one species during one part of their lives and a member of one or more different species in subsequent parts of their lives. As Wilkinson remarked after making a similar point, "If hybridization were common, the number of species necessitated by the formalism would be immense. These 'species' would all be monophyletic in Ridley's sense, but what price monophyly?" (437).

In answer to the set of objections based on founder populations, Ridley (1990) reaffirms that in each case "the new species has arisen in a *cladistic* sense" (447). But this is nothing more than to confuse methodology with ontology. Methodology is not ontology, any more than legislation is morality.

As for the proliferation of species that his cladistic concept entails, Ridley (1990) replies that "If the real cladistic structure has an 'immense' number of species, I do not see any reason to hush it up. There is a difference between awkward practical considerations and theoretical lucidity; cladism is interested in the latter" (449). But if the price of a so-called objective species concept is the many absurd consequences that we've seen above, then surely the sensible thing to do is to pursue another course.

And what about the claim of objectivity itself? Ridley (1990) tells us that "The danger lurking in non-cladistic definitions of species is subjectivity" (448), and it is clear from the context of this quotation that it is aimed against the use of similarity in the ontology of species. In chapter 2.4, I argued against the claim that similarity entails subjectivity. (I shall explore the topic further in chapter 5.3.) And here again it seems to me that the quest for objectivity by denying similarity falls into the same type of mistake. That such a so-called objective criterion for species delimitation, namely phylogenetic relations, should lead to such an outrageous species concept seems to me ample justification to rethink the possible objectivity of similarity.

The only similarity allowed for by strict cladists, interestingly, is the similarity between their different versions of the cladistic species concept. In an explicit "elaboration . . . of the Hennigian species concept . . . similar to that of Ridley (1989)" (409), but designed to overcome the pitfalls of both, D.J. Kornet (1993) has developed what she calls the *internodal species concept*. As we have seen above, Ridley thought of species as cladistic *sets* of organisms. Kornet has taken this to the extreme by applying formal set theory to the cladistic species concept. The result is an extremely technical species concept, the technicalities of which I will not enter into here. But much like Eigen's quasispecies concept examined in chapter 3.3, we can profitably ignore the technicalities and examine the basic features of the species concept.

Informally, Kornet tells us that according to the internodal species concept

> individual organisms are conspecific in virtue of their common membership of a part of the genealogical network between two permanent-splitting events or between a permanent split and an extinction [dead-end] event. [408]

That Kornet means by "set" what I mean by "set" (cf. chapter 1.3) is evident from her claim that her species concept is meant to comply with the "intuition of many biologists and others . . . that species are historical entities which originate at a particular moment, exist for some time and may end by becoming extinct or by producing new species"—in other words "species as individuals" as per Ghiselin and Hull (412). That species are supposed to originate at "a particular moment" is a claim I will return to later. That species on Kornet's view would not be accepted by Ghiselin and Hull as genuine individuals (as is the case with so many others who claim to follow in their footsteps) should be apparent alone from the use of set ontology. If it is not apparent it will become so as we go along.

At any rate, it is important to notice that the concept of "permanent split" is crucial to Kornet's species concept. As Kornet put it, "the only splits able to play the desired role as speciation events in the internodal concept are permanent splits, i.e. separations of the genealogical network into two or more chunks which are never thereafter reunited by interbreeding relations. After all, if non-permanent splits were taken as speciation events, then any momentary segregation of individuals in the network would be deemed to originate sets of new species. This problem would occur frequently" (408). Thus the problem of hybridization that plagues Ridley's version of the cladistic species concept, and made much of by Wilkinson, is avoided by Kornet's version.

But just what is it that determines, on Kornet's view, a permanent split? One might think, from her definition above, that it is the inability to interbreed, "inability" taken in the sense of Mayr's biological species concept. But this is not so. As Kornet put it, "The biological species concept is based on the ability to interbreed. . . . But potential relationships cannot knit organisms into a temporally continuous historical entity" (413). Instead Kornet focuses only on actual relationships. But not all actual relationships are used. Interestingly Kornet uses only two, which mark a further significant departure in her version of the cladistic species concept from that of either Hennig or Ridley. The first primitive relation that Kornet uses is the parenthood relation ("*a* is the parent of *b*"): "Only actual (i.e. not potential) parenthood relations are relevant, and they assure the cohesion and temporal continuity of the historical genealogical network" (413). The second primitive relation is the priority of birth relation ("*a* was born before *b*"): "this is needed if we are to distinguish between species in time, as between a daughter-species and its ancestor-species" (414). No other relations are used to determine branching points. Not potential interbreeding, as we have just seen. Not morphological similarity: "The morphological species concept does not allow an interpretation of species as historical entities" (412). (This is not necessarily true, as we shall see in chapter 5.2). And not ecological relations: indeed the ecological species concept is not even mentioned.

So only two primitive relations are used to determine internodal species. Indeed what motivates Kornet to employ only those two relations is the desideratum of "mutual exclusivity," in other words lack of overlap: "An ideal species concept should meet the various intuitions that we have about species. We envision species as mutually exclusive groups of organisms" (409). In other words, "Mutual exclusivity is often considered the most important requirement of the ideal species concept. It is generally felt unsatisfactory to have to admit that according to one's criteria an organism would have to be placed in two species at the same time" (410).

In terms of set theory, mutual exclusivity entails that any pair of members of a species should "possess the properties of reflexivity, symmetry, and transitivity" (410; cf. Fig. 1, 411). Reflexivity means that organism *a* is conspecific with itself. (Does this mean that there could be a species of one?) Symmetry means that if organism *a* is conspecific with organism *b*, then organism *b* is also conspecific with organism *a*. Transitivity means that if organism *a* is conspecific with organism *b* and organism *b* is conspecific with organism *c*, then organism *a* is conspecific with organism *c*. Thus from the viewpoint of the desideratum of mutual exclusivity, the morphological species concept fails because similarity, though it is clearly symmetric and arguably reflexive, is not transitive. Similarly the biological species concept fails because, though the ability to interbreed is symmetric, "the 'ability to interbreed' is not transitive," which is especially obvious in the case of ring species (411).[64]

But is mutual exclusivity *really* a desideratum of most evolutionary biologists? Certainly Darwin, as we have seen in chapter 2.3, held that evolution by natural selection entails gradualistic speciation, which for him (Darwin 1859) and virtually all biologists today entails "messy situations" in terms of branching speciation: "one species giving rise first to two or three varieties, these being slowly converted into species, which in their turn produce by equally slow steps other species, and so on, like the branching of a great tree from a single stem, till the group becomes large" (317). And except for the case of instantaneous speciation by polyploidy, modern Darwinism (i.e., the Modern Synthesis) is largely characterized by a rejection of saltational evolution (cf. the end of chapter 3.3). As Mayr

---

[64] Reflexivity, of course, is here entirely irrelevant, except in the case of self-fertilizing hermaphrodites. Presumably ecological relations are not even considered by Kornet because niches do not necessarily contain mutually exclusive sets of organisms, whether at any one time or over time.

(1991) put it, "Sudden new origins, . . . are not evolution. The diagnostic criterion of evolutionary transformation is gradualness" (18). But sudden new origins is exactly what Kornet's internodal species concept, guided by the criterion of mutual exclusivity, is all about. In fact Kornet's species concept entails the necessity of making the dividing line between two species so sharp that it is at the level of parent and offspring, such that an individual offspring—no matter how morphologically similar to its parent(s)—may be a member of a different species. Kornet makes this explicit in both words and diagrams. In words, she says "there is no reason to expect that such a split [i.e., a permanent split] in the network, . . . will always be accompanied by a morphological change of the organisms such that these species become visually distinguishable" (410). And in diagrams such as Fig. 4 (419) and Fig. 7 (425), we can see the actual lines demarcating internodal species drawn by Kornet between parents and offspring at the point of permanent splits. But not only need there be no morphological change between parent and offspring, there need not be any change in reproductive character either, since as we've seen the ability to interbreed is rejected by Kornet as a defining character. Ultimately, then, there need not be any genetic difference whatsoever.

Indeed Kornet's concept of permanent splits has many odd consequences. All that makes a split permanent on Kornet's view is the contingency of natural history. If the two halves resulting from a split should never again happen to have interbreeding relations between them, then the split is permanent and we have two species instead of one. And if, on the other hand, they should happen to interbreed at some future time, then the split was not really permanent and we have one species all along. This solution certainly avoids the problem of hybridization that plagues Ridley's version of the cladistic species concept, but at a much higher cost. First of all, speciation on Kornet's view is entirely retrospective. As Kornet put it, "The insistence that splits should be permanent in order to count as speciation events has the consequence that species of a genealogical network are defined in retrospect, i.e. looking back in history. Thus we can never, before the extinction of a network, exclude the possibility that parts of the network may still reunite into one species" (409). This is an approach to speciation that I rejected in §1n1 largely because it is teleological. But the problem with Kornet's version is even worse than that. Suppose, for the sake of argument, that a river divides the range of a species into two, that the two lineages remain in stasis for the next 10 million years, and that the split is permanent in Kornet's sense, its permanence sealed, say, by the dead-end extinction of both lineages caused by a new strain of virus. On Kornet's view the species we began with went extinct as soon as the river divided it, with the result that two new species were born.[65] All in all, then, Kornet would count three species in the above scenario, whereas most modern biologists (including even some cladists, loose cladists), I think it is safe to say, would count only one species. The problem of parsimony that plagues Ridley's

---

[65] Clearly Kornet also subscribes to Hennigian extinction, with the restriction that the splits must turn out historically to be permanent. This is evident not only from her definition of permanent splits but also from her diagrams.

species concept thus returns upon Kornet's attempt at an improved version with a vengeance. There is no satisfying the "intuitions" of evolutionary biologists here. On the contrary, it is an affront to their intuitions, theories, and practices. Almost all the biological relations that biologists find interesting and important when considering species are rejected by Kornet in an effort to impose an artificial desideratum of set-theoretic neatness.

There are other problems, of course, that we could add to the problems above. The species category is conceived to be "a set of sets" (423; cf. n60 above). The problem of numerical identity through infinite evolution, defined now by the absence of permanent splits, still exists, since Kornet's species concept is not character based. Hybrid zones between species, and hence multispecies (cf., e.g., Van Valen 1973: 72; Wiley 1981: 29; Templeton 1989: 10–12), are precluded by definition. Asexual organisms are excluded if "they lack both a sexually produced descendant and an ancestor which possesses such a descendant," because then they "do not belong to a genealogical network" (414). And, as Kornet herself points out as a "shortcoming" of her species concept, "The extermination of any isolated interbreeding group within a species introduces a permanent split into the network, which retrospectively originates from the time when the group became isolated" (430). Indeed with Kornet's internodal species concept we have moved very far away from anything that might be termed "biological."

All in all, the absurd consequences that follow from the cladistic species concept (variously conceived) merely serve to underscore once again the logical and therefore ontological priority of the horizontal over the vertical dimension. The ultimate lesson to be learned is that speciation should not be so radically decoupled from evolution. Exactly how speciation should be thought of, however, is a topic I will reserve for chapter 5.2.

Closely related to the cladistic family of species concepts is the plethora of so-called *phylogenetic species concepts*. The theorists who advocate these concepts all subscribe to the school of cladistic taxonomy, broadly conceived—in brief, the use of synapomorphies (shared derived characters), autapomorphies (unique derived characters), and plesiomorphies (ancestral characters) in determining maximum parsimony trees (i.e., winning cladograms). And like the theorists just examined they believe that a species concept should be consistent with phylogeny. Where they differ, however, is not always clear. The main difference, it seems to me, is that the advocates of phylogenetic species concepts want species to be more than just consistent with phylogeny; they want them also to be meaningful and useful in determining phylogeny. At any rate, it is common to divide phylogenetic species concepts into two groups (e.g., Baum and Donoghue 1995: 560; Hull 1997: 361) or three groups (e.g., Mayden 1997: 405) and even more (e.g., Davis 1995: 556). As we shall see, these divisions are rather arbitrary. The simplest division is between those phylogenetic species concepts that emphasize diagnostic characters and those that emphasize monophyly. This difference is often only one of emphasis, and some theorists have explicitly combined the two. The earliest of the diagnostic phylogenetic species concepts is commonly attributed to Donn Rosen. As we have seen in chapter 2.4, Rosen (1978) provided a nominalistic definition of "species" in favor of populations apomorphously and geographically defined. A year later, however, Rosen (1979) provided the following view:

> It seems evident . . . that if a "species" is merely a population or group of populations defined by one or more apomorphous characters, it is also the smallest natural aggregation of individuals with a specifiable geographic integrity that can be defined by any current set of analytical techniques. [277]

This certainly looks like a realist species concept apomorphously defined. Nevertheless it is rather vague and is rarely ever quoted in discussions on phylogenetic species concepts.

By far the most quoted and discussed phylogenetic species concept (indeed he coined the name) is that of Joel Cracraft. According to Cracraft (1983),

> *A species is the smallest diagnosable cluster of individual organisms within which there is a parental pattern of ancestry and descent.* [103; cf. Eldredge and Cracraft 1980: 92]

This species concept has a number of interesting features. One of these is the phrase "diagnosable cluster." Cracraft tells us that "Although most species will be defined by uniquely derived characters, this cannot be a component of a species definition (Rosen 1978, 1979), otherwise it would not be possible to recognize ancestral species, which must have primitive characters relative to their descendants (see Eldredge and Cracraft 1980, chapter 4). Species possess, therefore, only unique combinations of primitive and derived characters, that is, they simply must be diagnosable from all other species" (103). The words "smallest" and "diagnosable" involve features of Cracraft's species concept that I will discuss shortly, when I examine problems with his species concept. The remaining interesting feature is the phrase "parental pattern of ancestry and descent." Cracraft tells us that this phrase is necessary to avoid some undesirable consequences (consequences that we have seen are faced by the phenetic species concept). First, "diagnostic characters must be passed from generation to generation," and second, the phrase is necessary so as to avoid counting as distinct species "males, females, parts of life cycles, or morphs" (103).

Aside from its rising popularity, Cracraft's version of the phylogenetic species concept has many serious problems. The first of these is monophyly. There is no mention of monophyly in Cracraft's (1983) definition and no mention of it in his discussion. And indeed his focus on diagnosability would seem to preclude that species on his view need to be monophyletic. Accordingly this is how many have interpreted his view. For example, Melissa Luckow (1995) states that on Cracraft's view "The terms 'monophyletic' and 'paraphyletic' do not apply to species" (595). Similarly Mayden (1997): "For proponents of this concept, monophyly, paraphyly, and polyphyly apply only at a level of organization above species" (406). Seeing this as a failing, some followers of Cracraft have modified his species definition to include monophyly. For example, McKitrick and Zink

(1988), seeing no need to provide a species definition of their own, defend Cracraft's version but modify it so as to make it more consistent with the implicit lineage component in Cracraft's definition. Accordingly, they tell us that "we suggest that the grouping criteria of the PSC are diagnosability and monophyly, which are assessed by cladistic analysis of character variation. The smallest diagnostic cluster is given the rank of species" (2). The characterization of monophyly that they accept is "the one currently accepted by many evolutionary biologists, namely that monophyletic groups are those containing all the known descendants of a single common ancestor" (7).

A large part of the problem with determining whether Cracraft's phylogenetic species concept implicitly involves or should involve monophyly is that the concept of monophyly is ambiguous in cladistic literature. Hennig (1966) defined a "monophyletic group" as "a group of species descended from a single ('stem') species, and which includes all species descended from this stem species" (73). He also gave a second, ancillary definition: "a group of species in which every species is more closely related to every other species than to any species that is classified outside this group" (73). Given either definition, of course, it is logically impossible for a species itself to be monophyletic. The concept of monophyly, however, also has a much looser meaning both within cladism and outside it, namely monophyly in the sense of single origin. This is often what cladists mean by monophyly, as we have seen earlier and will see again.

Granting this ambiguity, it should become clear why some theorists claim that Cracraft's species concept does not involve monophyly while others attempt to bring it out. Keeping the ambiguity in mind, it is instructive to look at Cracraft's subsequent comments on monophyly in relation to his phylogenetic species concept. But before looking at those comments, it will be recalled from earlier in this section that Cracraft does not subscribe to Hennigian extinction, since he allows a species to survive following the budding off of a new species, as in peripatric speciation, providing that the parental species does not change in any of its diagnostic characters (cf. the diagram in Eldredge and Cracraft 1980: 91). And indeed this follows logically from Cracraft's (1983) species definition. So species according to Cracraft are not strictly monophyletic in some redefined sense following Hennig. However in the looser sense of monophyly, the sense of single origin, Cracraft clearly subscribes to monophyly. This follows not only from his implicit (Cracraft 1983: 102) and explicit (Eldredge and Cracraft 1980: 91, 245; Cracraft 1987: 332, 1989a: 39) adherence to the species-as-individuals view of Ghiselin and Hull, but also from his explicit claims about monophyly. For example, Cracraft (1989b) states that "In the majority of cases, phylogenetic species will be demonstrably monophyletic; they will never be nonmonophyletic, except through error" (35). Of course Cracraft, from what we have seen earlier, cannot here mean strict monophyly *sensu* Hennig and his followers. Instead he must mean monophyly in the sense of single origin. This is evident when he gives his reasons, which do not include any philosophical persuasiveness of the species-as-individuals view (cf. Cracraft 1987: 343). Instead for Cracraft, as for so many others of cladistic persuasion, anything but a monophyletic species (conceived in the sense of single origin) cannot be consistent with history reconstruction. As Cracraft (1989b) put it in the context of criticizing the biological species concept of Dobzhansky and Mayr, "Reproductive isolation is not an intrinsic attribute, but a relational concept, and thus does not constrain biological species to be strictly monophyletic. By definition, nonmonophyletic species imply history has been misrepresented" (39). And again, this time in the context of criticizing Paterson's recognition species concept, Cracraft (1989a) states that "it is possible for 'species' individuated on this basis to represent an assemblage of historically unrelated populations. . . . Such a species concept would not permit us to reconstruct the history of speciation accurately" (37).

I will return to the issue of monophyly a little later. Another claim often made is that Cracraft's phylogenetic species concept does not include asexual organisms. And indeed not only is there no discussion of asexual organisms in Cracraft's (1983) discussion, but he explicitly claims that "*all* species definitions must have some notion of reproductive cohesion *within* some definable cluster of individual organisms" (103). Again, some theorists have attempted to improve Cracraft's species concept, this time by explicitly including asexual species. Thus Nixon and Wheeler (1990), accepting the traditional view that asexual organisms do not form populations, define "species" as

> the smallest aggregation of populations (sexual) or lineages (asexual) diagnosable by a unique combination of character states in comparable individuals (semaphorants). [218]

According to Echelle (1990), "The PSC [Cracraft's phylogenetic species concept, which he defends] is applicable to non-Mendelian forms, but only if the diagnosability criterion is not extended to mutants at the molecular level." Thus he advocates morphological over allozymic data, all the while admitting that this is not only "a loss of some objectivity" but a "sacrifice," since "one apparently will not be able to use the same criteria in assigning taxonomic rank" (111).

In spite of what he wrote in his 1983 article, however, Cracraft later provides hints of a change in view. For example, in the context of criticizing Mayr's biological species concept, Cracraft (1987) states that "the BSC lacks generality inasmuch as it cannot be applied to asexual organisms . . . . it is that the BSC, *in principle*, cannot be applied to an important component of biological diversity" (339; cf. 344n5, Cracraft 1989a: 40).

Although it has increasingly seemed attractive to many in biology (cf. the papers in Claridge *et al.* 1997), there are serious problems with Cracraft's phylogenetic species concept (and related versions). Perhaps the most expressed criticism is its high degree of refinement, namely that it is a splitter species concept *par excellence*. For a start, Cracraft's species concept is not polytypic; it does not allow for subspecies. As Cracraft (1983) put it, "Subspecies cannot have ontological status as evolutionary units under a phylogenetic species concept" (104; cf. Rosen 1979: 277). This, of course, is because of the criterion of diagnosability. Thus, "In general, the major effect will be to elevate some subspecies to species" (106). What this means is that adherence to the phylogenetic species concept results in more entities recognized as species taxa than recognized by the biological species concept. In the case of birds, Cracraft (1997: 331) estimates that the phylogenetic species concept would roughly double the 9,000 or so species currently recognized in accordance with the biological species concept. This in itself is not bad, and in fact Cracraft (1997) claims, what does indeed seem to be a rising concern, that "A comparison of the BSC with the PSC demonstrates that phylogenetic species, not polytypic biological species, are the most appropriate units for conservation" (335).

But the issue does not stop there. Many have argued that the phylogenetic species concept is too fine-grained, that it results in species distinctions that go flat against sound biological sense. For example, Mayr (1992), in reference to Rosen's (1979) version of the phylogenetic species concept, points out that "In the highly variable poeciliid fishes of Central America, it would mean that the population of almost every tributary of every river would have to be given species rank" (227). Ghiselin (1997) has given the problem an interesting twist by reviving the monogenist versus polygenist controversy that raged in America during the civil war (cf. Gould 1981: 39–72):

> According to Cracraft's definition, *Homo sapiens* is not a species. . . . Politically, the repercussions of abandoning the biological species concept in favor of smaller units might be most unfortunate, and I urge every morally responsible reader to consider it carefully. . . . The rules of nomenclature are such that, were we to have species be only parts of biological species, the name ["human being"] would cease to apply to all of us. . . . Most people would be outraged by someone even suggesting the possibility that they are not a human being, and life might become downright miserable if one put oneself in a position of having to explain why one gave even the appearance of being a racist. [116–117]

But the problem does not stop there. Elisabeth Vrba (1995) has provided the following *reductio:*

> My grandmother was born with the mutant phenotype of an abnormal left lower rib morphology. By chance assortment of alleles it happens that all her descendants so far bear this phenotype. Thus, we are a "smallest diagnosable cluster of individual organisms within which there is a parental pattern of ancestry and descent." Thus, I gather that under the Phylogenetic Concept we are a species. [11]

Similarly, Mallet (1995) has argued that "with detailed morphology or modern molecular techniques, one can find apomorphies for almost every individual" (298). Mishler and Brandon (1987), however, take this line of criticism the furthest: "It is not sufficient to say that a species is the smallest diagnosable cluster (Cracraft, 1983) or even monophyletic group, because such groups occur at all levels, even *within* organisms (e.g., cell lineages)" (408).

In reply, Echelle (1990) has argued, as we have seen above, that molecular techniques should not be used for discriminating asexual species. But as Avise and Ball (1990) point out, such statements "overlook the huge size and extreme variability of eukaryotic genomes. Evidence from molecular biology demonstrates enormous polymorphism within most taxa. . . . in some taxa nearly every organism can be distinguished with the limited genetic information already at hand" (47).[66] In reply to the criticism of Mallet (1995) and Avise and Ball (1990), Cracraft (1997) states that it is "irrelevant because the PSC is not about the diagnosability of individuals but of populations. Such a criticism forgets that species concepts are populational concepts that are used to delimit basal taxa. The concept of apomorphy, moreover, has no meaning at the level of an individual but only at the level of taxa. Delimiting species taxa is a problem for systematics not population biology" (330). However, although this reply does obviate some of the criticisms, it does not answer all of them. The problem still remains that Cracraft's phylogenetic species concept is a splitter concept and that his formulation of it, contrary to his intentions, invites not only arbitrary but seemingly ridiculous results. Until the concept of diagnosability is clearly defined and the definition defended from other possibilities, this problem will remain to haunt Cracraft's phylogenetic species concept.

On the other side of the coin, a further problem concerns sibling (cryptic) species. According to Claridge *et al.* (1997), "A major disadvantage of phylogenetic species concepts that is not often recognized is the improbability that they will reveal the existence in many groups of organisms of complexes of sibling species. . . . this is a major weakness" (10). Claridge *et al.* credit this weakness to lack of "incentive." But the problem is more interesting than that. Cracraft (1983: 104) explicitly states that reproductive isolation is not a diagnostic character according to his species concept. "Thus," he says, "even if two sister-taxa broadly hybridize, both can still be considered to be species if each is diagnosable as a discrete taxon (of course, it may not be possible to assign hybrids to one or the

---

[66] I should briefly discuss here Avise and Ball's (1990) *genealogical concordance species concept*, in which they "combine what we perceive to be the better elements of the PSC and BSC" (46). On their view "population subdivisions concordantly identified by multiple independent genetic traits should constitute the population units worthy of recognition as phylogenetic taxa" (52). Thus on their view species are phylogenetic units with two or more apomorphies (cf. Mayden 1997: 398). Furthermore, they recognize that genetic concordance (concordance between genes at multiple loci) can result either from genetic (intrinsic) or geographic (extrinsic) reproductive barriers, where "the latter can be ephemeral due to breakdowns of geographic barriers" (58). Because of the latter ephemerality, they advocate that "the biological and taxonomic category 'species' continue to refer to groups of actually or potentially interbreeding populations isolated by *intrinsic* RBs [reproductive barriers] from other such groups" (58). Moreover, they advocate the recognition of subspecies in sexually reproducing organisms, since these can evolve their own genetic concordances while in allopatry but without any change in intrinsic reproductive barriers. Thus subspecies "*are groups of actually or potentially interbreeeding populations phylogenetically distinguishable from, but reproductively compatible with, other such groups*" (60). Although an interesting take on species, their concept is a hybrid concept and accordingly suffers from many of the faults of its parental concepts.

other species)" (105; cf. 1987: 342). Cracraft (1989a) is even more explicit as to his reasons: "Reproductive isolation . . . is not an intrinsic attribute of populations; it is instead a relational concept. Populations do not evolve reproductive isolation. . . . Reproductive isolation is best interpreted as an epiphenomenon or effect of differentiation" (37; cf. 51n9, 1989b: 34–35). Since, as we shall see in chapter 5.2, Cracraft considers species on his view to be epiphenomena, it is curious to see him use this epiphenomena criticism against reproductive isolation as an attribute of species. At any rate, the point remains that sibling species do not seem to be affirmed in Cracraft's species concept. Since the recognition of sibling species is important in many areas of biology, especially biological control (cf. chapters 1.1 and 2.4), it is interesting that Cracraft's species concept shares this difficulty with the species concept of phenetic taxonomists (cf. chapter 3.3).

A further problem relates to Cracraft's use of monophyly. As we have seen, he allows for a species to be paraphyletic. But does that mean he allows for an interspecific hybrid species? Cracraft does not say, but Echelle (1990: 111) accepts this possibility. However, Echelle does not discuss the issue of multiple origins, and we know from Cracraft's emphasis on single origins that he would not allow for the possibility. Thus his species concept invites the same problems as Ridley's (1989) rejection of multiple origins discussed above, in particular multiple origins by repeated hybridization events between the same parental species. In chapter 5.2, I will examine this issue more closely, as it relates to real-life examples. I will also there examine Cracraft's claim that his phylogenetic species concept is a pattern rather than a process species concept such that this makes it a viable candidate for a universal species concept. Although I agree with the pattern species concept approach, I will argue that a concept of monophyly, in the loose sense of single origin, is not only *not* necessary for history reconstruction (contrary to what many theorists maintain), but that it must be scrapped if we are to have a truly viable species concept worthy of the Modern Synthesis.

Related to this issue is the question of the temporal dimensionality of species on Cracraft's view. Unlike the species-as-individuals view and cladistic as well as punctuated equilibria species concepts, it is not clear where Cracraft stands on this issue. Cracraft (1997) tells us that the phylogenetic species concept "is not strictly a cladistic concept" (326) and that phylogenetic species are "terminal taxa" (332). This is very curious. As Panchen (1992) points out, the phrase "terminal taxa" refers in cladistics to "the tips of the branches" (7). Thus defined, a terminal taxon sounds like a horizontal entity, unlike the branches themselves which are vertical entities. But this flies flat in the face of so many of Cracraft's earlier statements on phylogenetic species. For a start, as we have seen above, he subscribes to the view of Ghiselin and Hull that species are individuals. Moreover, there is his diagram in Eldredge and Cracraft (1980), which makes it obvious that species on his view are vertical entities. Indeed, below the diagram he says "Thus species are individuals, with their own internal cohesion, and give rise to descendant individuals rather like a parental *Hydra* buds off young individuals" (91). If Cracraft's view has changed since then, he certainly needs to tell us, as this is no small matter.

A final issue is that phylogenetic species, on Cracraft's view, do not evolve, but that is an issue I will reserve for chapter 5.2. I now want to examine a very different concept of phylogenetic species, namely the phylogenetic species concept of Brent Mishler and Robert Brandon (1987). They provide the following definition of "species":

> A species is the least inclusive taxon recognized in a classification, into which organisms are grouped because of evidence of monophyly (usually, but not restricted to, the presence of synapomorphies), that is ranked as a species because it is the smallest "important" lineage deemed worthy of formal recognition, where important refers to the action of those processes that are dominant in producing and maintaining lineages in a particular case. [406]

By "monophyly" they mean "a grouping that had a single origin and contains (as far as can be empirically determined) all descendants of that origin" (405–406). By "single origin" they mean an "ancestral individual [which] could be a single organism, a kin group, or a local population. . . . it would never be a whole species because we share the widespread view that new species come about only via splitting, not by any amount of anagenetic change" (409).

Although this might look like a cladistic species concept, it really is not. True enough, they employ monophyly as a grouping criterion, but they are pluralistic with regard to species ranking criteria. As they put it, "The ranking concept to be used in each case should be based on the causal agent judged to be most important in producing and maintaining distinct lineages in the group in question" (406). This is very different than Ridley (1989), for example. Whereas Ridley, as we have seen, thought that in general there is congruence between the different causal processes that maintain lineages, Mishler and Brandon do not believe this to be the case. On their view "there is no reason to believe that reproductive processes and selective processes pick out the same units in nature" (401). Instead they claim that groups of organisms defined as species "with respect to one process are often not congruent with groups defined with respect to a second process" (401). Thus, on their view, their pluralism "reflects a fact of nature" (407) and "This diversity of causes for evolutionary divergence reinforces the need for a pluralistic ranking concept" (409).

In chapter 5.3, I will dispute their claim about the lack of congruence. For the present, it should be noted that it is odd that they should put so much emphasis on plurality and the lack of congruence when, on the other hand, they fully admit that "In the great majority of cases, little to nothing is actually known about any of these biological processes [i.e., the different processes that keep lineages together]" (406).

In addition to the above, there are some further interesting features of their species concept that should be brought out. For one, they distance themselves from Cracraft's phylogenetic species concept because the grouping criterion of diagnosability "does not rule out the possibility of paraphyletic species" (408). On their view "the paraphyletic species should be broken up into smaller monophyletic species" (410).[67] In addition, although they allow for the possibility of hybrid speciation, they do not allow for multiple origins. As they put it, "if similar hybrids are produced elsewhere in the ranges of the two original species, or if hybrids are produced in the *same* locality but discontinuously in time (i.e., if the first hybrid population goes extinct *before* the new hybrids are produced), then the separate hybrid populations would have to be considered as separate monophyletic groups and could not be taken together and named as a new species" (410; cf. Donoghue 1985: 179). But this, of course, would involve them in all the absurdities that, as we saw above, are faced by the cladistic species concepts. Populations resulting from repeated hybridization events between the same parental species would not be conspecific. Budding would result in the extinction of the parental species, since they reject paraphyletic species. And because they only recognize cladogenetic speciation, as we saw above, a species that undergoes "infinite evolution," to use Ridley's (1989) term, would remain numerically the same species throughout.

Thus far it might seem that their species concept is in many ways very similar to the species-as-individuals view of Ghiselin and Hull. And in fact Mishler and Brandon explicitly subscribe to this view, although they believe that "it is time to move beyond the simple class-individual distinction to a more detailed consideration of properties held by biological entities" (398). Accordingly they list and discuss "four major sub-concepts of individuality [which] can be recognized" (398–399). The first is "spatial boundedness." By this they mean that "entities at all taxonomic levels . . . are spatially restricted" (399). The second is "temporal boundaries." By this they mean that "individuality involves temporal restriction of an entity. A taxon must have a single beginning and potentially have a single end in order to count as an individual under this criterion" (399). The third is "integration." By this they mean not only "active integration among parts of an entity," but also whether "the presence or activity of one part of an entity matter[s] to another part" (399–400). The fourth is "cohesion," which they caution should not be confused with integration. By "cohesion" they mean "situations where an entity behaves as a whole with respect to some process. In such a situation, the presence or activity of one part of an entity need not directly affect another, yet all parts of the entity respond uniformly to some specific process." They list as two examples "developmental canalization in biological systems, and processes of density-independent natural selection" (400).

Interestingly, where Mishler and Brandon differ most from Ghiselin and Hull is in their claim that "the 'species as individual' concept developed by Ghiselin and Hull cannot be applied in its simplistic form to most species taxa as currently

---

[67] In accordance with this view they explicitly follow Donoghue (1985: 179), who suggested the term "metaspecies" for supposed species that are not presently known to be monophyletic or paraphyletic. Whether known or not known, neither accepts paraphyletic species as genuine species. Cf. Olmstead (1995) for a defense of paraphyletic species— which he calls *plesiospecies*—as genuine species within the context of phylogenetic species concepts.

delimited" (400). With regard to integration, they point out that "species taxa as currently delimited often do not meet this criterion of individuality (even though they may meet one or both of the two criteria listed above [i.e., spatial boundedness and temporal restriction])" (400). And with regard to cohesion, they state that "species taxa as currently delimited may show cohesion as defined in this way, or integration, or both, or neither" (400). Thus not all of the four criteria that they discussed are each necessary and jointly sufficient for species individuality on their view. Instead what they have is a partial cluster concept of individuality, with only spatial boundedness and temporal restriction being necessary. This, of course, is a far cry from the individuality thesis of Ghiselin and Hull. Many of the groups that Mishler and Brandon would count as species individuals would not be so on Ghiselin and Hull's view, evident alone in the fact that Mishler and Brandon allow for asexual species (407). Moreover, on Mishler and Brandon's view some species will be process entities (those with integration) while others will be pattern entities (those without integration). Again, as I have pointed out before, I do not think that the two halves of each distinction are on a par, and in chapter 5.2, I shall argue for a species concept where species are primarily horizontal entities and are exclusively pattern entities.

Oddly, David Baum (1998), in defending Mishler and Brandon (1987), claims that the species-as-individuals thesis "leads" to the view of Mishler and Brandon, in that the former "implies ontological pluralism, the idea that for each type of connection among organisms, a different type of species individual and corresponding lineage may exist in parallel. These distinct types of species need not be coextensive (i.e., contain the same set of organisms) and, even when they are, they still are not the *same* individual" (651). Baum concludes that "after three decades of consideration, it is about time that systematists came to terms with all the unsettling implications of the species-as-individuals thesis" (651).

Indeed a number of systematists, in addition to Baum, have accepted the approach of Mishler and Donoghue (cf. Donoghue 1985; de Queiroz and Donoghue 1988; Baum and Shaw 1995). Some have modified it by replacing the definition of monophyly above, which stems from Hennig's first definition of monophyly, with a version of Hennig's second definition, modified to be applicable to species. Thus we have Baum's (1992) definition of monophyly as "exclusivity," by which he means "a group whose members are all more closely related to each other than they are to any individuals outside the group" (1).[68] In all of this, in-

---

[68]   Baum and Shaw (1995) employ this definition of exclusivity and the concept of gene coalescence to develop what they call the *genealogical species concept*. On this view a species will be monophyletic, in the sense of exclusivity above, and the species barrier (speciation) will be the boundary between reticulate and phylogenetic (splitting) relationships. Accordingly they define a genealogical species as "a taxon whose relationships with other taxa are predominantly divergent but whose parts (the organisms ascribable to the taxon) are related to each other by a predominantly reticulate genealogy" (291). To determine the speciation divide they borrow coalescence theory from population genetics, a recent theory which "views the gene as a minimal stretch of DNA whose nucleotides share an identical history" (Doyle 1995: 576). On this view each gene has a gene tree, such that

cluding the pluralism of Mishler and Brandon, there is not the slightest hint of species pluralism leading to species nominalism. As we have seen in chapter 2.4, however, not only is there more than a strong hint of species nominalism in the pluralism of Mishler and Donoghue (1982), but I argued that species pluralism entails species nominalism. What an odd consequence, then, if I am right and Baum is right. It would mean that the species-as-individuals thesis entails species nominalism, precisely the position the former was out to avoid in the first place.

At any rate, although the phylogenetic species concept, conceived individually or as a family, has often been claimed to be the usurper of the fifty-or-so-year reign of the biological species concept (e.g. Donoghue 1985: 173; Echelle 1990: 109; Cracraft 1997: 325–326), the many dissensions within this family, and the many conceptual problems that they share, do not bode well, I suggest, for their future. They might currently be riding a wave of popularity, but trends are not laws, and I do not think their trendiness will last.

Finally, a more basic physicalist possibility is to think of species as *lineages*. Bradley Wilson (1995) and Kevin de Queiroz (1999) have both argued that this change in emphasis provides an acceptable solution to the species problem, whether lineages are conceived as organism lineages or as population lineages (Wilson 1995: 349, de Queiroz 1999: 50). Both, moreover, argue that not only do many biologists explicitly use the term "lineage" when talking about evolution, but many even explicitly use the term in their species definitions (as we've amply seen in the present work). Moreover still, both pay homage to the species-as-individuals thesis of Ghiselin and Hull, but think their own emphasis on lineages marks an improvement in that view, even though they de-emphasize the importance of horizontal cohesion (as we've seen, horizontal cohesion is a *sine qua non* of species individuality according to Ghiselin and Hull).

---

particular copies of a gene descend from—"coalesce" to—a single ancestral copy, so that the alleles possessed by two individual organisms that trace back to a common ancestral gene may coalesce more recently than either organism with a third organism possessing the allele (cf. Baum and Donoghue 1995: 566). Thus on the view of Baum and Shaw the gene trees within a genealogical species will not be congruent, since because of reticulation they will conflict with one another, while the gene trees of different genealogical species will be congruent, since there is no more reticulation (cf. Luckow 1995: 594). One consequence of this view (a consequence I take to be a requirement of a viable species concept) is that it allows for some fuzziness between closely related species, since the gene copies of each species might not yet all be monophyletic (Baum and Shaw 1995: 296–297). A negative consequence, however, is that it does not allow for asexual species, since reticulation applies only to sexual organisms. Another feature of their view which is unlike other phylogenetic species concepts is that they conceive of species as horizontal entities, such that they "equate species with synchronic individuals, limiting species to single slices of time" (Baum 1998: 645; cf. Baum and Shaw 1995: 300). Although, of course, I take horizontal species concepts to be privileged over vertical species concepts (cf. chapter 2.3/4), Baum and Shaw's concept of horizontal species is much too extreme. At any rate, Luckow (1995: 594) argues that the coalescence methodology is not congruent with the monophyly desideratum, while Doyle (1995: 578–579) provides a number of theoretical and practical problems with the gene tree approach.

But apart from these and a few other points, their views diverge in important ways. For one, they don't use the term "lineage" in quite the same way. Wilson defines a lineage as "a sequence of reproducing entities, causally related to one another via reproduction" (341). De Queiroz defines a lineage as "a series of entities forming a single line of direct ancestry and descent" (50). The difference between these two definitions is subtle, but becomes apparent when their authors reconceptualize species as lineages. Wilson tells us that "species can be thought of as groups of lineages, either organism lineages or population lineages" (349). De Queiroz, on the other hand, does not talk about species as *groups* of lineages. Instead, in accordance with what he calls the *general lineage concept of species*, he tells us that "species are not equivalent to entire population lineages, but rather to segments of such lineages" (53).[69] Accordingly Wilson would possibly include a branched lineage, symbolized by the letter "Y," as a single lineage, whereas de Queiroz (80n2) explicitly distinguishes his view from Wilson's, in that he would include as lineages the stem itself, each of the upper branches, the stem and the left upper branch, and the stem and the right upper branch, but not the "Y" as a whole. Instead for de Queiroz the "Y" as a whole would symbolize an example of what he calls a clade (cf. his diagrams, 51).

A further difference is in their conceptions of what breaks up a lineage (or lineages) into a species. Wilson mentions "gene flow," "stable selection pressures," and "homeostatic epigenetic programs" (354). De Queiroz's list is more comprehensive. He mentions "potential interbreeding or its converse, intrinsic reproductive isolation," "common fertilization or specific mate recognition systems," "occupation of a unique niche or adaptive zone," "potential for phenotypic cohesion," "monophyly," and "distinguishability" (60). Whatever this list should be, one serious problem is that these various criteria need not be congruent. A closely related problem is that, granting the common origin of life on earth, virtually all organisms on this planet are, in a sense, parts of one huge lineage. If this one huge lineage is to be divided into species, then, much more is going to have to be taken into account than just individual lineages. But if much more has to be taken into account, it seems to me to only prejudice the issue by not including any of that something more in the ontology of species. In other words, if the relation or relation complex of lineage descent is by no means sufficient to group organisms into species, then why include only that relation and no others in species ontology? Without this prejudice, of course, species cannot simply *be* lineages or groups of lineages.

This problem, however, only faces Wilson's treatment of the matter. De Queiroz avoids it partly by using (as we have seen) a different concept of lineage. But more importantly, he makes a distinction between *species concepts* and *species criteria* (analogous to the distinction between a disease and its symptoms) and

---

[69]   De Queiroz makes it clear that by the term "population" he merely means "an organizational level above that of the organism, rather than in the specific sense of a reproductive community of sexual organisms" (52). Wilson's concept of population is similar (348), in that it includes asexual organisms but other than that is somewhat vague.

argues that "All [the different species] definitions [given by biologists] have a common primary necessary property—being a segment of a population-level lineage—but each has a different secondary property—reproductive isolation, occupation of a distinct adaptive zone, monophyly, and so on" (63). Thus, unknown to them, "Virtually all modern biologists have the same general concept of species" (79)—"the common thread" of the segmented lineage—and have only been differing in species criteria, in how to segment lineages. Accordingly "the alternative species definitions are reconciled, and the species problem is thereby solved" (64). Even the debate between monism and pluralism is solved: "Monism accounts for the common theme underlying all concepts of species—that is, the general lineage concept itself; it reflects the unity of ideas about species. Pluralism accounts for the numerous variations on that common theme; it reflects the diversity of ideas about species. There is no conflict . . . ." (73).

This is a massive, and certainly revolutionary, reconceptualization of the species problem. But does it work? I suggest not. First, arguing that there is a "common thread" to definitions of species by modern biologists, in particular that of segmented lineage, is not unlike the claim that all religions say basically the same thing. Only those who want to believe it will be convinced. A more sober examination finds no underlying unity. Darwin is a good point of departure. De Queiroz claims that Darwin's *Origin* contains an "early version, or at least a precursor, of the general lineage species concept . . . . Darwin represented species as dashed and dotted lines, or collections of such lines, forming the branches of what would now be called a phylogenetic tree" (76). As I have shown in chapter 2.3 and more fully in Stamos (1999), this is just not the case. The same is often true when we look closely at more recent definitions of species given by biologists. Granted, as we have seen, many include the concept of lineage. But many do not. For instance, the morphological species concept, examined in chapter 2.4, does not, and the same is true for the cluster concepts, whether phenotypic or genotypic, examined in chapter 3.3. For the ecological species concepts examined in chapter 3.4, most do include the concept of lineage, but one has to be careful here. Van Valen (1976), for example, includes in his definition the qualification "or a closely related set of lineages" following his opening clause that a species is a lineage (70). This is significant, because later in his article he makes it clear that he wants to allow for "multiple origins" in the case of repeated hybridization (73). As such, his species concept does not conform even minimally to de Queiroz's reconceptualization (nor to that of Wilson, as we shall see below). The species concepts that do conform, as we have seen, are the ones mainly to be found in chapter 4. But they certainly do not constitute the vast majority, let alone the whole, of modern competing species concepts.

What is also problematic, and why I don't think de Queiroz's solution to the species problem will ever take hold, is that in order to make it work he has to conflate away a number of distinctions that have grown increasingly prominent in the literature on the species problem. One is the temporal distinction between horizontal and vertical species. According to de Queiroz, "a lineage (at the popu-

lation level) is a population extended through time, whereas a population (in itself) is a short segment—a more or less instantaneous cross section—of a lineage
.... the two categories of definitions do not describe different concepts of species; they merely describe time-limited and time-extended versions of the same
species concept" (54). Another important distinction that de Queiroz conflates is
between patterns and processes. On his view "processes and their products are
intimately related, so that an emphasis on one or the other does not reflect a fundamental difference regarding ideas about the nature of species" (54). In the next
chapter, I shall develop a solution to the species problem that takes these two
distinctions very seriously. Such an approach has the advantage of not having to
reinterpret what biologists actually mean (as opposed to what they think they
mean) when they talk about species and the species problem.

But there is a further problem with the view of de Queiroz, and also Wilson,
and this has to do with what is common in their definitions of lineage. Because
they in some sense equate a species with a lineage, they cannot allow for a species to have multiple origins. Wilson, in explicit reference to Kitcher's unisexual
lizard example that I examined earlier in the previous section (Wilson focuses
only on the second scenario, in which the first hybrid population is completely
wiped out before another is created), finds that he must reject the conclusion that
the two hybrid populations of *C. tesselatus* should be thought to be conspecific.
This is because on Wilson's view "the genealogical structure of a species is of
fundamental importance" (350). But why should that structure be so important?
Claiming that his answer is "based on current views about evolutionary processes," Wilson replies that "One requirement for a group of organisms to function
as an evolutionary unit is that there be genealogical continuity within that group"
(350). Thus, because the two lineages of hybrid lizard in Kitcher's second scenario are "temporally discontinuous," *C. tesselatus* "would not constitute a species on the lineage conception, since that group of lineages could not function as
a single evolutionary unit" (351). The same situation is even more pronounced in
the actual example of three allopatric sexual populations of *Senecio cambrensis*,
determined by Ashton and Abbott (1992) to have arisen by three separate events
of allopolyploidy between the same two parental species. Interestingly, these
botanists offer not the slightest hesitation in considering the three polyploid populations as one species. (I examine this example in detail in chapter 5.2.) Wilson,
because the three polyploid populations lack genealogical continuity and are temporally discontinuous, would have to argue, contrary to perhaps the majority of
botanists, that they are three species, not one. And de Queiroz, given his more
restrictive concept of a lineage as "a *single* line of *direct* ancestry and descent"
(italics mine), would have to do the same.

But surely Wilson, and by implication de Queiroz, is operating on a false
premise here (recall my discussion in chapter 2.4 on Mary Williams' reply to
Bunge). To me, at least, it seems false that modern evolutionary theory *requires*
that an evolutionary unit have genealogical continuity within itself. It is not true
for languages (cf. the discussion on language evolution in §4 above). It is not true
for theories (cf. §4n52 above). It is not true even for eyes (cf. Mayr 1988a: 409:

"eyes evolved in the animal kingdom at least 40 times independently").[70] Why, then, would it have to be true for species? (Consider, furthermore, Ruse's recombinant DNA example discussed in §4n45.) It is not even true, as we've repeatedly seen earlier, that modern evolutionary theory requires species to evolve! Darwin certainly did not think so, since his view not only entailed extinction but allowed for stasis as well. Clearly modern evolutionary theory encompasses both of these in addition to evolution.

That Wilson's and de Queiroz's approach not only multiplies names beyond necessity but entails counterscientific results (e.g., Ashton and Abbott 1992) should really be of no surprise, since in taking only one kind of relation as important (the series relation of lineage) their all-too-narrow focus cannot help but obscure other relations that might be equally if not more important in species ontology.

In sum, although the various physicalist solutions to the species problem help to bring the ontology of species "down to earth," so to speak, something is clearly missing that would make their view more acceptable, something that they sometimes hint at, sometimes imply, sometimes foreshadow, but ultimately fail to grasp. It is that something that I hope to bring out in the following chapter.

---

[70] Paul Griffiths (1994) has attempted to apply a cladistic ontology (as opposed to the traditional functionalist ontology) to organism traits, but this application has even odder consequences than it does for species. For example, to be truly monophyletic such a view would have to entail, as we have seen in the case of cladistic species, that an organism trait in a strictly monophyletic species retains its numerical identity through "infinite evolution" (providing there is no branching). Thus an arm, for example, that evolves in a strictly monophyletic species into a wing would be numerically the same kind of trait from start to finish. (Griffiths indeed thinks of clades as "natural kinds"; cf. 210; also Griffiths 1999.) All things considered, I think a much more sensible view would be that organism traits, just like fitness, are supervenient, and hence not at all "cladistic kinds."

Chapter 5

Species as Relations

[N]either the term class nor the term individual expresses the ontology of bio-logical species satisfactorily. Rather, it would seem necessary to introduce a new term, so far not in use in the terminological repertory of philosophy.

—Ernst Mayr (1987a: 159)

5.1 The Origin of an Idea

So far we have examined three basic paradigms for the ontology of biological species, namely as unreal (species nominalism), as abstract entities (elementary, cluster, and ecological classes), and as physical entities (individuals, sets, clades, and lineages). Aside from the nominalism category, which is highly problematic although it remains the default position, must we restrict our search to within either the abstract category or the physicalist category, which themselves are also highly problematic? Or may we, or indeed must we, as suggested in the quotation by Mayr above, search beyond into some other category?

It was with these thoughts in mind that the possibility of a fourth paradigm occurred to me during the summer of 1993, during which I was engaged in read-ing Bertrand Russell's analytical writings on metaphysics in preparation for a graduate course on Russell at the University of Toronto.[1] In traditional metaphys-

---

[1] I should perhaps at this point respond to the possible charge that it is somewhat pre-tentious of me to engage in this work in any sort of autobiography. In fairness, however, it cannot be doubted that the autobiography of discovery and/or development has always been of interest in both science and philosophy. In science we readily think of the fabled apple of Newton, as well as the real-life story of the discovery of the genetic material by Watson and Crick. And of course there is a whole industry of scholars devoted to extract-ing from Darwin's writings the precise development of his theory of evolution by natural

ics, Russell pointed out, everything falls into one or the other of two broad categories—universals and particulars—and if something does not fall into either category, then it's not real. Having already read a number of writings on the modern species problem, and having achieved (or so I thought) a reasonably good feel for the field, it seemed to me that all of the proposed solutions to the species problem were simply more elaborate variations on an old theme. Species nominalism, of course, with its various reasons for denying that species are real, accorded with universals nominalism (so long as it was thought that the reality of species had to be as classes). Each of the various class views of species, of course, accorded with universals realism. And finally, each of the physicalist views of species were merely attempts to recategorize species as particulars of one sort or another. In other words, all of the modern solutions to the species problem accorded with the various categories in the traditional problem of universals—all of the traditional categories, that is, except for one.

What awoke me in reading Russell was that on his view the universals category is dichotomous: universals according to him are of two kinds, on the one hand *qualities/properties* (Russell sometimes simply says predicates),[2] and on the

---

selection. In philosophy one readily thinks of the narratives of personal philosophical development given by, for example, Russell, Popper, and Quine. The literature on the modern species problem, interestingly, is much more modest in this regard. Only occasionally can one find excursions into autobiography, and then they are very brief, explaining what led one to develop a certain species concept, or why one converted to a different species concept. But for their modesty these personal narratives are nevertheless just as interesting. Ghiselin (1997), for example, tells us that the individuality thesis came to him during the mid-1960s while working on a postdoctoral fellowship under Mayr at Harvard (x). But apparently the seeds were already sown while working on his doctoral dissertation. He tells us that while trying to develop a more accurate phylogenetic tree for a subclass of gastropods, he was acutely aware that he would have to defend his dissertation against the problem of knowability. He tells us that accordingly he read voraciously both recent systematic literature and related philosophical literature, which eventually led him to appreciate and resolve the equivocation between biology and philosophy over the term "individual" (13) and to turn species nominalism on its head (14). Hull (1987), who earlier subscribed to a cluster class theory for species (Hull 1965), and who refereed Ghiselin's 1966 paper but remained unconvinced, tells us that by 1974 he had converted to Ghiselin's view, though mainly because of reading J.J.C. Smart's claim that in biology there are no unique laws of nature (183). Eldredge (1985b) tells us that he did not convert to Ghiselin's view from reading Ghiselin's 1974 paper but only after he realized its implications for hierarchy theory (124–125 and 138n3). It would no doubt be interesting to continue this list and extend it as far as possible to other authors of species concepts examined in this book. But such an exercise will not be engaged in here. Rather, I only give these examples to justify my own excursion into autobiography. In my own case this has the added advantage of showing the successive steps and self-corrections that led to the final version of my species concept. It explains why I did not stop at those earlier steps, and thus why I think others should not either, and it helps to further justify why I arrived at the species concept which is now my settled view.

[2] Examples of his use of the terms "qualities" or "properties" for universals may be found in Russell (1912: 54, 1914: 60, 1944: 684). Examples of his use of the alternate term

other hand *relations*. Now, all those who have held that species are universals (classes) have held that conception in the sense of qualities or properties.[3] No one, as far as I know, including Russell,[4] ever focused on the other universals

---

"predicate" may be found in Russell (1913: 81, 90–92, 1940: 343, 1944: 686).

[3] On the elementary class view, a species as a class is constituted in terms of a set of one or more defining properties. On the cluster class view this set necessarily involves more than one property and one or more minimum quorum conditions. The ecological class view also involves a set of more than one defining property, with the main difference being that the defining set of properties is ecological and therefore external to the member organisms.

[4] Russell seems to have held, at various times, each of the three basic paradigms I discussed above. Often he speaks of species simply as universals (cf. Russell 1912: 61, 1913: 95, 1940: 24). But sometimes he is more specific, as when he says "we may mention a defining property, as when we speak of 'mankind'" (Russell 1919: 12; cf. Russell 1914: 53, 1940: 256, 1944: 684). (For Aristotle, of course, a species is not a property but an entirely different category altogether.) Given Russell's longstanding (though increasingly qualified) universals realism, that would make him a species-as-universals (property) realist. But Russell also speaks of species as classes. He sometimes explicitly calls mankind, for example, a class (cf. Russell 1919: 136, 186). But then that would make him a species nominalist, for Russell long argued against the view that classes are part of the ontological furniture of the universe. His view instead was that classes are logical fictions. This is a view he developed quite early in his career, presaged in his 1905 "On Denoting" and found fully formed in his and Whitehead's *Principia Mathematica*, completed in 1910 (cf. Russell 1959: 57–65). Russell's basic argument (which I examined in chapter 3.5) was that a class "is only an expression, . . . a convenient way of talking about the values of the variable for which the [propositional] function is true" (Russell 1959: 62). Since both propositions and propositional functions were long held by Russell to be ontologically nothing, in the sense that a complete description of the universe would not include them (cf. Russell 1918: 223–230, 1940: 183, 189), it should come as no surprise that at one point he went so far as to say "'mankind' means nothing, though it can occur in significant sentences" (Russell 1944: 692). What I find most interesting, however, is that for a brief period in the early part of his career Russell expressly thought of species as physical entities. In regard to Darwin's *Origin* and the evolution of species, Russell (1914) wrote: "The doctrine of natural kinds, . . . was suddenly swept away for ever out of the biological world" (22). A few pages later, again on the topic of biological evolution, he wrote: "What biology has rendered probable is that the diverse species arose by adaptation from a less differentiated ancestry. This fact is in itself exceedingly interesting, but it is not the kind of fact from which philosophical consequences flow. Philosophy is general, and takes an impartial interest in all that exists. The changes suffered by minute portions of matter on the earth's surface are very important to us as active sentient beings, but to us as philosophers they have no greater interest than other changes in portions of matter elsewhere" (26). Incidentally, that Russell here exaggerates against the importance of Darwinism for genuine philosophy is evidenced by context and comparison. The context of the above passage is a critique of the concept of evolutionary progress or cosmic purpose, particularly as expressed by Bergson. Elsewhere, in for instance his celebrated 1927 essay "Why I am Not a Christian," Russell certainly thinks Darwinian evolution has important consequences for philosophy, since he uses it in a disproof of the argument from design. Needless to say, then, when all of the above passages are considered, Russell did not have a settled view on

category. In other words, no one ever emphasized relations in the ontology of species. This seemed to me a possible fourth paradigm, one with potentially fruitful consequences, which ought to be explored.

Why, one might wonder, has this possible fourth paradigm gone hitherto unexplored? Indeed why, one might further wonder, has it not even been noticed as a possibility? One reason might be because it is by no means obvious. But that, of course, may be said for many of the proposed solutions to the species problem. The real reason, I believe, has to do with the short shrift that relations as a category have received in the history of thought. Although Russell perhaps more than anyone attempted to remedy this deficiency, he also went far in providing a historical analysis for this relations denial in traditional metaphysics. Interestingly, Russell traced the root cause to a linguistic prejudice, namely the traditional belief, founded on the influence of Aristotelian logic, that all facts can be expressed in subject-predicate form. As Russell (1914) put it, "Traditional logic, since it holds that all propositions have the subject-predicate form, is unable to admit the reality of relations: all relations, it maintains, must be reduced to properties of the apparently related terms" (56). What this view further entailed, since the subject-predicate form of proposition conforms to the substance-attribute metaphysics, is that, as Russell (1959) put it, "every fact consists of a substance having a property" (48). It was this fundamental belief, on Russell's analysis, that served as the foundation for each of the wild idealistic metaphysical systems of Spinoza, Leibniz, Hegel, and Bradley, in which so much of what is commonly thought to be real is held to be an illusion (including relations).[5] But it was not only the idealists who were misled by the subject-predicate logic. As we saw in chapter 2.2, empiricists such as Occam and Locke also tended to subscribe to the traditional logic, resulting for them in an ontology of particulars which denied reality to both universals traditionally conceived (including species) as well as relations (which is all the more troublesome for empiricism, since it regards far less of experience as illusory than does idealism).

It is only when the traditional logic is given up, or modified and expanded to include subject-subject propositions—and Russell more than anyone is responsible for the development of the new logic (which of course includes far more than just subject-subject propositions)—that the world of common sense and science follows. As Russell (1914) put it, "The belief or unconscious conviction that all propositions are of the subject-predicate form—in other words: that every fact consists in some thing having some quality—has rendered most philosophers

---

the ontology of biological species, probably because the species problem for him was not important. Indeed the species problem was not really a problem for philosophers in Russell's time, since philosophy of science was dominated by philosophy of physics.

[5]   Naturally it would take me too far afield to examine how each of these metaphysical systems is founded, along with relations denial, on the traditional subject-predicate logic. For Leibniz, who served as the foundation for Russell's insights on this topic, cf. Russell (1900: 4–11, 1946: 572–573, 575); for Spinoza, cf. Russell (1946: 553–554, 559–560); for Hegel, cf. Russell (1914: 48, 1946: 702–703); and for Bradley, cf. Russell (1959: 42–48).

incapable of giving any account of the world of science and daily life" (54–55). Put most simply, with subject-subject propositions a great portion of the universe of facts may now be expressed linguistically. Hitherto those facts either had to be denied altogether or warped and twisted to fit some other ontological category.

It is my belief, then, that the whole nature of the debate in the modern species problem is but an extension of a prejudice of history. Species are still conceived either as universals, in the sense of classes defined by properties, or as concrete particulars of one sort or another. Relations, on the other hand, have an ontology quite different from both classes and concrete particulars, an ontology that may indeed better fit what biologists say about species. Granting, as so many have, that it is a contradiction in terms to say that a relation can exist without relata, I would go even further and say that the relata of a relation are *parts* of the relation. But in the case where the relata are physical things, this does not mean that the relation itself is a physical thing. A relation is not just its relata but its relata *plus something more*, and that something more is *insubstantial*. But it is not insubstantial in the way that classes or property universals are insubstantial. As argued in chapter 3.5, a class may in a sense be real even if it has no members. And on many theories of universals a universal may be real even though it has no instances. So the *something more* in a relation, by virtue of the fact that a relation cannot be real without relata, is insubstantial in a way very different from that of either classes or universals traditionally conceived. As such, then, relations seem to me to constitute a sort of hybrid category, neither abstract nor concrete, but a fusion of both.[6] With this category, then, contra Mayr, there may indeed be no need to go beyond the repertoire of philosophy in our search for the ontological category which best fits the nature of biological species. Philosophy may have had the answer all along, only it was ignored because of historical prejudices.

While reading further along in Russell during the summer of 1993, I came across the distinction between internal and external relations, a distinction at the core of the philosophical controversy over monism versus pluralism and which reached its zenith around the turn of the century. Russell (e.g., 1959), quoting a much earlier work of his, defines the doctrine of internal relations as follows: "Every relation is grounded in the natures of the related terms" (43). From this definition we may define the term "internal relation" as denoting a relation that is grounded in the natures of its relata, while we may define the term "external relation" as denoting a relation that is not grounded in the natures of its relata.[7] From

---

[6] Interestingly, at one point Hull (1977) says "doubt might remain whether or not they [species] should be interpreted as individuals. Perhaps they belong in some hybrid category like individualistic classes, as Leigh Van Valen has suggested, or complex particulars, as Fred Suppe . . . has proposed, but the evaluation of such suggestions must wait for further elucidation of these notions" (87).

[7] Russell's discussion on internal and external relations virtually always forms a part of his refutation of the theory of internal relations held by Russell's older and highly influential contemporary F.H. Bradley, with whom Russell was in fundamental disagreement. Bradley held that all statements of relation express intrinsic properties of the relata, so that all relations are internal and ultimately unreal. (If all relations are internal, to state the

these definitions we may proceed a step further, important from the viewpoint of biological evolution, and say that a relation is internal if a change in the relation entails an intrinsic change in at least one of its relata, whereas a relation is external if a change in the relation does not entail an intrinsic change in any of its relata. Thus *above* would be an example of an external relation, since if the fact that *a* is above *b* changes to *b* is above *a*, there need not be any intrinsic change in either *a* or *b*. On the other hand, *similarity* would be an example of an internal relation (cf. Armstrong 1989: 43), since if a similarity relation between *a* and *b* changes it must be because either *a* or *b* (or both) have intrinsically changed in the relevant respect.[8]

The distinction between internal and external relations seemed to me an important distinction, one that would help to distinguish for my view those relations which should be taken as constitutive of species from those that should not. External relations would seem to be clearly of the latter kind, since if species are to evolve the change must be grounded in the intrinsic natures of the member organisms. Whether organism *a* moves from the left to the right of organism *b* would not seem to make a difference to their species. A species does not evolve simply as a result of its member organisms moving around. Instead it evolves when its member organisms undergo descent with modification, to use Darwin's

---

argument all too briefly, then there are no two perfectly independent relata, no ontological pluralism, but instead one and only one subject, an ontological monism, which Bradley called the Absolute. Accordingly all propositions are really of the subject-predicate form and predicate some property of the Absolute. Cf. Russell 1910: 139–146.) In arguing against what he called the "axiom" or "doctrine of internal relations," by means of what he called the "doctrine of external relations," Russell had much to say in defense of the reality of external relations to the almost complete exclusion of internal relations. In attempting to refute the axiom of internal relations, then, Russell appears to have gone overboard. To my knowledge his only explicit acknowledgment that there may be internal relations in addition to external relations is to be found only much later in Russell (1959), wherein he says "With some relations this view [that some relations are internal] is plausible. Take, for example, love or hate" (42). In this respect G.E. Moore's treatment of the issue is much more even. According to Moore (1922), "*some* relations are 'internal,' others, no less certainly, are not, but are 'purely external'" (80). One interesting example that Moore gives is based on mereology. According to Moore (88), though the relation of a part to its whole may sometimes be external (since some parts may exist unchanged when no longer part of the whole), the relation of a whole to its parts is necessarily internal (since any change in a part changes the whole). Interestingly, what this entails for the various physicalist views of species (individuals, sets, etc.) is somewhat of a *reductio*, for there is clearly a sense in which a species may remain the same even though some of its members (parts) come and go (cf. chapter 4.3). Thinking of species as relations, as we shall see, easily captures this sense, as well as species change via membership change.

[8]   As we shall see in the following section, Russell was an ardent similarity (resemblance) realist in terms of universals realism. And yet it is extremely odd that he never acknowledged (at least as far as I can tell) that similarity is an internal relation, especially since it is highly implicit in Wittgenstein's *Tractatus* (1922: 4.014–4.0141). Apparently Russell's battles with the monists (cf. n7 above) is responsible for this lack of admission.

phrase, which entails genetic changes between organisms in a genealogical sequence. Thus, whether organism *a* is more or less similar in certain respects to organism *b* can make a real difference to a species, in particular when one is the descendant of the other and when the similarity is genetically based (more on this later). Such change is the stuff of biological evolution.

Thus, I began to experiment by defining the species category in terms of internal relations. A few years ago an eminent scholar who I greatly admire and respect (though it might not be appropriate to mention his name) wrote in his latest book that "David Stamos (personal communication) informs me that he has been exploring the thesis that a species is a particular kind of relational property." I am grateful that this sentence is worded in such as way as to imply that the view he ascribes to me was no more than an exploration on my part and not my settled view. However, I wish that he would have relayed more accurately the view I was in fact exploring at the time of our early correspondence. Sometime between late 1993 and mid-1994 Robert Haynes (an eminent molecular biologist who agreed to be on my Ph.D. committee) sent to this scholar some of my writings which contained my early speculations on the subject. I do not know which of my writings were sent, but they had to be from among three pieces, and may indeed have been all three. The earliest piece would be my Ph.D. dissertation proposal, which I submitted in September of 1993. In that paper I proposed to argue, ultimately, "that species are internal relations," and as internal relations I included not only similarity but other relations such as mating relations and ontogenetic relations.

The second possible piece would be a paper titled "Russelling Species," which I submitted to Robert Tully on January 7 of 1994 for his Russell course at the University of Toronto. In that paper I attempted to develop my view "that species words denote a type of relation" and that "if species are a type of relation, then they must be internal relations." And I did indeed elsewhere in that paper write of "the species relation," but when fleshing out my view I made it clear that I thought of species as "delimitable (sufficiently discernible) complexes of internal relations" and that the relations I included were not only similarity relations but what I called "the appropriate polymorphic relations," among which I primarily included mating relations, ontogenetic relations, and sociomorphic relations (by which I meant primarily the internal relations between members of different castes in social insects).

The third possible piece would be a paper titled "Species as Relations," which I first submitted for publication in August of 1994 and which was never published (a rejection for which I have since been very thankful). In that paper I introduced my thesis, following a preliminary discussion, as follows:

> A further ontological category for biological species, then, hitherto overlooked, suggests itself and begs to be explored, namely that species *are* relations . . . . In saying that species are relations, I mean that when biologists correctly delimit species and when the rest of us correctly use species words we are all in effect referring neither to entities abstract or concrete nor to their members or parts, but to individual organisms and the various relations between them that make them

ontological unities, what may be called *relation complexes*. I shall argue that these relation complexes are made up of internal relations, specifically similarity plus what I shall call polymorphic relations (ontogenetic relations, sexual relations, social relations, etc.).

After reading whatever were the manuscripts that Haynes had sent him, the scholar wrote a letter to Haynes, dated November 5, 1994, in which he indeed twice ascribed to me, and then proceeded to reject, the thesis that "species are relations." In my first letter to him, dated February 22, 1995, I attempted to clarify my position. After briefly summarizing and criticizing the views of others on the topic of species, I wrote that I thought of species as "relation complexes," that it seemed to me that "Different species . . . may be thought of as being constituted not only by their respective organisms but also by some of the various relations between them, with the provision that the relations must be internal relations." Again I included not only similarity relations but mating relations, ontogenetic relations (which I now called cyclomorphic relations), and sociomorphic relations.

What should be clear from the above is that even during my early view, during which my ideas were developing, I did not entertain the thesis that a species is *a* relation or a *type* of relation—although I certainly sometimes wrote loosely in ways that would strongly suggest one of these—let alone that "a species is a particular kind of relational property," which is the view that the scholar supposes me to have been exploring. In all of my early pieces and in my correspondence with him I ultimately made it clear that I thought of species as *relation complexes*, composed both of organisms and various internal relations between them.

This experience was, in fact, one of many (along with following through on the paraphrases of secondary sources only to find that they were misleading if not just plain incorrect when compared with the primary sources) that made me realize just how easy it is to misrepresent someone else's view—even for a first-rate scholar such as my correspondent—and how important it is to supplement one's representation of their view with choice quotations from their actual writings: a program that I very mindfully implemented, I might add, in the present work.

At any rate, what about my earlier view that a species is a relation complex composed of its member organisms and all the particular internal relations between them? This is not, after all, such a strange view. It will be recalled from chapter 4.5 that Kent Holsinger (1984) expressed the view of many biologists when he wrote that "a [species] taxon is not just a collection of individual organisms. It is composed of individual organisms that interact with one another, are related to one another in particular ways, and participate in biological processes in similar ways" (296).

To begin, then, perhaps the most obvious candidates to include in the ontology of species are interbreeding relations, given the fact that the biological species concept is still the most widely accepted species concept today. (Interbreeding relations, of course, are absolutely meaningless for my purposes unless they entail horizontal gene flow and the production of fertile progeny.) Indeed a sexual species might be viewed as a complex web of interbreeding relations which in most cases segregates one sexual species from another and which allows each to have its own evolutionary path. It seems to me that we must take these relations seriously, to borrow from Edward Wilson (1992), "if we are to avoid chaos in general discussions of evolution" (56). And certainly interbreeding is an internal relation, for it is grounded in the natures of the two organisms. Moreover, from the perspective of evolution interbreeding relations are clearly internal since a change in the nature of those relations may well result in an evolutionary change in the species. Nevertheless, interbreeding relations alone are insufficient to group organisms into species and to delimit one species from another. For a start, many organisms in a sexual species are either sterile or (for whatever reason) never participate in the production of progeny. But more importantly, as we've seen in previous chapters, many sexual species have subpopulations which have very little if any interbreeding between them. Thus for sexual species the relations which constitute their ontology must include other relations as well. (Potential interbreeding relations, of course, cannot count, since potential relations are not relations.)

Given that interbreeding relations are insufficient for the ontology of sexual species, and given the fact, as we have seen in many places earlier, that many biologists accept asexual species as genuine species, it would seem that one should also include similarity relations. And after all, these relations do seem to connect organisms in a way that interbreeding relations either don't or can't. As such, Edward Wiley's view, examined in chapter 4.4/5, would be highly instructive here. Many biologists, of course, as well as many philosophers, have great difficulty in taking similarity seriously, for a variety of reasons, but we have seen that many of the problems with their own views on species ontology stem from this aversion.

In addition to interbreeding and similarity relations, there are many other relations which it would seem should be included in the ontology of species on the present view. For example, in many species elaborate mating rituals are highly characteristic of those species. Consider Edward Wilson's (1992) description of male jumping spiders:

> Male jumping spiders (family Salticidae) recognize females of their own species by sight. . . . The males posture and dance before the females. Those of one species supplement the yellow and black vertical striping of their fangs by raising black-tipped yellow forelegs above their heads in a gesture resembling surrender. Those of others variously bob up and down, weave side to side, or raise their forelegs and wave them side to side like semaphore flags. When biologists give females a choice between males, the spiders use the colors and movements to select mates of their own species. [58]

Such relations, of course, are not only a prelude to interbreeding relations and are species specific, but they are also presumably genetically based and consequently internal. Thus a change in those relations over time would constitute an evolutionary change in the species.

Another kind of relation that one might consider is nurturing relations. For example, as Margulis and Schwartz (1988) point out, "Mammals nourish their young with milk secretions produced in mammary glands of the mother" (249). That nurturing relations are internal is indicated by the fact that a change in these relations may indeed result in a change in at least one of the relata. But more importantly there is a genetic component to these relations, since the relevant morphological properties and behaviors of the organisms are genetically prescribed. And of course from an evolutionary point of view the relations for which mammals get their name marked an important step in evolutionary history and obviously have continued to evolve in species-specific ways.

Indeed from the above it looks as if we are moving toward an ontology of species that is in accord with the kind of species descriptions that one typically finds in biology books, descriptions that include not only the physical properties of organisms (and hence the role of similarity) but also their typical behavior. Accordingly one could add relations such as competitive relations and even more widely ecological relations. For example, it will be recalled from chapter 4.1n8 that although lions and tigers have different social behaviors it is also true that the former prefer open savannas while the latter prefer forests. It is partly these different ecological relations that have kept them from hybridizing even in Roman times. Indeed when one flips through any biology book that includes descriptions of species, the relations that figure in those descriptions are often what may be broadly termed ecological relations. For instance, in an extensive study on the relation between evolutionary escalation and predation, Geerat Vermeij (1987) points out, among many examples, that "The rotifer *Brachionis* develops spines with a demonstrated antipredatory function in the presence of the predaceous rotifer *Asplanchna* . . . . Species of the cladoceran crustacean genus *Daphnia* form crests in the presence of predaceous insects. The crests render the potential victims more difficult for the predators to manipulate . . . . Some trees increase the concentration of herbivore-repelling substances in leaves and branches that are attacked by rabbits or damaged artificially" (39–40). One could also, of course, in an entirely different book, read about the relations of resource utilization, responses to temperature change, and on and on. Moreover, in another book one might read about symbiotic relationships, which can range anywhere from occasional parasitism or mutualism to full-fledged interdependency. To cite a notable example from Lynn Margulis (1991), "In spite of sociobiological dicta to the contrary, organisms behave to increase the fitness of symbionts with which they have *very few genes* in common (e.g., the cow licking her calf ensures the continuity of her entodiniomorph rumen ciliates)" (11).[9] Indeed, as we saw in chapter

---

[9]   Interestingly, Margulis (1993) sees this as creating a problem not so much for species distinctions as for individual organism distinctions (the very relata in my species concept!): "By tradition, species binomials—biological names—refer nearly always to the most conspicuous member of the consortium: the member most likely to be encountered by the first scientist. . . . Of course, the cow (*Bos taurus*), rather than its cellulose-digesting, entodiniomorph ciliates, contributes the name . . . . We people tend to see and

3.4, even a species individualist such as Wiley (1978) had to admit that "Species cannot be divorced from their environment any more than they can be divorced from their gene pools or their morphologies" (87). Perhaps, then, ecological relations should also be included in the ontology of both sexual and asexual species.

This is especially important in the case of asexual species, for usually there is little else besides the existence of morphological gaps to help determine their boundaries. Many biologists, of course, think that there are no such things as asexual species, or that the topic is not important, since "The vast majority of species are sexual" (Wilson 1992: 49). But it seems to me an important desideratum to include asexual species, for reasons examined in chapter 4.4. One apparent advantage for my view, then, is that since I take similarity seriously, I may admit the reality of taxonomically good asexual species. Although, of course, some asexual forms such as dandelions display a lot of intergrading smeariness (cf. Maynard Smith 1975: 225, Ridley 1989: 8), most asexual forms, as Templeton (1989: 8) and others have stressed, are taxonomically good species. And as we have seen in chapter 4.4, even Ghiselin and Hull grant that such forms (they call them lineages, not species) may evolve.

Nevertheless, the inclusion of ecological relations in the ontology of species opens up a serious problem, namely that we are now including relations not only between organisms of the same species but between different species and even abiota as well. Originally one of the great attractions to me for employing only internal relations was that unlike external relations they require one to take into account certain properties of the relata, in particular the ones that allow for intrinsic changes in species, which of course is fundamental from the viewpoint of evolution. In adding ecological relations to species ontology, however, what used to appear to be external relations might now appear to be internal relations. Indeed the distinction between the two kinds of relations appears to have become quite blurred.

There is a further problem, however, that the inclusion of ecological relations in species ontology makes especially apparent, although this further problem by no means depends on that inclusion. The problem now is that of pluralism. The species concept developed thus far is highly heterogeneous. Although every species includes organisms, the kinds of relations that are also included are highly varied. Some species will include habitat relations to trees, for example, others not. Some species will include interbreeding relations, others not. And on and on. Interestingly Holsinger (1984), the botanist whose view quoted above helped to inspire the concept of species examined here, himself took a pluralistic view on species ontology. As he put it, "taxa participate in a variety of processes and no unique characterization of these entities is possible with respect to all of the processes in which they participate" (305). The problem now, of course, as we have seen in chapter 2.4, is that species pluralism all too easily leads to species nomin-

---

name only the most superficial or largest members of the consortium; we then act upon the self-deceptive construction that this consortium is an independent 'individual'" (216). For my reply to the problem that this might pose for my species concept, cf. §3 below.

alism. Thus what began as a realist species concept now seems to slide down a slippery slope into a *reductio ad absurdum.*

As with many slippery slopes, however, there is a way to stop the slide downward. In the case of the species concept being developed in this chapter, it is time to see how this can be done. The solution is surprisingly simple and yet unique, but most important of all it accords, I suggest, with the theories and practices of biologists better than any species concept hitherto examined.

5.2 Species as Biosimilarity Complexes
In the summer of 1996, I happened to read an article by Arda Denkel titled "Real Resemblances" (Denkel 1989). Denkel defends a version of what is known as Resemblance Nominalism (more fully developed in Denkel 1996). For Denkel similarity relations are real, in an extra-mental sense (hence the similarity realism), but they each are particulars, not universals, and hence are not repeatable or have instances (hence the universals nominalism). Moreover, on Denkel's view it is from the vast array of particular similarity relations that we observe between objects in the world, and not any identities between those objects, that we actually base our concepts of properties and kinds. As Denkel (1989) put it,

> The main contention of what has been characterized as the RT [Resemblance Thesis] is that the 'common aspects' observed among things in nature, such as properties or kinds, are a matter of resemblance rather than identity. It is owing to such resemblances in different degrees that we are able to speak about 'common' properties and natures and thus classify the world accordingly. From the point of view of the RT, Aristotelian Realism is a simplification, a summary of the actually more complex relation which holds between particulars; on this latter account, all resemblances, whatever their degree, are reduced to identities. What one needs to do in order to realize that the so-called identical properties are in fact merely similar is to examine them more closely. . . . Aristotelian Realism enjoys the advantage of allowing a practical and easy classification of the world, but it also invites the nominalist revolt. It assumes certain existences which it claims to be identical in different particulars. [37]

Although I have some difficulties with Denkel's view, I will reserve them for the next section, where I will also attempt to deal with other specific and more general difficulties, including the reality and nature of similarity relations. For the present it is useful to assume for the sake of argument the extra-mental reality of similarity relations. This is where my revised solution to the species problem comes in. While Denkel applies his thesis to kinds such as properties, he nowhere applies it to kinds such as species. (Nor do any other resemblance nominalists, as far as I've since been able to tell.) Almost immediately this seemed to me a golden opportunity for theory development. What if we conceive of species as similarity complexes, composed of their member organisms and all the various similarity relations between them? Of course the first objection that comes to mind, or should come to mind, is that all organisms in the world have between them various similarity relations. How, then, is this vast array of similarity relations to be pared down into the much smaller complexes that basically accord with what biologists today commonly recognize as species? Given the species concept that I had been working on at the time, described in the previous section, the answer came to me almost as suddenly as the objection. Taking the horizontal dimension as primary, nature itself is partly composed of various causal relations that more or less divide the similarity relations between organisms into species. Hence the birth of what I have since come to call the *biosimilarity species concept:*

> *A species is a primarily horizontal, all the while dynamic, phenotypic similarity complex of organisms objectively and maximally delimited by causal relations, in the case of sexual organisms mainly interbreeding, ecological, ontogenetic, and possibly social and sociomorphic relations, and in the case of asexual organisms mainly ecological, possibly gene transfer, and possibly social (e.g., colony formation) relations.*

One of the many virtues of this species concept (although many theorists, as we have already seen, will consider it a vice) is that it takes similarity seriously, as part of the ontology of species. In the previous chapter especially, we have seen what happens when one attempts to exclude similarity completely. The case of strict cladists such as Ridley and Kornet is particularly instructive, and we have seen how loose cladists such as Wiley and Cracraft find themselves forced to let similarity in through the back door, so to speak.

But what of those who take a primarily horizontal view of species? Many of them will also remain recalcitrant, and Mayr will probably serve as their icon. Indeed Mayr's rejection of similarity relations in the ontology of species taxa is well known. In this rejection Mayr most often refers to sibling (cryptic) species, what Kitcher (1984b), along with pronounced sexual dimorphism, calls "Mayr's pincers" (625). As Mayr (1988a) put it, "The discovery of the high frequency of morphologically indistinguishable species (sibling species) has demonstrated the invalidity of the morphological species better than anything else" (316). Holsinger (1984) tells us that in a personal communication Mayr told him that in defining the species category "the morphological yardstick is better regarded not as part of the definition, but as part of the evidence that the definition has been satisfied" (300n5). Indeed Mayr (1988a) further tells us that "as a first approach in a preliminary analysis of the diversity of a fauna and flora, it is sometimes necessary to recognize provisional species based exclusively on morphological criteria. . . . such tentative arrangements will be confirmed or rejected or at least modified as soon as additional information becomes available." He then immediately adds that "It must be emphasized that there is a complete difference between basing one's species concept on morphology and using morphological evidence as inference for the application of a biological species concept" (316).

In this claim Mayr is simply following Simpson. In an often cited passage on the distinction between "definition and the evidence that the definition is met," Simpson (1961), drawing an analogy between species and monozygotic ("identical") twins for the importance of monophyly in evolutionary taxonomy, says "We *define* such twins as two individuals developed from one zygote. No one has ever seen this occur in humans, but we recognize when the definition is met by *evidence* of similarities sufficient to sustain the inference. The individuals in question are not twins because they are similar but, quite the contrary, are similar because they are twins. Precisely so, individuals do not belong in the same taxon because they are similar, but they are similar because they belong to the same taxon" (69). Accordingly Mayr (1988a), after referring to Simpson above, states that "We do not combine two populations into one species because they are similar; rather, we conclude that they are similar because they belong to the same species" (329; cf. Mayr 1987a: 146–147).

Simpson's claim, of course, if ever construed as an argument against similarity as constitutive of species (or higher taxa for that matter), is only as strong as the analogy between species and organisms. As I believe I have shown in chapter 4, that analogy is poor and in many senses breaks down. It cannot be used, therefore, to argue that similarity is evidence of but not constitutive of species. Indeed I think that biologists (and not just biologists) need to take similarity more seriously, not merely in the sense of evidence but ontologically as well. (I will return to the question of monophyly a little later.)

In the case of Mayr, the fundamental mistake that he and many others make is in their inference that since the use of interbreeding relations works so well while that of similarity relations alone often leads to mistakes in the delimitation of species taxa, then only interbreeding relations are important and similarity relations must not be included in the ontology of those taxa. It's a bit like Plato teaching that the senses are never to be trusted because a straight stick appears bent in the water. At any rate, given Mayr's pincers, with the exception of his rejection of overall similarity (the "morphological species") his conclusion need not follow. One could just as easily conclude that similarity relations are constitutive of species taxa but that they need to be restricted or bounded by interbreeding relations, so that the latter are evidence of but are not constitutive of those taxa.

At this point one might wonder about the kind of organism similarity that is being taken seriously here. Organisms, as we know, have both a phenotype and a genotype. But in the definition of the biosimilarity species concept given above, it explicitly mentions only phenotypic similarity. But why not genotypic, or more simply genetic, similarity? Indeed it might be said that a certain degree of genetic similarity is necessary for successful interbreeding. (A high degree of genetic similarity, of course, can also count against successful interbreeding, in the sense of inbreeding depression.) And yet it should never be forgotten that the passing on of one's genes in successful interbreeding is largely a phenotypic activity (a fact, no doubt, for which most of us are exceedingly grateful). But more to the point, my reason for restricting my definition of the biosimilarity species concept to phenotypic similarity is simply this: it is to avoid the possibility of having a species of one. If we allow that the relevant kind of similarity is genotypic, in other words genes in common, defined in terms of DNA or RNA, then we are faced with the possibility that a single multicellular organism, given the genes in common between its cells, could be a species unto itself. The problem now is that, as we have seen in chapter 4.4n48, it is agreed by modern biologists that an organism is not a unit of evolution. Ghiselin, however, seems to be an exception. As we have seen in chapter 4.4, according to Ghiselin (1981) "A species can consist of just one organism—especially if he and she is a self-fertilizing hermaphrodite" (306; cf. Ghiselin 1987a: 137, 1997: 86). But if a species can consist of just one organism, and a species is a unit of evolution, then it follows that a single organism can be a unit of evolution. But this is not evolution in the modern sense. An organism can change in the sense of ontogeny, but such change is not evolution, since (for one thing) the genotype of the organism remains virtually the same from the birth of the organism through to its death. Quite simply, then, organism change is not evolution in the modern sense of species evolution. Species evolution requires phylogeny, not just ontogeny, and that is something that at the absolute minimum requires at least two organisms. Since relations require at least two relata, the biosimilarity species concept meets one of the basic requirements of a modern species concept. But to avoid the problem of allowing for a species of one, the relata must be organisms, more specifically their phenotypes, not their genotypes, for the reasons already given.[10]

What this further means is that the biosimilarity species concept conforms with a holistic biology that sees the organism as something more than, rather than

---

[10] Indeed at this point we may note a serious incongruence between Mayr's view that phenotypic similarity is not part of the ontology of species taxa and his concept of evolution. To begin, Mayr (1988a) dutifully affirms that species are units of evolution: "Species are the product of evolution, and, owing to the genetic turnover in populations, all species are evolving all the time. . . . Every deme evolves and so does every subspecies as well as every higher taxon (aggregate of species). Therefore, the fact of evolving is not in the slightest diagnostic for the species. Alfred Emerson . . . had 'evolving' in his species definition, but in 1942 I rejected this inclusion as obviously not being a defining characteristic of species" (321). Mayr's reasoning here, of course, is confused. Just because species may not be the only entities that evolve, it does not follow that "evolving" should not be included in the definition of species. Surely the *capacity* to evolve must at least be included (if only implicitly) for a species concept to be a contender in the modern debate. At any rate, Mayr agrees that species evolve. He then elsewhere (Mayr 1988a) argues as follows: "the traditional definition of evolution adopted by the geneticists ('changes in gene frequencies') is highly misleading. . . . All macroevolutionary phenomena and processes such as adaptation, convergence, rate of evolution, and shift of adaptive zones relate to phenotypes and can be studied without reference to their genetic basis. Indeed, up to the present time, everything we know about macroevolution has been learned by the study of phenotypes; evolutionary genetics has made virtually no direct contribution so far" (403). And again: "organic evolution is a hierarchical process, and nothing demonstrates this better than the difference between the gene frequency variation phenomena within populations and the phenotypic changes in macroevolution" (417). Given the above we must wonder: If phenotypic changes play such a central role in macroevolution, then is this not a highly implicit appeal to genetically based changes in phenotypic similarity relations? And if this latter is the case, why then are not these similarity relations considered to be part of the ontology of species taxa? Surely there is a serious incongruence here, a tension between two opposing views which do not easily, if at all, harmonize with one another.

as something reducible to, its genes. And indeed phenotypic similarity is arguably *real* similarity, since it cannot be reduced to genotypic similarity because of the supervenience of amino acids on codons (cf. chapters 2.2 and 3.2). Neither can it be reduced to, contrary to what Occam thought (cf. chapter 2.2), bodily characters in common, since, as a simple inspection of our noses will prove, characters vary, and so are not a simple matter of all or nothing, but instead involve a degree of similarity themselves. Indeed as Kurt Fristrup (1992) pointed out in his review of the concept of character in modern biology, "By recognizing likeness, we define a character, but its importance lies in emphasizing variations: . . . In fact, characters are never defined unless an observed or probable variation exists" (45). Genotypic or genetic similarity, on the other hand, is reductive: it reduces to nucleic acid sequences, which, like subatomic particles, may within each of their various kinds be all but numerically identical, contrary to Leibniz's law. The fact is that genetic similarity, in spite of its reductionistic attractions, is rarely ever used to define species themselves. (And given the reasons provided in chapter 3.2/3, as well as above, I doubt that it will ever be more than rare.) Genetic similarity, however, has proven quite useful for inferring phylogenetic relationships between species, as for example in inferring when the most recent species that led to modern humans and chimps diverged.[11] And as we shall see a little later, there is nothing about the biosimilarity species concept that is inherently incompatible

---

[11]   One of the now most common and powerful techniques is what is known as DNA-DNA hybridization. It is known that the two DNA strands in a double helix can be split by heating them to a certain temperature, and that they will reassociate when the temperature is brought back down. In DNA-DNA hybridization, then, to put it all too briefly, a portion of a single strand of DNA (normally around 500 nucleic acid bases) is taken from an organism of one species and combined with a comparable DNA strand of the same length from an organism of a different species. The degree of genetic similarity is determined by the degree of the overall bond between the two strands. The more base pair matches (T-A, A-T, G-C, C-G), the stronger the overall bond, while the more mismatches (T-T, A-A, G-G, C-C, T-G, G-T, A-C, C-T, none of which bond), the weaker the overall bond. Thus, the greater the overall bond, the greater the temperature required to split the hybrid double helix, while the weaker the overall bond, the lower the temperature needed to do the trick. Thus, the temperature required to split a hybrid double helix indicates the recency of common ancestry: the higher the temperature, the stronger the overall bond and therefore the closer the phylogenetic relationship. Nevertheless, there are serious problems with this approach (which alone should render obvious its inapplicability in delimiting vertical species). For one, as Mayr and Ashlock (1991: 174) point out, the virtual universality of mosaic evolution (cf. chapter 2.4) means that one will get very different results depending on the pieces of DNA used. If the pieces of DNA one chooses happened to have changed very little in, say, the past 2 million years, then the two species will appear much more recently related than if the pieces of DNA one chooses happened to change very much during that same period of time. Second, the DNA-DNA hybridization results are combined with the concept of the "molecular clock" (the theory that the rate of random mutations is fairly constant and that such mutations occur within a specifiable rate) in order to estimate the period of divergence. But as Panchen (1992) points out about the concept of molecular clock, "there is still no agreement about its validity" (209).

with this kind of research.

At this point one might wonder about the meaning of the term "phenotype" that is being used above. Certainly there is no problem in extending the traditional meaning of the term to include genetically prescribed behavior, since this can have its interorganism similarities and dissimilarities too. Moreover, this would have the added advantage of enriching the nature of species descriptions (which of course can at best be no more than rough approximations) from the viewpoint of the biosimilarity species concept. Nevertheless, I am hesitant to extend the meaning of the term to the extent that Richard Dawkins does. Dawkins (1982) extends the meaning of "phenotype" beyond the physical expression of an organism's genotype, beyond even its behavioral expression, to include "functionally important consequences of gene differences, outside the bodies in which the genes sit" (292). Thus Dawkins allows for scenarios in which, for example, "A mutant gene arose which, when present in the body of a male mouse, had phenotypic expression in the bodies of female mice with whom he came in contact" (230). Dawkins here refers to the so-called Bruce Effect, the automatic abortion of a recently fertilized (not yet implanted) egg in a female mouse caused in part by a chemically-active odor from a second male mouse (and of course in part by the corresponding biochemical mechanisms of the female), thus rapidly returning the female to estrus and allowing the second male to impregnate her. Dawkins interprets this as a phenotypic expression of the second male's genotype: "Genes in male mice have phenotypic expression in female bodies" (229–230). Again: "Abortion in female mice, according to this hypothesis, is a phenotypic effect of a gene in male mice" (231). Dawkins tells us that his hypothesis or concept of the extended phenotype is "more elegant and parsimonious" (232). I won't comment on the elegant part, but it is difficult to see how it is more parsimonious. It seems to me that a more parsimonious interpretation of the Bruce Effect would be one that is restricted to single-body genetics. Accordingly, one possible line of interpretation, which Dawkins is well aware of, is that the Bruce Effect is nothing more than an adaptive strategy on the part of the female, in spite of the apparent advantage to the genes of the second male (thus simply a matter of good fortune rather than adaptive selection). As Malte Andersson (1994) points out, "pregnancy blocking might reduce female investment in offspring that risk being killed by the new male" (378). But even if one wants to include the male signal or pheromone as an adaptive strategy on the part of the male, there is no explanatory need to extend the male phenotype to anything beyond the behavior of the male. To extend it beyond that behavior as Dawkins does is both unnecessary and luxurious, certainly not parsimonious. Moreover, it has the unwelcome consequence of blurring the distinction between individual organisms. Indeed for many biologists this consideration alone will be reason enough for rejecting Dawkins' thesis. All things considered, then, it seems to me desirable to somewhat limit "the long reach of the gene."

There remains much more to be done, of course, in the way of unfolding and developing the biosimilarity species concept, so as to reveal its many applications and possible virtues. (I will reserve real problems for the next section.) At this point we may consider the use of similarity by biologists in practice, when they are actually delimiting species. What we will see is that the biosimilarity species concept is really not so radical after all, contrary perhaps to its first impression.

Peter Stevens (1992) has pointed out that "The history of the circumscription of taxonomic species in botany is perhaps aptly interpreted in pragmatic and instrumentalist terms . . . . Whatever taxonomists say that their species concepts are, it is what the taxonomists are perceived to *do* that matters" (310–311). And what botanists do, interestingly, along with many other biologists who delimit species, is employ similarity to a large degree, and in a way highly compatible with the biosimilarity species concept. Mayr, as we have seen, claims that similarity should only be used as a preliminary step in delimiting species, that similarity should be increasingly disregarded as more information of a strictly biological nature is obtained. But this is not how most biologists who delimit species in fact operate. Two recent studies on published papers devoted to species support this judgment.

Melissa Luckow (1995) examined papers on species (mostly empirical papers, though some were theoretical) in the journals *Systematic Botany* and *Systematic Zoology/Biology* for the period 1989–1993 (a total of 114 papers from the former journal and 16 papers from the latter journal). Her first observation is that "most of the papers were not explicit about which species concept was being used" (598). For most of the papers, then, she inferred species concepts from the diagnoses. One of her conclusions from this analysis is that while one-third of the biologists in *Systematic Zoology/Biology* subscribed to the biological species concept, only 7% of the authors in *Systematic Botany* did so. Moreover, in 75% of the botanical papers (collected presumably from both journals) "qualitative or quantitative differences in morphological characters were used to distinguish species" (599). Does this mean that most of the authors subscribed to a morphological species concept? Not necessarily. As Luckow immediately points out following the above quotation, "Other data may be used in exploring modes of speciation but these data are seldom used alone to delimit species." What this indicates, then, is that similarity plays a much more important role, in practice if not also in theory, than sources such as Mayr would have us believe.

Lucinda McDade (1995) surveyed 104 botanical monographs, accounting for a total of 1,790 species, published in three botany journals, namely *Systematic Botany Monographs* (1984–1993), *Systematic Botany* (1988–1993), and *Opera Botanica* (1984–1993). Like Luckow, McDade found that "The majority of monographers . . . did not discuss the concepts or criteria used to delimit species" (612). And again, like Luckow, McDade found that "Most of the monographs were based on morphological data, but a number incorporated other sources of evidence (e.g., chromosomes, ultrastructural characters, molecular data, common garden experiments)" (608). Interestingly, of the twenty-eight monographs that did discuss species concepts and criteria for delimitation, twenty-one stated that they used morphological differences to distinguish species, while fifteen explicitly subscribed to a morphological species concept. Of further interest is that of these twenty-eight monographs, nine explicitly used reproductive isolation as a distinguishing characteristic (seven in explicit addition to morphology), two others used ecological coherence (both in explicit addition to morphology), while one used distinctive habitats (in explicit addition to both reproductive isolation and ecological coherence). (This breakdown is derived from McDade's Appendix 1, 619–622.)

Although both of the above studies focused mostly on the work of botanists, what they reveal, I believe, is a much larger picture, namely that phenotypic similarity is taken much more seriously by biologists devoted to the actual practice of species delimitation than it is by biologists devoted to the theory of species delimitation (in other words biologists who spend time on the species problem). In fact what both Luckow and McDade have found is that the work of biologists devoted to the practice of species delimitation is little affected by the controversies circulating in the modern literature on the species problem. Luckow points out, for example, that "the idea of species-as-individuals is marginal to empirical science" (596). McDade is even more explicit, and her conclusions are highly important in this regard:

> It is clear . . . that monographers largely rely on observable *patterns* of differentiation while assuming that these patterns are due to a variety of biological *processes* that underlie the morphological integrity of species.
>
> The ideas that species should be monophyletic . . . were expressed infrequently. The monographic work of systematists thus appears to have been largely unaffected by recent contributions to the species controversy. It is difficult to determine the reasons for this (authors rarely explain why they did not pursue a particular course). However, a number of monographers who discussed species concepts explained their focus on morphological differentiation by pointing to problems with the practical application of other concepts. This suggests that many monographers find the species controversy to have little practical bearing on their work. [613]

This is *not* to say, however, that monographers should not better familiarize themselves with the current controversies so as to attain a higher level of consistency in their taxonomic practice. Indeed both Luckow and McDade provide examples of inconsistencies (e.g., claiming to use a monophyletic species concept while using phenetic techniques to delimit species). Both authors, moreover, include in their discussions a list of recommendations for practicing systematists, for example that "describing species concepts and criteria used to delimit species is a key component of monographic papers and [I] would urge authors of monographs (and other works focused on the species level) to include at least a brief statement on these topics" (McDade 1995: 612), and that "Scientists naming species should be familiar with and evaluate the assumptions inherent in each species concept. . . . The species concepts I have discussed rest on conflicting assumptions, and with real data these assumptions will lead to different species circumscriptions" (Luckow 1995: 600).

Clearly, then, a work such as the present book has not only theoretical but practical value. And as for the present chapter, it would seem that not only is the biosimilarity species concept less radical and farfetched than it might first appear, at least in relation to the practices of monographers, but given those practices there is a definite field for the possibility that the biosimilarity species concept could actually change the way that biologists think of and even delimit species.

Be that as it may, it will be helpful at this point to examine how the biosimilarity species concept works in theory, given some simple cases, as well as some of the ways in which it might arbitrate disagreements in actual practice.

Perhaps a good place to begin is with "Mayr's pincers." Beginning with sibling species, the biosimilarity species concept incorporates them quite easily. The demarcating causal relations, of course, are the interbreeding relations within each sibling species, between which there are none, and possibly also differentiating ecological and behavioral relations.[12] The *Anopheles* example, referred to in chapter 1.1, offers a prime example, and it is useful here to continue with Wilson's (1992) description:

> The first such "characters" detected were the size and shape of the rafts of eggs laid by the female mosquitoes on the water surface. Those of two species were found to produce no clusters at all, instead laying each egg separately. . . . More characters were quickly added: color of the eggs, gross structure of the chromosomes, hibernation versus continuous breeding in winter, and geographical distribution. Most important of all, some of the species distinguished by these traits were found to feed on human blood and thus to carry the malarial parasites. [45]

And indeed the general consensus seems to be that sibling species turn out not to be so cryptic after all, which of course would have to be the case if they are ever to be discovered. As Mayr and Ashlock (1992) put it, "once discovered and thoroughly studied, they [sibling species] are usually found to have at least a few previously overlooked morphological differences" (92; cf. Sokal 1973: 371), such

---

[12]   Indeed it will be seen in this and the next section that for the biosimilarity species concept the operation of the internal isolating mechanisms of Dobzhansky and Mayr are still of prime importance, but they are relegated to the position of being a subset of the causal relations that delimit species. (I will reserve the possible problem of incongruence, in particular between interbreeding and ecological relations, for the next section.) I should add that the causal relations that delimit species according to the biosimilarity species concept are characterized by their direct ability to maintain the phenotypic unity of a species (the inclusion of indirect causal relations results in an infinite regress), such that changes in those relations are the proximal cause of evolution and speciation (this is necessary for the species concept to be an evolutionary one). In other words there must be a hereditary component to at least one of the relata in each of the causal relations. Accordingly, mere geographical separateness, or even geographical barriers (e.g., the interposition of a river within the range of a species), are excluded; in fact they do not play a direct but only a preliminary role in species evolution (cf. Mayr 1942: 247). As Cain (1993) put it, "geographical separation is probably never merely that. It will entail differences in the environment, physical and especially biotic, which will exert selection to cause genetic divergence" (195). I am indebted to my ex-fiancée, Elizabeth Pereira, for forcing me to make explicit the above distinction.

that "it is generally possible in a group of sibling species to identify old type specimens to the particular species to which they belong" (93).

No more troublesome for the biosimilarity species concept is the case of sexually dimorphic species, even those that are markedly differentiated. When discussing niches (chapter 3.4), I focused on sexually dimorphic woodpeckers, where the extreme dimorphism is in bill morphology. A more striking example is with mallard ducks, where unless one saw the males and females in pairs one would swear they were different species (indeed Linnaeus classified them as separate species; cf. Mayr and Short 1970: 88). An even more striking example is with fur seals, where the male is not only somewhat different in structure but is several times larger than the female (cf. Weisz 1967: 166). In each of these cases, in spite of the dimorphism there is no reason to reject the importance of similarity relations. In every case the males form one distinguishable similarity complex (subphenome) while the females form another. In these cases what joins one similarity complex with another into a single species is the interbreeding relations between them. But the importance of the interbreeding relations certainly does not justify the exclusion of the similarity relations. On the contrary, the similarity relations are just as important, for without them not only could we not delimit these species, but neither could they recognize their conspecific mates.

Polymorphisms (subspecies, varieties, races, breeds) can also be dealt with in much the same way. Allopatric polymorphisms are somewhat troublesome (as they are for many other species concepts), and will be dealt with in the following section. For the present, we may note that not all polymorphisms are allopatric. Some polymorphisms, such as the red and blue forms of the butterfly *Heliconia erato*, originally thought to be two distinct species, develop from identical larvae which feed together in the same cluster on a tree (cf. Bates 1862: 557). Similarly described originally to be two separate species, the two very different skin patterns of the California king snake, *Lampropeltis getulus*, are often found in the same litter and interbreed freely without any mixture of the two forms (cf. Futuyma 1986: 92–93). Similarly, the bridled (a white ring surrounding the eyes) and unbridled forms of the Common Guillemots, *Uria aalge*, which also involve slight differences in skull structure and in the shape of the tail feathers, nest together and interbreed at random (cf. Maynard Smith 1975: 219–220). Such species may easily be dealt with in the same way as sexual dimorphism, namely as two or more subphenomes connected by interbreeding relations. Maynard Smith's reflections on the third example above are especially interesting: "the relevant point here is that even when two forms are distinct, and the difference extends to several features, this is not regarded as a satisfactory reason for placing them in different species, if it is known that they breed together freely in the wild" (220). That the several differences are not satisfactory, it may be added, is because interbreeding is a causal relation essential not only for the perpetuation but for the evolution of the species.

Equally interesting from the perspective of the biosimilarity species concept, many species are strikingly dissimilar in different stages of their life-cycle, in particular species which undergo metamorphosis (what may be called ontogenetic dimorphism), such as Lepidoptera (butterflies and moths from caterpillars). In such cases the species will consist of two overall similarity complexes (each corresponding to one of the cyclical stages) connected by what may be called ontogenetic (or cyclomorphic) relations, the relation complex that obtains when every relatum in the adult subphenome is spatiotemporally continuous with one and only one relatum in the conspecific larval subphenome. This causal relation is especially interesting from the perspective of evolution since species evolution may even be defined in terms of changes in ontogenetic relations, as Maynard Smith (1975) argues: "although evolutionary changes are usually described in terms of the differences between successive adults, i.e. as phylogenetic changes, the differences between those adults were the consequence of differences between the paths of development which gave rise to them, i.e. of ontogenetic changes; phylogenetic changes are the result of changes in ontogeny" (310).

Biosimilarity species become more complex, but not inherently more difficult, with those cases which involve more than two forms. Social insects (termites, ants, bees, and wasps) provide a classic example, not only because each colony involves markedly different forms (castes) interrelated in complex associations (cf. Weisz 1967: 177), but also because in many cases the various forms within a species are produced not by different kinds of genotypes but instead by different nurturing relations. As Wilson (1992) points out, "Depending on the food and chemical stimuli she receives as a larva, a female ant becomes a queen or a soldier or a minor worker" (86). Equally fascinating are the highly polymorphic colonial coelenterates, such as the extraordinary *Physalia physalis* (the Portuguese man-of-war), consisting of various functionally interdependent forms of sessile and free-swimming medusae, the former vegetatively, the latter sexually reproductive. In both examples the species consists of the group of overall similarity complexes (each subphenome corresponding to one of the different body types) connected by the appropriate polymorphic relations, what may in the above cases be called sociomorphic relations. Sociomorphic relations are the causal relations that obtain when relata in each subphenome are reproductively dependent upon relata either in their own or other subphenomes (or possibly both) and are functionally interdependent with relata in all of the other subphenomes.

The above cases which involve species with both sexually and asexually reproducing forms raise the interesting question about species that cycle back and forth between these two modes of reproduction, in particular cyclical parthenogenesis (virgin birth). Cain (1954), for example, describes the following cases:

> In many waterfleas and greenfly, parthenogenesis is used during favourable conditions (especially in the summer). The young, all females, are either brought forth alive, or are hatched from soft-shelled 'summer eggs.' On the approach of less suitable conditions some males are produced which then fertilize the females in the normal way. These females lay resistant tough-shelled 'winter eggs' which can last through periods of adversity. [98–99]

Cain's description here is somewhat reminiscent of ecophenotypic switches, which as we've seen in the case of snails (chapter 4.3) involves a radical change in morphology depending on environmental conditions. With cyclical parthenogenesis, however, the response to changed environmental conditions is not a radical change in morphology but rather a radical change in reproductive mode. Both cases are easily handled by the biosimilarity species concept. The case of ecophenotypic switches, although it involves two different similarity complexes, is nevertheless a united similarity complex simply because of the fact that the switches are merely cases of species change, not species evolution (a distinction that I shall return to and shall elucidate a little later on), a distinction firmly embraced by the biosimilarity species concept. As for cyclical parthenogenesis, since such cases fall well within the horizontal dimension there is no problem considering each species as a similarity complex bounded by a combination of uniparental and interbreeding relations.

The case of cyclical parthenogenesis, of course, raises the question concerning species that reproduce consistently by parthenogenesis. It is believed that almost all such species, because of their morphological similarity to sexual forms, arose from sexual forms relatively recently. Among the likely exceptions are parthenogenetic species of Bdelloid rotifers, the distinct morphologies of which suggest a long history of obligate parthenogenesis (cf. Maynard Smith 1975: 4–5; Futuyma 1986: 280). In spite of being obligately asexual, however, many species of rotifers are exceptionally "good" species; in fact the asexual species of rotifers are generally better in this respect than the sexual species of rotifers (cf. Templeton 1989: 8).

Indeed the question now concerns asexual species in general, which includes obligate parthenogenesis, obligate hermaphroditism, and vegetative reproduction (budding and fission), and which may be extended to include obligate sib mating. Here, of course, one may simply delimit asexual species according to distinct morphological gaps. Since, as Templeton (1989) pointed out (providing it is true), "the asexual world is for the most part just as well (or even better) subdivided into easily defined biological taxa as is the sexual world" (8), the biosimilarity species concept has no problem with asexual species. Since the gaps cannot be due to reproductive isolation as with sexual species, it is legitimate to infer that they are due to ecological relations. But besides these there are often other causal relations that may be used to delimit asexual species. In the case of fungus-like organisms such as cellular slime molds (Acrasiomycota; cf. Margulis and Schwartz 1988: 132–133), one might also use relations of colony formation. Although species of slime molds are composed of free-living unicellular amoebas that feed on bacteria and typically reproduce by fission, under adverse conditions they aggregate into a pseudoplasmodium, a mobile mass which has the appearance of a slug. (Indeed at this stage alone they could superficially be misclassified with shell-less snails, the true slugs.) As conditions become more adverse, the migration of the slug stops and it produces a stalk with a fruiting structure on top from which it produces spores, beginning the whole cycle all over again. Such colony relations clearly unite the micromorphologies of the unicellular slime molds with their social macromorphologies into one species. In the case of bacteria, there are also bacterial slime molds (Myxobacteria; cf. Margulis and Schwartz 1988: 62–63), but because as we have seen in chapter 4.4, namely that some bacteriologists such as Sonea would go so far as to argue that the whole bacterial world is a single species because of the degree and scope of gene transfer between bacteria, I would be inclined to agree (provided that what they say is true), from the viewpoint of the biosimilarity species concept, that we should consider all bacteria a single species.

At this point, I will provide two final examples of how the biosimilarity species concept would work in practice. Both are particularly important since they show how this concept would adjudicate on two of the most divisive issues in the theory and practice of modern taxonomy. The first addresses the issue of multiple origins versus monophyly, the second addresses the related issue of hybrid species such as the red wolf, a case of particular concern to conservation biology.

Beginning with the first issue, it is important to note that many modern evolutionary biologists, when describing a species, naturally allow for that species to have multiple origins. Dobzhansky (1937), for example, in his chapter on polyploidy, tells us that "At least in one case, that of *Galeopsis tetrahit*, an existing species has been experimentally resynthesized from its putative ancestors" (207). Much more recently, Ashton and Abbott (1992), faced with strong evidence for at least two recent (this century) occurrences in Wales and Edinburgh of plant allopolyploidy resulting from the same two parental species, offer not the slightest hesitation in calling the resulting allopolyploids the same species—namely *Senecio cambrensis* (I will return to this example a little later)—and in leaving open the possibility of future origins. On the other hand, many other authors, influenced either by the now dominant school of cladism or by the equally dominant view of species as individuals (or by both, for after all they go together rather well), refuse to name such complexes a single species, but instead divide them into two or more species, one for each singular hybrid origin. An example of this is Frost and Wright (1988), who focus in their paper on the taxonomy of parthenogenetic lizards of the genus *Cnemidophorus*. Frost and Wright explicitly follow Ghiselin and Hull's thesis of species as individuals but they prefer Wiley's version of cladism since it allows for species of asexual/uniparental organisms (201). Accordingly on their view, "Regardless of similarity by *any* measure, historical groups derived from different origins do not constitute one entity and should not be forced into one binomial" (205). Thus any putative species of *Cnemidophorus* that has two or more origin (hybrid) events (even if the parental species in each case are the same) is on their view not a real species but a "nominal" species, a possible example being *C. velox* (207).

In chapter 4, I inveighed against the class/individual distinction upon which the individuality thesis rests, and I have also shown that loose cladists such as Wiley bring in ontological similarity at the species level through the back door. But I will not rehash those arguments here and show how Frost and Wright's combination of the theses of Ghiselin and Hull on the one hand and Wiley on the other is incoherent. Instead I will focus on two claims made by Frost and Wright (1988). First, they claim that "The origin of any historical entity is the *outcome* of some process. To say that some processes are valid . . . but others not . . . clearly is poor metaphysics" (203). I accept that a modern species concept should admit a plurality of evolutionary processes for the origin and delimitation of species, a point that I will return to later again. But they also claim that "Taxonomy must not be inconsistent with phylogeny" (204), by which they mean, primarily but not exclusively, that a taxon (species and higher) must be monophyletic (in their sense, strictly monophyletic, since they exclude paraphyly; cf. 205). In all of this, one senses a confusion between, or a conflation of, causal processes in evolution and monophyly. But it must be emphatically stated that *monophyly (however defined) is not a causal process.* As Kluge (1990) points out, "monophyly achieves logically consistent phylogenetic hypotheses and taxonomies," but he sees "no purpose in considering monophylysis to be a natural process." Instead he calls it a "convention" (427n1), a characterization that fully accords with the observation of McDade (1995: 613) discussed above. That monophyly is a convention rather than a causal process should be evident alone from the fact that it does not produce anything, unlike true causal processes such as ontogenesis or natural selection. Certainly it is a mistake to think of it as a mode of speciation. Instead it is a convention for arranging organisms into groups, ideally without any ontological appeal to similarity. But of course as with all conventions, it may easily result in splinter schools as numerous as the denominations that resulted from the Protestant Reformation. And this, indeed, is much the present case with cladistic and phylogenetic species concepts, not all of which I have cared to examine in this book (though I think my examination is sufficient). Of course as with all conventions, the current emphasis on monophyly need not be expected to last, its proclivity for logical neatness notwithstanding. At any rate, since monophyly is not a causal process but rather a convention, it cannot play a role in the biosimilarity species concept. Consequently, the biosimilarity species concept favors approaches like that of Ashton and Abbott (1992) and rejects approaches like that of Frost and Wright (1988). The advantages of this view are many, not the least of which is a greater parsimony with regard to the number of species under consideration. Moreover, it results in species delimitations that are much more in accord with the actual causal processes in nature that are responsible for the existence of the species delimited. And of course it results in a decrease in concern over whether a particular group of organisms form a real species or not (since monophyly, however defined, is exceedingly difficult and often impossible to determine in practice; cf. Mayr and Ashlock 1991: 256; Luckow 1995: 597; McDade 1995: 613), a decrease in concern that is consistent with taking the horizontal dimension as primary for species reality.

The case of the red wolf is equally interesting but of graver concern. As Wayne and Gittleman (1995) point out, the red wolf (*Canis rufus*) used to be widespread throughout the American southeast. By the late 1970s, however, it was almost extinct, its numbers reduced to a single small population in southern Texas-Lousiana, due mainly to many years of hunting and habitat destruction. It was saved from the brink of extinction by the U.S. Fish and Wildlife Service, which managed to capture fourteen red wolves thought to be pure (i.e., not wolf-coyote hybrids). With these wolves it established a captive breeding population, which has been used for ongoing reintroduction back into the wild, particularly in select national parks. The program has been hailed as the flagship of American conservation biology, a shining example of the successful application of the Endangered Species Act of 1973. However, due to the cost of the program (the budget for the late 1990s was roughly $4.5 million), due to the increasing numbers of species that need similar attention, and due especially to recent genetic evidence, the species status of the red wolf has been called into question, and accordingly its priority and worthiness for protection.

The problem is that the red wolf does not seem to constitute a species by any of the currently prevailing species concepts. It is highly doubtful that it would satisfy the criteria of Mayr's biological species concept or Paterson's recognition concept, since when its population numbers have been low it has been known to breed extensively with both gray wolves (*C. lupus*) and coyotes (*C. latrans*). Under the morphological species concept it has tended to do a little better. Although often thought to be merely either a subspecies of the gray wolf or a gray wolf-coyote hybrid, one prominent researcher, Ronald Nowak, argued that it is a good morphological species, since its cranial features are intermediate between those of gray wolves and coyotes. Indeed based on older and fossil skulls he concluded that the red wolf is the ancestral species of the two. Wayne and Gittleman, however, have concluded that the red wolf is a hybrid taxon after all. Their extensive genetic analyses (both nuclear and mitochondrial DNA) have led them to conclude that red wolves, based not only on modern specimens but on old pelts (taken from when their numbers were high), have no unique genetic traits. From the viewpoint of the phylogenetic species concept, then, diagnostically conceived, which is what Wayne and Gittleman were initially trying to apply, the red wolf is not a genuine species (irrespective of whether its hybrid origin happened more than once), since it has no diagnostic genes.

In spite of all of this, however, Wayne and Gittleman argue that the red wolf is worth protecting. Even though on their view the three reputed species have crossed throughout various periods of their history, so that "we feel the red wolf never truly developed into a separate species" (38), they also think it might be "the last, albeit impure, repository of genes from a now extinct gray wolf subspecies," so that "it should certainly be preserved." Moreover, they add "ecological concerns," namely that red wolves "are once again important predators of many wild animals" and "may also play a role in some habitats that its smaller kin, the coyote, cannot entirely fill" (39).

While I agree that the latter considerations are important, I cannot agree that the red wolf deserves the same consideration as good or true species. From my own perspective, the red wolf does not much qualify as a "phenotypic similarity complex of organisms objectively and maximally delimited by causal relations." This is both an evolutionary and taxonomic perspective, and from it the red wolf just does not wash as a species. And certainly there are many good species from this viewpoint that are either threatened or endangered and that are in need of protection. Given the need to catalogue the diversity of life, in the face of mass extinction #6, and the sadly limited conservation resources presently at our disposal, priority should go above all to the species-level of diversity. Criteria for determining target groups that are either too fine-grained (such as phylogenetic species concepts) or too coarse (such as morphological species concepts) work against the many needs and limitations of conservation biology. Placing questionable ecological roles above these considerations, as Wayne and Gittleman seem to do in the case of the red wolf, can only result in conservation policies that are as shifting and indefinite as the purported objects in their view.

At any rate, having seen how the biosimilarity species concept works in theory, it is still important to develop the concept a little further. What I have in mind is two exercises that will complete this section. The first is to individuate the biosimilarity species concept from those species concepts which it might seem to imitate or at least bear a strong resemblance to. The second is to show where the biosimilarity species concept stands in relation to a number of fundamental distinctions found in both the biological and philosophical literature. Both of these exercises will help to further clarify the biosimilarity species concept as well as bring to light more of its theoretical virtues.

To begin with the first exercise, it should by now be abundantly obvious that the biosimilarity species concept is not a morphological species concept, based on overall similarity and gaps, as for example the species concept of Cronquist (cf. chapter 2.4), even though both species concepts involve phenotypic similarity between organisms and have between them a number of other similarities, such as conceiving of species as being primarily horizontal. Although a morphological species concept might be attractive for botany, as well as microbiology, it has little attraction for zoology, as cases of radical sexual dimorphism alone amply illustrate. Accordingly it cannot hope to come close to being a universal species concept, worthy of the Modern Synthesis.

For many of the same reasons the biosimilarity species concept should not be confused with the phenetic species concept of the numerical taxonomists (cf. chapter 3.3), which after all is really only an ultrasophisticated version of the morphological species concept. As we have seen, both cannot handle many cases which biologists naturally take to be separate species, such as sibling species, and cases which they naturally take to be single species, such as caste polymorphism, ontogenetic dimorphism, and sexual dimorphism. Indeed it is sometimes almost a *reductio ad absurdum* to see how authors who subscribe to a phenetic species concept try to grapple with these relatively simple species situations. For example, in the case of highly differentiated sexually dimorphic species Sokal and Sneath (1963) suggest that "In such cases sexual forms should be treated as if they were stages in the life cycle and thus male characters compared only with other male characters, and so on. A similar procedure should be followed in social insects when differentiation among castes is considerable" (88). The crass male chauvinism in this suggestion is particularly striking! Clearly similarity must be added to in some nonphenetic way.[13] Aside from such practical difficul-

---

[13]   Such obvious failings, of course, could not be expected to go long without attempts at

ties, it should be noted that the biosimilarity species concept has even less in common with the species concept of the numerical taxonomists than it does with the morphological species concept. Not only does the biosimilarity species concept have no use for state spaces, but more importantly it does not share the same concept of similarity. Although numerical taxonomists repeatedly refer to and employ what they call "overall similarity" (a concept that I will return to in the following section), similarity is not for them a *real* relation, but instead is reducible to properties in common. For example, Sokal and Sneath (1963) state that "overall similarity . . . is a function of the proportion of attributes shared by two entities" (18), where attributes logically reduce to a matter of "'Yes' or 'No,' 'Possessed' or 'Not Possessed'" (63), each of which can be represented by a "binary digit" (64).[14] This approach, of course, is necessary if a character state space is to be used in defining species, since one must have a limited and manageable number of axes if a state space is to be operational. In reducing similarity to properties in common, numerical taxonomists are following in a tradition that we have seen goes back at least to Occam (cf. chapter 2.2). In fact it goes back even further, back all the way to Pythagoras, for whom mathematics was the paradigm and quality is accordingly reducible to quantity. This is a tradition which, following modern metaphysicians such as Russell and Denkel, I take to be unempirical and misconceived in the extreme. But I will save the issue of whether similarity can be reduced away for the following section. Suffice it to say at this point that the biosimilarity species concept is not an offshoot of the phenetic species concept.

Another species concept that the biosimilarity species concept might seem to closely resemble (Ereshefsky, pers. comm.) is the phylogenetic species concept of Joel Cracraft. As we have seen in chapter 4.5, Cracraft uses the phrase "smallest diagnosable cluster," which might be taken to imply that similarity between organisms is, in some sense, constitutive of species taxa; in fact Cracraft (1989a) uses the phrase "phenotypic cohesion" (43). Moreover, the concept of phenotype employed by Cracraft (1983) explicitly includes behavior in addition to morphology (97), reproductive isolation is not taken to be constitutive of species distinctness (104–105), his species concept includes asexual species (Cracraft 1987: 339), and it is explicitly an *effect* species concept aimed at being universal by

---

repair. For example, Sokal and Crovello (1970) state that "When assembling similar individuals, dimorphisms and polymorphisms may give rise to practical difficulties, and relational criteria based on knowledge of the biology of the organisms involved may be invoked. Thus, knowing that given caterpillars give rise to given butterflies, we shall associate them, and in cases of marked sexual dimorphism we would wish to associate males and females that appear to form sexual pairs" (37). Such attempts at repair are certainly laudable, but it should be noticed that the price of such repairs is that one is left with something that can no longer be called a phenetic species concept.

[14]   From the later writings of numerical taxonomists, there is every indication that this has remained their view. For example, Sokal (1986) writes of "overall similarity" (423), the "ordination approach" (436), "the occupation of phenetic space by the OTUs being ordinated" (436), and the ideal of "the greatest number of shared character states" (425).

allowing for a plurality of evolutionary and speciation processes (more on this a little later).

But beyond these similarities the two species concepts are radically different. For a start, on Cracraft's species concept the constitutive similarities are both genotypic and phenotypic (Cracraft 1983: 97, 107, 1989a: 36–37, 1989b: 30–31, 36). Moreover, the constitutive similarities consist of "unique combinations of primitive and derived characters" (Cracraft 1983: 103), which elsewhere Cracraft (1989a) refers to as "shared similarity in character data that can be used to distinguish them from other such clusters ['clusters called species']" (36). The biosimilarity species concept, on the other hand, does not limit itself to the cladistic concept of characters. This difference leads to more overtly profound differences, namely in the dimensionality of species and the role of monophyly. As we have seen in chapter 4.5, Cracraft mainly thinks of species as individuals, as vertical entities and as monophyletic. And although he does not accept Hennigian extinction and therefore allows for paraphyletic species, anything else, on his view, as we have seen in his criticisms of Mayr and Paterson, implies that history has been misrepresented. I will very soon argue, when I return to the topic of multiple origins, that this view is simply, and deeply, mistaken. At this point it need only be emphasized that by virtue of conceiving of species as individuals and as loosely monophyletic, the phylogenetic species concept of Cracraft, at least in its earlier formulations, is a primarily vertical species concept, whereas the biosimilarity species concept is primarily horizontal. This difference is profound, illustrated alone by the fact that the latter allows for multiple origins whereas the former does not.

Perhaps a closer comparison is between the biosimilarity species concept and the cohesion species concept of Alan Templeton (cf. chapter 3.4). Templeton's definition, it will be recalled, explicitly uses the phrase "phenotypic cohesion," and he views that cohesion as being maintained by a partially disjunctive plurality of cohesion mechanisms.

Nevertheless, just as with Cracraft's species concept, the differences outweigh the similarities. For a start, Templeton, as we have seen, uses the term "evolutionary lineage," which on his view is what a species is. This, however, implies monophyly, which the biosimilarity species concept disavows; moreover, it implies a primarily vertical conception of species, whereas the biosimilarity species concept conceives of species as primarily horizontal entities, allowing, once again, for multiple origins. In addition, as we have seen, Templeton goes far beyond "phenotypic cohesion," in that "both process and pattern are part of the cohesion species concept" (Templeton 1998: 33). The biosimilarity species concept, on the other hand, is exclusively a pattern concept, with the relevant causal relations being delimitative, not constitutive. All of these differences are profound ones, and they not only distinguish the two species concepts from one another but they also indicate, as argued in various places in this book, some of the advantages of the biosimilarity species concept over the cohesion species concept. For example: a horizontal species concept is superior to a vertical one; monophyly is a convention, not a causal relation or mechanism; pattern should be distinguished from process; etc.

Comparison of the biosimilarity species concept with the phylogenetic and cohesion species concepts might suggest to some readers that the biosimilarity species concept is really a pluralistic species concept, in one or more of the senses of this label examined and rejected in chapter 2.4. But nothing could be further from the truth. Pluralistic species concepts, as we have seen, all agree that modern biology requires different kinds of species concepts for different kinds of research programs and different kinds of explanatory situations. The biosimilarity species concept, on the other hand, not only does not share the fundamental premises of species pluralism, but its use of a variety of causal relations to delimit species does not at all mean that it uses a variety of species concepts. Nor does this mean that it subscribes to a variety of ontologies for species. All species, on the biosimilarity species concept, share the same ontology: they are intricate complexes of phenotypic similarity relations supervening on a partially disjunctive base of causal relations. Accordingly, the biosimilarity species concept is a monistic species concept, a competitor against other monistic species concepts, and together with them a competitor against pluralistic and nominalistic species concepts.

Finally, one might compare the biosimilarity species concept defined and developed in this section with the reconstruction of Darwin's mature species concept defined and developed in chapter 2.3. Again, however, while there are some striking similarities, such as the priority of the horizontal dimension, the use of similarity as constitutive of taxa, and the use of causal relations to delimit these taxa, there are also some striking differences, though perhaps not as striking as with the previous species concepts. To begin, Darwin's emphasis on common descent (monophyly in the loose sense) is rejected by the biosimilarity species concept. The latter would place, for example, all domestic dogs into one species, contrary to Darwin's view. This also leads to what is the most important difference, namely that the biosimilarity species concept does not discriminate between adaptive and nonadaptive characters. This is a fundamental difference that distances the biosimilarity species concept not only from Darwin but from many other biologists. As Peter Stevens (1980) pointed out,

> Botanists have long ranked characters according to real or presumed functional and/or adaptive significance. Darwin noted that the less a character was concerned with special habits, the more use it was in classification. Since then, characters apparently of adaptive or functional value to the plant have often been considered useful at lower levels in the hierarchy; characters apparently lacking such values have been considered to be of more use at higher levels. [344]

What I have shown in chapter 2.3 (and more fully in Stamos 1999), however, is that Darwin also used adaptive characters at the species level (and nonadaptive characters for higher levels) and that this was fully in accordance with his theory of evolution by natural selection. The biosimilarity species concept, on the other hand, does not share Darwin's (and not just Darwin's) emphasis on adaptations and, by implication, natural selection. It was this emphasis that made Darwin so resistant (albeit not dogmatically) to instantaneous speciation. The biosimilarity species concept involves no such resistance. A new species can arise by polyploidy (or by other means) even if initially it has no new adaptations. Indeed it may even be to some extent maladaptive. If outcompeted and driven to extinction by another species, it was still a species (cf. Wiley 1978: 87). Natural selection is one of, and may even be (as so many think it is, myself included), the main mechanism of speciation, but it is not the only one and a modern species concept ought to be able to accommodate this fact. Interestingly, Stevens (1980) goes on to point out that most authors have found that determining which characters are adaptive and which not is far from being an easy task, that often such characterizations are subjective and subject to change, and that some zoologists use adaptive characters to delimit higher taxa. All things considered, it seems to me better not to have such distinctions in a species concept. Accordingly on the biosimilarity species concept vestigial characters such as hen's teeth and horse's toes (to use two famous examples; cf. Gould 1983) are just as much a part of the similarity complexes in these two cases as are beaks and hooves, and neither should be eliminated from the respective species descriptions.

In closing this section, I want to show where the biosimilarity species concept stands in relation to a number of fundamental distinctions found in both the biological and philosophical literature. This will help to clarify even further the nature of the biosimilarity species concept and bring out even more of its theoretical virtues.

The first distinction is a purely philosophical one. It is the distinction that motivated the discussion in the previous section, namely the distinction between internal and external relations. As argued in the previous section, if a species concept is to be developed using relations, and if that species concept is to conform to a basic desideratum of a modern species concept, namely that it must allow for species evolution, then the constitutive relations must be internal relations, not external ones. When causal relations were used, we saw in the previous section that this led to a blurring of the distinction between internal and external relations. In using only similarity relations, however, as in the present section, this problem is obviated. Quite simply, similarity is the internal relation *par excellence* (cf. Armstrong 1989: 43), since it is clearly rooted in the respects or natures of its relata. We have seen further in the present section that for the biosimilarity species concept to be an evolutionary species concept, in the modern sense of evolution, the relata must be organism phenotypes, not organism genotypes. This specification has the further advantage of allowing for an important distinction in modern evolutionary theory, namely the distinction between species change and species evolution. It is important that a species concept capture this distinction, since not all species change is species evolution. If all humans, for example, were to shave their heads, this would be a change in the species, but it would hardly be an instance of evolution. To be an instance of evolution, the change would have to be genetically based, in other words descent with modification, where the modification is a directional change in successive genotypes within a population.

To give a more realistic example, a species that undergoes an ecophenotypic switch (cf. chapter 4.3) certainly undergoes species change, so much so that it can even fool biologists into thinking that it is an instance of species evolution, but that species change is certainly not evolution, since the change is not the result of changes in genotypes; indeed the species need not have changed genetically at all. Moreover, a species that doubles its population within a few generations is again an example of species change, an example that could have a profound effect in terms of its MVP (cf. chapters 3.4n51 and 4.1n2), but it is no more an example of species evolution than the previous example. The biosimilarity species concept perfectly conforms to this distinction (unlike, for example, many cluster class theories, such as the species concept of the numerical taxonomists). A species on the biosimilarity species concept is a dynamic entity, constantly in flux, changing as the phenotypes of its member organisms change, but none of that change is evolution unless it is an expression of cumulative change in the genotypes of those organisms.

A further important distinction is a distinction made much of throughout the present work, namely the distinction between the horizontal and vertical dimensions for species. There is an amazing recalcitrance in many theorists to admit this distinction, but as I have shown earlier (cf. chapter 2.3n15 and 2.4n20), it is a distinction that pervades the literature. What is even less appreciated is that, as I argued in chapter 2.4, the horizontal dimension is logically and therefore ontologically prior to the vertical dimension (cf. Stamos 2002). The implications of this for the issue of species reality, as we have seen in chapter 2.4, are enormous. And we have seen in chapter 4.3/4/5 what happens with realist species concepts that take the vertical dimension as primary. The absurd and awkward consequences are indeed overwhelming. With the biosimilarity species concept, on the other hand, since it takes species to be primarily horizontal, these consequences are avoided, and species realism is robustly affirmed.

In taking the horizontal dimension as primary for species reality, it is important to avoid a number of misunderstandings. First, it is not to say, as pointed out a number of times earlier, that a species on this view cannot have a vertical reality. Instead, it is only to say that the reality of a species does not depend on whether it has a vertical reality, only that it must have a horizontal reality (however defined).

Second, although on the biosimilarity species concept a species is a primarily horizontal entity, the biosimilarity species concept is nevertheless not a *chronospecies* concept. This is a concept that comes from paleontology. The biosimilarity species concept, on the other hand, is a neontological species concept, and it is important that the two are not confused with each other. Chronospecies are normally defined either as successive species of a phyletic lineage delimited fortuitously by gaps in the fossil record (cf. Eldredge 1995: 108) or as successive species of a phyletic lineage delimited in spite of a relatively good fossil record (cf. Damuth 1992: 108). Either way chronospecies are held to be arbitrary, ultimately nothing more than subjectively delimited segments of a continuum, mere matters of taxonomic convenience. And of course from a vertical perspective, chronospecies do indeed appear arbitrary. On the other hand, from a horizontal perspective, vertical species appear just as arbitrary! Since for species reality I take the horizontal dimension to be logically and therefore ontologically prior to the vertical dimension, the side I take on this issue should be clear. Given a phyletic lineage in the fossil strata that spans, say, 10 million years, the question "How many horizontal species, on your view, are there?" is an illegitimate question. What makes it illegitimate is that the question is being asked from the vertical perspective. From that perspective one might just as well answer that there are infinitely many species as that there are none. The answer makes no sense because it is being asked from the wrong perspective, wrong because the vertical dimension is neither logically nor ontologically the primary dimension for species reality. The concept of chronospecies is a product of that wrong perspective, and that is what makes it an illegitimate, nonsensical concept. It is not the horizontal dimension and the horizontal perspective that make it so. Accordingly, when it comes to the question of species reality, and the question of how many species there are, the only legitimate perspective is the horizontal perspective. From that perspective meaningful questions get meaningful answers. (I will return once again to paleontology in the next section.)

The third misunderstanding is perhaps the deepest of all. It is to say that because the biosimilarity species concept is a horizontal species concept, species on its view cannot be historical entities; accordingly a species concept that does not conceive of species as historical entities is antievolutionary. This misunderstanding is bound to be so pervasive that it is important to dig it up from its roots. Once again, the problem is a confusion based on the wrong perspective. For a start, consider a perfectly good set of statements made fairly recently by Jonathan Losos (1996) in a short introductory paper heading a special issue of *Systematic Biology*:

> The revolution in comparative biology that occurred over the past 15 years stemmed from two related developments. In the early 1980s, a number of workers argued that macroevolutionary phenomena can be interpreted only in an explicitly historical context . . . . Shortly thereafter, workers realized that as a result of shared ancestry, species are not statistically independent entities; consequently, statistical analyses of comparative data are invalid unless phylogenetic information is incorporated . . . . The result is that workers in all fields of biology are now aware that phylogenetic information must be incorporated into any comparative study that investigates causal hypotheses (i.e., studies that are not purely descriptive). The number of journal articles that incorporate phylogenies has increased substantially, not only in those journals devoted to evolutionary issues but also in journals such as *Animal Behavior*, *Ecology*, and *Development*. [259]

Accepting the truth of statements like these, many biologists and philosophers have gone further and concluded that the acceptance of these facts necessitates a vertical species concept. Paul Griffiths (1994), for example, tells us that "Functional classifications of traits are superficial . . . . They ascribe a function to a trait, but tell us little else about it. Phylogenetic classifications, on the other hand, allow us to infer a great deal of detail about a trait" (216). Accordingly, he tells us, "I adopt the historical view of species developed by M. Ghiselin, D. Hull and others, and the particular version developed in the writings of cladistic systematists. . . . see Ridley 1989" (206). Time and again one can find authors make the move from the importance of history reconstruction to a vertical species concept. And time and again in the literature one can similarly find a conflation of vertical entities with historical entities. As we have seen in chapter 4.3, Eldredge (1985a) claims that "punctuated equilibria puts the icing on the cake in the argument that species are real historical entities, comparable in a formal manner to individual organisms" (122). And as we have seen in chapter 4.2, Ghiselin and Hull argue that since classes are spatiotemporally unrestricted entities they are ahistorical entities, while since individuals are spatiotemporally restricted entities they are historical entities. Accordingly, species must be vertical entities (individuals) if they are to be historical entities. To give yet one more example, Frost and Wright (1988) tell us that "only individuals (=entities) exist independently of definition and have histories. Thus, only individuals should be included in biological taxonomies" (201).

But can a primarily horizontal entity (*entitas* = any existing thing) also be a historical entity? I believe that it can. All we need is to ask ourselves: What is it that makes an entity a historical entity? Certainly if an entity is a vertical entity, it is a historical entity. But while being vertical is sufficient for being a historical entity, it is by no means necessary. All that an entity needs to be a historical entity is to have a particular history. But that history does not have to be part of the being of that entity for it to be a historical entity. Therein lies the confusion. The fact of the matter is that horizontal species are just as amenable to historical analysis as are vertical species. And after all, is it not the determination of historical relationships (along with rates and mechanisms) that is really the matter of interest to those of us who value the evolutionary point of view? From such a point of view, making historical-vertical identity claims is really beside the point.

To see that this is so, consider once again the matter of multiple origins. Cracraft, as we have seen in chapter 4.5, claims that "By definition, nonmonophyletic species imply history has been misrepresented." This claim is not only false, but it provides us with a paradigm example of a definition that is nothing more than a persuasive definition. At any rate, it is a pervasive claim. Wiley (1978) also claims that para- and polyphyletic groups "misrepresent history" (81), and one can find the same or similar claims in many others. But given the primacy of the horizontal dimension for species reality, it follows that species may have multiple origins. And those origins, wherever they exist, are a matter of history and history reconstruction, and ought not to be ignored. One problem, however, is that multiple origins cannot be determined using cladistic methods. But why must history reconstruction be confined to only one method? If actual history is more complex than any one method can handle, then what is needed is a plurality of methods, not an imposing restriction on history. And again, if determining history is what we're really concerned about, then our allegiance should not be confined to any one method of determining history. In the case of multiple origins, such histories are no less interesting than branching histories, even though the former cannot be determined by cladistic methods (of whatever variation). As McDade put it, "current phylogenetic methods are inappropriate for taxa with complex reticulating histories, and yet the history of such groups is at least as interesting as those of divergently evolving groups" (616–617).

As a case in point, consider the paper by Ashton and Abbott (1992). In examining living populations of *Senecio cambrensis*, Ashton and Abbott determined that this species stems from at least two and possibly three occurrences of allopolyploidy speciation (each from the same parental species), all of these occurrences happening within this century, the earliest near Wrexham in North Wales sometime between 1910 and 1948, the second somewhat later (early 1960s?) in Mochdre, also in North Wales, approximately 40 kilometers west of Wrexham, and the third more recently (early 1970s?) in the Leith area of Edinburgh, Scotland, which is approximately 300 kilometers north of Wrexham. This is an interesting history. But equally if not more interesting is how Ashton and Abbott determined it. First, all three populations of *S. cambrensis* are highly similar, both genotypically and phenotypically; consequently Ashton and Abbott had not the slightest hesitation (and correctly, on my view) in classifying the three populations as conspecific. As we have seen earlier, however, many biologists are divided on this issue. A case in point is an exchange not too many years ago in *Taxon*. According to Robert Dressler (1990), "In the special case of allopolyploids, it is well known that they may be polyphyletic and still meet any reasonable criteria for species status" (448). In reply, Lidén and Oxelman (1990) claim that "A 'thing' (here the singularity in the evolutionary model) cannot be polyphyletic. In the special case of allopolyploidy it is of course possible that two events may give very similar results, but if we were not there to see it happen, the only way to infer polyphyly is by character analysis" (449). As for their first claim, I think enough has been said to show why I think they are wrong. But their second claim is also not true. Ashton and Abbott used much more than character analysis (however broadly defined). First, they accepted that the three highly similar populations, commonly classified as *S. cambrensis* ($2N = 60$), are allopolyploids, based on repeated colchicine experiments with the supposed parental species, *S. squalidus* ($2N = 40$) and *S. vulgaris* ($2N = 20$). Second, they used what can only fall under the category of biogeography. They knew that *S. squalidus* is an introduced species, having escaped from the Oxford Botanic Garden around 1910. Of the three populations of *S. cambrensis*, the population at Wrexham, first discovered in 1948 and the earliest known of the three populations, is closest to Oxford, being approximately 175 kilometers northwest of Oxford; the population at Mochdre, first discovered in 1966, is a little farther from Oxford but close to Wrexham; and the population at Edinburgh, first discovered in 1982, is the farthest away from both Oxford and Wrexham. Third, all three populations are geographically separate from each other; in other words, members of *S. cambrensis* are not found halfway, for instance, between the Wrexham and Mochdre populations. Fourth and finally, an electrophoretic survey of isozyme variation (an isozyme is a variant of a given enzyme in individuals of a species) showed that the population at

Edinburgh had the least genetic diversity of the three populations (consonant with a later origin), while the populations at Wrexham and Mochdre showed more genetic diversity (consonant with an earlier origin) as well as some significant diversity between them (although it is quite possible that the population at Mochdre is not the result of a separate origin but of a founder population from Wrexham). All the arrows of evidence, then, converge on a single explanation, namely that there were at least two and quite possibly three separate origins of *S. cambrensis* in Britain during this century. Interesting also is that given the fact that hybrids ($2N = 30$) of *S. squalidus* × *S. vulgaris* have been found in various locations in Britain where no members of *S. cambrensis* are to be found, Ashton and Abbott concluded that "it is possible that new origins of *S. cambrensis* will occur elsewhere in Britain in future years" (31). At any rate, the moral of the story is that history reconstruction in species biology countenances a wide variety of methods and that the biosimilarity species concept is fully consistent with this fact.[15]

Having finished with the above three common misunderstandings in taking the horizontal dimension as primary for species reality, we may now examine a further important distinction, briefly discussed in chapter 1.2/4, namely the distinction between species as process entities and species as pattern entities. As we saw in chapter 2.4, Levin (1979) pointed out that "Similar products need not derive from the same processes. For this reason, we should avoid promulgating species interpretations founded upon a common underlying process or interaction. These species are formulated by edict. Species interpretations based on the products of evolution are not shackled with implicit or explicit assumptions of causation" (384). In spite of this insight (i.e., that pattern supervenes on process), Levin took the species nominalist route, but there is certainly no necessity to do so, as evidenced by other theorists who think of species both as pattern entities and as real. Joel Cracraft (1983), for instance, proffered a species concept that is both realist and monist, and yet his view is that "a species concept is best formulated from the perspective of the *results* of evolution rather than from one emphasizing the processes thought to produce those results" (102). Logically there is no reason why one cannot combine a *realist* with a *result* species concept (cf. chapter 1.2) and have it be a *monist* one at that.

Indeed, Cracraft has some interesting things to say about species as pattern entities. Importantly, Cracraft (1989a) uses the process/pattern distinction to dichotomously classify competing species concepts. More specifically, he places the species concepts of Mayr (cf. chapter 4.1), Paterson (cf. chapter 4.1), Van Valen (cf. chapter 3.4), and Wiley (cf. chapter 3.4) in the species-as-process-entities category and the species concepts of Sneath and Sokal (cf. chapter 3.3), Cronquist (cf. chapter 2.4), and Cracraft (as we have just seen) in the species-as-pattern-entities category. Each species concept in the former category, Cracraft points out, "can be interpreted as being derived from a particular view of the evo-

---

[15] For further examples of phylogeny reconstruction in cases of multiple origins in hybrid plant species, and with the same style of species ascription as Ashton and Abbott (1992), cf. Werth *et al.* (1985), Sytsma (1990), Song and Osborn (1992), and Wyatt *et al.* (1992).

lutionary process" (34–35). While this is true for the most part, it is important to be as clear as possible on what the precise difference is between these two very different categories for species concepts. I would state it as follows: On the view of species as pattern entities, a species is purely an effect or result of causal processes in nature; those processes themselves are not part of what a species is. On the view of species as process entities, on the other hand, the causal processes specified are indeed part of what a species is. This is the crucial difference, and as we shall see below its implications are profound. (In the very least, it avoids Harrison's 1998: 25 rejoinder that isolating mechanisms are effects.) At any rate, one could, of course, add more examples to Cracraft's classification above, with some species concepts not clearly falling into either category (even the classification of species concepts has its "messy situations"), the most notable exception being Templeton's cohesion species concept (cf. chapter 3.4), since it conceives of species as process entities but it is based on many rather than on one or a few evolutionary processes.[16] But Cracraft's basic point still remains. As he put it, "this view of species [i.e., species as pattern entities] appears to be 'theory neutral,' at least in the sense that no specific theory of speciation (or evolution) has been invoked to justify or derive the definition" (34). In other words, species concepts which conceive of species as pattern entities do not stand or fall with theories of evolutionary mechanisms and speciation processes. Moreover, species concepts which conceive of species as process entities generally do not do justice to an understanding of the possible variety of processes and mechanisms (known and possibly yet to be discovered) responsible for species evolution and the diversity of life. As Cracraft (1983) points out, "Biologists have now come to believe that these processes are highly variable and often depend upon the group being studied" (102). Accordingly, species concepts which conceive of species as process entities might in actual fact present an obstacle to advanced understanding in this area. We have seen, for example, that Paterson's recognition species concept does not countenance speciation by polyploidy (cf. chapter 4.1), and one may reasonably conjecture that the latter processes would never have been minutely studied had Paterson's species concept, or something like it, been the prevailing view throughout the present century. Vrba (1995), for example, who heavily favors Paterson's species concept, claims that "all species should be the same kind of ontological entity and that all speciation should denote one particular kind of process" (3). While it may indeed be desirable that all species should be of the same ontological kind (I will address this shortly), it by no means follows that all speciation should be thought to be of only one kind. As Cain (1993) noted, "It is a good working rule that biological phenomena are more complicated than you

---

[16] The problem with Templeton's combination of both approaches (a process species ontology with many rather than only one or a few constitutive processes and mechanisms) is that in taking whatever are the processes and mechanisms that account for phenotypic cohesion as part of what a species is, it leaves the ontology of particular species indefinite, and accordingly the species category pluralistic. Pattern species concepts do not have this problem.

think" (197). Why should this be any different for speciation processes? It is logically possible, and probably good biology, to combine a monistic species concept
with a pluralism of speciation processes. But the fact remains that process species
concepts generally fail in this regard, while pattern species concepts have no
problem whatsoever. As Kluge (1990) put it, "my definition [remarkably similar
to Simpson's and Wiley's] requires that pattern be distinguished from process,
and in that regard I believe our understanding of the nature of diversity will improve. Any definition that confounds the two will surely lead to misunderstanding" (424; cf. Budd and Mishler 1990: 171).

There are more reasons, however, for preferring a species concept which conceives of species as pattern entities. For one, Cracraft (1989a) points out that process species concepts "are more distantly removed from species-as-observables in
that the concepts obtain some of their justification from particular views of the
processes of speciation and/or evolution" (34). Similarly, Luckow (1995) points
out that process species concepts are largely untestable (592), the reason being
that "For most species on earth, the process of speciation has not been directly
observed" (591).

Another important reason for preferring a pattern species concept is that it
allows for a universal species concept where all species are of the same ontological kind. As Luckow (1995) points out, "a universal and definitive species concept is most important to those systematists who actually circumscribe and name
species" (589). To this she adds that "if species are viewed as the endpoints or
results of various processes rather than as participants, a universal species concept is still viable" (590). A universal species concept, of course, is perhaps the
biggest lacuna remaining in the Modern Synthesis, the biggest barrier to its com-
pletion as a genuinely new paradigm. As Cracraft (1987) put it, "the 'synthetic
theory of evolution' has incorporated very disparate concepts of species, so much
so that it undermines the very notion of a 'synthesis'" (334).

Some writers on the species problem, however, have thought that the quest for
a universal species concept is unrealistic, even impossible. Hull (1997), for example, sees species concepts as trying to fulfill one or more of three basic goals,
namely universality (generality), applicability (operationality), and theoretical
significance (specifically evolutionary theory). The ideal species concept should
satisfy all three. But this, says Hull, is impossible because they are mutually exclusive: "Any species concept, no matter which one we choose, will have some
shortcoming or other. Either it is only narrowly applicable, or if applicable in
theory, not in practice, and so on" (357). Again: "Most importantly, if a species
concept is theoretically significant, it is hard to apply, and if it is easily applicable, too often it is theoretically trivial" (358).

Hull's view has a familiar ring to it. For example, in 1933 Ernest Rutherford
claimed that those who search for an energy source in atomic transformations are
"talking moonshine," while in 1963 J.B.S. Haldane claimed that the genetic
modification of humans "must surely be millennia away" (cf. Haynes 1990: 163).
Just as suspect as Hull's pessimism, of course, is an optimism that supplies a
simplistic solution to a long and seemingly intractable problem. Mayden's (1997)

attempt at a "denouement" of the species problem, which he thinks "should finally put the species problem to rest" (384), is a good example. Adding to Hull's (1997) three criteria the desideratum of monism over pluralism, Mayden attempts to place "at least 22 concepts that have been developed to characterize diversity" (389) into a hierarchy, with the evolutionary species concept of Simpson on top as the primary species concept and "most of the other species concepts" (419) at some level as secondary species concepts. In itself this approach not only fails to address the many problems with Simpson's species concept (cf. chapter 3.4), but it fails to acknowledge that most of the species concepts which Mayden classifies as secondary are not congruent in their division of organisms into species. As such, the hierarchy quickly falls.

What both of the above extremes illustrate yet once again is the failure to take seriously two of the most important distinctions in the species problem, namely the logical and therefore ontological priority of the horizontal over the vertical dimension for species reality and the numerous advantages of conceiving of species as pattern entities rather than as process entities.

In addition to conceiving of species as primarily horizontal entities, the biosimilarity species concept is quite clearly also a concept of species as pattern entities, since it takes similarity, and no other relations, as constitutive of species. It is superior to other pattern species concepts, however, on a number of counts. Its superiority to pattern species concepts such as Cronquist's consists mainly in the fact that it allows for both sibling (cryptic) species and polytypic/polymorphic species. The phenetic species concept of Sokal and Sneath has the same failings, plus the added one that it does not take organisms as constitutive of species (without which much of what modern biologists say makes very little sense). Mallet's (1995: 296–297) and other genotypic cluster definitions suffer from similar difficulties.

The biosimilarity species concept is also superior to the pattern species concept of Cracraft, but for very different reasons. For one, the biosimilarity species concept affirms the logical and therefore the ontological priority of the horizontal over the vertical dimension for species reality. For another, the biosimilarity species concept affirms that species evolve, what I take to be an absolute necessity for a modern species concept aiming at universality. Cracraft (1989a), on the other hand, holds that species "are epiphenomena, developed or evolved—in the true historical sense of the term—from lower-level processes." Thus, "they do not themselves evolve" (48). To my mind (and I suspect to that of most others), this alone eliminates Cracraft's species concept as a contender in the modern debate. But one should not thereby dismiss other pattern species concepts by a sort of guilt by association. Biosimilarity species *evolve*. They evolve as the similarity relations that constitute them undergo a directional change (providing that the similarity relations are genetically based).

Equally important, the biosimilarity species concept allows for speciation to be primarily a process rather than an event. This is essential, for as we have seen in chapter 2.4 messy situations are not only part of the reality of evolution but part of its evidence. So many biologists accept this that one can say it is accepted by virtually all of them (cf. Mayr 1982: 600; Otte and Endler 1989: xii; Templeton 1989: 24; Ghiselin 1997: 100). This is one of the main advantages of taking similarity seriously, as part of what a species is. Species concepts that avoid similarity and conceive of species as being constituted only by their component organisms cannot consistently conceive of speciation as a process but must conceive of it as an event. This is an extreme consequence, but it has been accepted, for instance, by Kornet (1993), for whom "species are historical entities which originate at a particular moment" (cf. chapter 4.5). To me, as well as to most biologists, this is unacceptable, and I suggest to them that they stop and consider the many consequences of not taking similarity seriously. Only a species concept that takes similarity as constitutive of species can possibly deliver a species concept that is both a species concept that takes speciation to be a process and a species concept that is truly universal.

Conceiving of species as both pattern and primarily horizontal entities has a further important consequence for speciation, namely that speciation is necessarily coupled to evolution. Many will find this consequence absurd, but the absurdity is really only apparent, as the following will show. Viewed from the horizontal dimension, most will agree that a species can evolve from being monotypic to being polytypic, all the while retaining its numerical identity. In evolving thus, most will also agree that it has undergone *part* of the process of speciation (cladogenesis). But if the same monotypic species should transform by the same amount (*same* in the sense of any of its geographic isolates in the first scenario) while still remaining monotypic, why then should this not also be considered part of a speciation process, in this case anagenesis? To allow that the change in the first scenario is speciation but not the change in the second scenario seems to me nothing more than a mere prejudice, a prejudice which has its roots, I suspect, in the *relational* aspect of Mayr's biological species concept, and which whatever its roots is contrary to the deepest insights of Darwin. Of course once this prejudice is removed, speciation is necessarily coupled to evolution, and the horizontal dimension for species ontology takes its rightful place of priority. In all of this there is no "underestimation" of evolution, contrary to what Ridley claimed (cf. chapter 4.5). Indeed it is the evolution in the horizontal dimension that is the very stuff of speciation. A common mistake is to think of the horizontal dimension as static, so that speciation and species need a vertical dimension. But the horizontal dimension is dynamic, gradually passing with every day, and as such provides everything (in the sense of time) that is needed for species to be fully real.

At any rate, a further reason, and I shall leave it as the final reason, for preferring a pattern to a process species concept is the former's better ability at capturing biodiversity. It is interesting to see how different species concepts arrive at different answers here. For example, according to Mayr and Short (1970), from the viewpoint of the polytypic biological species concept there are "607 species of birds breeding in North America," of which "no less than 315 North American subspecies were described initially as full species" (105). From the viewpoint of, for example, Cracraft's (1983) species concept, "the major effect [of which] will be to elevate some subspecies to species" (106), there are really something like 922 species of birds in North America (cf. Cracraft 1997: 331). (Interestingly, both monophyletic and phenetic species concepts tend to exaggerate the number of species.) This might make a practical difference from the viewpoint of endangered species laws, particular if they do not make any provisions for subspecies. Important also for biodiversity studies is whether a competing species concept includes asexual species. Clearly from the viewpoint of adequately assessing the world's biodiversity, and its decline, a pattern species concept is to be preferred over a process one, particularly one that takes similarity as constitutive of taxa. But on the other hand, whatever way one cuts it the fact remains that mass extinction #6, caused by none other than *Homo sapiens*, is occurring at a rate unprecedented in the history of life on earth. And if, as Edward Wilson (1992: 333) and so many others contend, the main and most important way of saving biodiversity is by saving whole ecosystems, then a debate over competing species concepts might seem pretty much beside the point. However, this is not necessarily so. Eisner *et al.* (1995) provide four strong reasons why conservation biology should retain a focus on individual species, among which is the fact that particular species are more discrete and objective than particular ecosystems, as well as the fact that many species are indicator species (e.g., frog species), such that a major decline in their number may indicate a serious stress to an ecosystem before it is obvious.

So much for the distinction between species as process entities and species as pattern entities. Another distinction that pervades the modern literature on the species problem is, of course, the distinction between species as classes and species as individuals. And as one can gather from chapter 4, this is definitely *not* a distinction that has guided the development of the biosimilarity species concept. In fact, not only do I think that the class/individual distinction is, ontologically speaking, a false dichotomy, but I think it has done far more harm than good. Guided by an analogy with individual organisms—interestingly an analogy that Darwin, as we have seen in chapter 4.4, initially toyed with but then quickly rejected—it has led to a proliferation of different vertical species concepts, each claiming to best capture the species-as-individuals view. But more damaging is that it has intimidated many biologists into rejecting what seems to others ridiculously obvious, such as that there are asexual species, that the extinction of a species is not necessarily forever, that species may have multiple origins, that a species that undergoes "infinite evolution" is not numerically the same species, that the primary reality of a species is horizontal, and that species do not need stasis to be real. Indeed it has even led some authors, as we have seen above with Cracraft, to deny that species are entities that can evolve. Fortunately, the biosimilarity species concept is not motivated by the class/individual distinction. Instead it accords with the intuitions of biologists when confined to the practice of biology. Much like languages, biosimilarity species evolve, they allow for multiple origins, and their extinction is not necessarily forever.[17] Accordingly they are nei-

---

[17]   Whether similarity relations are repeatable is an issue I will reserve for the next sec-

ther classes nor individuals, but they are nevertheless real. Indeed a further ontological category is needed to capture this reality, perhaps something like Wiley's
category for historical entities (cf. chapter 4.4).[18] That such a category has yet to
be fully elucidated only serves to highlight the fact that metaphysics is no more a
finished or completed field than its natural science counterparts.

5.3 Problems with Species as Relations

The time has now come to focus on various problems with the biosimilarity species concept. In the previous section we have examined a few of the problems,
but the main focus of that section was on elucidation of the biosimilarity species
concept. In the present section, instead, the main focus will be on problems. Although by no means meant to be complete, the present section will progress from
the more practical problems to the more theoretical, from the more overtly biological to the more overtly metaphysical, ending with a discussion on the concept
of similarity itself. Many of the following problems, of course, apply not only to
the biosimilarity species concept but to many other species concepts as well. The
later discussion on relations, in particular, applies not only to the biosimilarity
species concept but to other species concepts that take relations to be real, and not
just similarity. For ease of discussion I will introduce each problem with a question.

    What about paleontology? As pointed out earlier, the biosimilarity species
concept is basically a horizontal species concept with maximum application to
neontology. It has very little application to paleontology. But as we have seen in
chapter 2.4, some paleontologists think that paleontology is better off without a
species concept anyway. I tend to agree. Even though Shaw's (1969) example of
conodonts might not have been the best to illustrate his point, it is a point that can
nevertheless be further defended. "Paleontology," as defined by Abercrombie *et
al.* (1990), is the "Study of fossils and evolutionary relationships and ecologies of
organisms which formed them" (419). This definition is about as good as any. In
terms of evolutionary relationships, aside from phylogeny reconstruction paleontologists focus largely on rates of evolution and extinction. Rates of evolution
are usually divided into character rates (often called morphological rates) and
taxonomic rates, the former referring to the rates of change in particular characters or character complexes in fossil lineages and the latter to the rates of replacement of taxa (species and higher) through the fossil record. As a paleontologist,

---

tion. It may be, however, that *precise* repeatability is no more necessary than in the case of
languages. If true, then repeatability need not be necessary for the logical possibility of the
reappearance of an extinct species.

[18] Cf. Mishler and Donoghue (1982), who explicitly think that Wiley's third category
applies in many cases to commonly recognized species taxa. As they put it, "These units
are not strictly 'individuals' or 'classes,' but clearly they can function in evolutionary theory and phylogeny construction" (127). My proposal is to go even further, to apply something like Wiley's third category to *all* species taxa. The main advantage here is that it
rescues the species category from pluralism and the concomitant danger of nominalism.

Shaw (1969) embraced the former kind of study and rejected the latter. Defenders of the theory of punctuated equilibria (cf. chapter 4.3), on the other hand, reject this restriction and argue that paleontology, to be done properly, requires a species concept. Which is the more reasonable view? What needs to be remembered is that when it comes to *rates* of evolution, all that paleontologists can study is fossil morphologies through strata. Epistemologically, the study of character rates has priority over the study of taxonomic rates. Why? In their useful review of contemporary methods of measuring morphological rates of evolution, Fenster and Sorhannus (1991) point out that "Measurement of taxonomic rates of evolution are reflective of morphological rates of evolution, since in the fossil record most taxonomy is based on morphology" (376). Moreover, they point out that measurement of taxonomic rates "are subject to the vagaries of systematic methodology" (376). If all of this is true, then it follows that the study of change in character rates, the kind of paleontology that Shaw (1969) prescribes, is the more basic and stable kind of study than the other. Indeed it may even be that character paleontology, what Shaw calls analytic paleontology, is the only kind of paleontology that is really needed. The rest is metaphysics (cf. chapter 4.3). And given the logical and therefore ontological priority of the horizontal dimension for species reality, this conclusion receives only stronger support.

This is not to say, of course, that paleontologists cannot or should not make species distinctions at any given horizontal dimension in fossil strata. But given that the causal relations which delimit species in the horizontal dimension are not preserved in the fossil record, that most of the morphology of fossil organisms is not preserved, and that accordingly little of their former ecology and behavior can be safely inferred, such species distinctions can never be as secure as they are in neontology. For example, it is well known that Neanderthals and anatomically modern humans coexisted in various parts of Europe and elsewhere for many thousands of years (cf. Feder and Park 1993: 286–295). Should Neanderthal fossils, then, be named *Homo neanderthalensis* or *Homo sapiens neanderthalensis?* The choice is arbitrary. We don't have access to the causal relations, for any given level in the fossil record, that would decide the issue. From the viewpoint of the biosimilarity species concept, then, the species designations, whatever they are, are based on similarity alone (and on only partial phenotypes at that), without the delimiting causal relations. They are therefore not biosimilarity species. Nonetheless what we can focus on, and what paleontologists can have a meaningful dialogue on, are the characters that distinguish each of the two forms, where they are to be found and when, and their rates of change (and disappearance) through time. The species or subspecies designations, however, whatever we decide, are little more than conventions, mere conveniences that aid the truly meaningful exchange of ideas. This is not to deny, of course, that paleontology has no other uses. Take for example the question concerning when the common ancestors of modern humans and chimps diverged. Paleontological evidence has converged with DNA and other genetic evidence to give an answer of 4 to 6 million years before the present (cf. Feder and Park 1993: 104, 169–176). In helping to answer this interesting and important question, and others like it, the significance should not be lost that no species concept peculiar to paleontology is required, so it is no argument against the biosimilarity species concept that it does not readily apply to paleontology.

As for paleoecology, it is amazing how much information about the former ecological relations of fossil organisms can be inferred, for example from dentition (cf. Weishampel 1995) and fossil dung (cf. Chin 1995), but surely none of this allows a degree of resolution sufficient to make species distinctions. Much more promising is the existence of "bone beds," localized fossil remains tragically deposited over a period of time as short as a few months or even hours and representative of an entire population or species. As Scott Sampson (1995) put it, "the entombed animals most likely represent a single species, perhaps even a single population—truly a snapshot in time" (37). From such bone beds it is possible to infer that somewhat different forms are really the two halves of a sexually dimorphic species (especially if the fossils from each are roughly equal in number), or that a form previously thought to be a species itself is really the juvenile form of another. And indeed for the latter kind of case one does not even need bone beds. For example, my friend Thomas Carr of the Royal Ontario Museum has recently argued (Carr 1999), based on well-represented ontogenetic changes in *Albertosaurus libratus* (an earlier or primitive relative of *Tyrannosaurus rex*), that contrary to previous opinion the fossil named *Nanotyrannus lancensis* is misnamed, that its craniofacial features strongly suggest that it is not a pygmy tyrannosaur (*Nanotyrannus* having been made a new genus) but instead a juvenile *Tyrannosaurus rex*; similarly for a fossil named *Maleevosaurus novojilovi* he has argued that it is really a juvenile *Tyrannosaurus bataar* (a sister species of *T. rex*). In all of the above, species that are both nontypological and horizontal are being inferred in paleontology based on inferred relations that perfectly accord with the biosimilarity species concept. Nevertheless, those relations can at best be only inferred, never observed. The biosimilarity species concept is still fundamentally a neontological species concept, although it is interesting that the biology upon which it is based can and actually is being applied, albeit in a very limited extent, to the distant past. (As Carr put it to me, he likes to call it the attempt to put the "ontology" back into "paleontology.") But when these same paleontologists take the further step of identifying their species vertically over vast stretches of time, they are stretching their implicit species concept beyond the limits of its validity. The new trend may be to rearrange dinosaur exhibits according to cladistic principles, following the 1995 reopening of the Dinosaur Halls at the American Museum of Natural History in New York City, wherein the old Halls have been replaced by the Hall of Saurischian Dinosaurs and the Hall of Ornithischian Dinosaurs, so that "Rather than juxtaposing creatures that lived at the same time, we have grouped them according to their evolutionary relationships" (Gaffney *et al.* 1995: 33). Nevertheless, nothing is gained from such practices, and much is regrettably lost, from the viewpoint of the biology and logic of species.

What about allopatric populations that are highly similar? This question, of course, at least as far as the biosimilarity species concept is concerned, applies only to the horizontal dimension and presents a problem for many species concepts. Many biologists think that the classification of allopatric populations is more arbitrary than not. Cain (1954), for example, says "We . . . can only guess" (73). Maynard Smith (1975) goes almost as far:

> There must, however, be some degree of difference between two geographically isolated populations which will justify placing them in different species. Since all degrees of difference between such populations may exist, from barely recognizable statistical differences to clear-cut differences of the same order of magnitude as separate different species inhabiting the same area, the decision in any particular case is to some extent an arbitrary one.
>
> This difficulty arises because, with geographically isolated populations, it is impossible to apply the test, 'Do the two forms interbreed freely in the wild?' [226–227]

The problem, however, is not the same as in paleontology, where the additional problem is whether allochronic populations or individuals are conspecific. In both cases, of course, whether one is dealing with allochronic or allopatric situations, it is a matter of inference whether the forms in question are conspecific. But if in the paleontological situation it is a matter of *mere* inference, the same cannot be said for neontology. As Mayr (1987b) put it, "such inferences are always based on numerous factual pieces of evidence. They are not simply wild guesses" (219). In neontology the amount of available evidence is so much greater than that of paleontology that the difference is in many orders of magnitude. There is a much greater access to morphological similarities and dissimilarities, including chromosomal evidence, behavioral similarities and dissimilarities, including periods of gestation and reproductive behavior, and ecological similarities and dissimilarities. As for the latter, Van Valen (1976), surprisingly, claims that "it is arbitrary whether otherwise similar populations on isolated islands are called different species. . . . the decision seems to be one of taste rather than biology." Nevertheless, he claims that he would not make such a splitting unless it could be determined that "their evolution is controlled by pressures sufficiently different that the evolution of these populations is also separate. I believe that this criterion is one of the most commonly applied to such situations in practice, at least implicitly" (71). But if this is truly the case, then the decision is indeed largely a matter of biology—and most importantly for the modern species problem, evolutionary biology. To this we could add considerations from biological control, such as immunology. If from the viewpoint of biological control the two populations are in effect the same, then we have a further reason to think of them as conspecific. If, on the other hand, there is some difference from that viewpoint, then this alone does not mean that they should be considered different species (since even within "good" species there is usually some variation from the viewpoint of biological control), but it is further evidence of evolutionary divergence and should be treated accordingly. In the absence of evidence from biological control (pro or contra), Rosen (1978: 31) suggests that allopatric populations that are highly similar should initially be considered as different species, since it is better to err on the side of safety from the viewpoint of biological control. I would tend to agree, as long as we are willing to change our species distinctions in the light of new evidence. Using this approach is not in the least inconsistent with the biosimilarity species concept.

What about ring species? As discussed in chapter 2.4, ring species are often thought of as living examples of geographic speciation, but they are problematic for species concepts based on reproductive isolation.[19] So would a ring species, from the viewpoint of the biosimilarity species concept, be considered a single species or a multitude of species? Since the reproductive relations of a "good" ring species delimit one continuous phenotypic similarity complex, and since it is phenotypic similarity, on my view, that is constitutive of species, not reproductive isolation, it seems to me that I would have to consider a ring species as a single species. Since to divide a "good" ring species into two or more species is completely arbitrary, as Cain (1954: 142) observed, this solution seems to me a further positive consequence of the biosimilarity species concept, although it is mitigated by the relative rarity of ring species. Fales' (1982: 84) rejoinder that one could turn a ring species into two or more species simply by killing some of the intermediates ignores not only that nature itself produces species relatively instantly, as in polyploid speciation, but that in evolutionary history extinction plays an essential role in nature's production of species. The downside of considering a ring species as a single species, however, is that it is problematic to consider a ring species as an evolving unit. It is, of course, logically possible that a ring species could evolve as a unit, but given its geographic divergence this is highly unlikely. If gene flow were high, this would not be unlikely, but because a ring species is typically constituted by overlapping rings of delimitable subspecies, gene flow between the subspecies must be low, otherwise the whole would not have the appearance of overlapping rings of subspecies. A case in point is the lungless salamander *Ensatina eschscholtzii* studied by Wake *et al.* (1989). This ring species, which circles around the Central Valley of California, is commonly

---

[19]   Although both Dobzhansky (1937: 315) and Mayr (1942: 180–185, 1970: 291–293) think of ring species (called "rings of races" by the former and "circular overlap" by the latter) as prime examples of geographic speciation, both avoid giving a clear answer on whether a ring species should be classified as a single species. The problem is that although a ring species constitutes by their definition a single gene pool, and hence a single species, the terminal populations are reproductively isolated from one another and so on their view constitute two separate species. It is no wonder, then, that they each avoided addressing the issue. Ghiselin (1997: 94–95), on the other hand, offers no such avoidance. Using the example of *Canis familiaris* (domestic dogs), where Chihuahuas cannot mate with Great Danes but they are nevertheless considered to be conspecific because of their reproductive links with other breeds, Ghiselin holds that a ring species likewise constitutes a reproductive whole so that each ring species should count as a single species. Ghiselin's solution is consistent with his shift in emphasis from what keeps species apart to what holds a species together (cf. chapter 4.2). Nevertheless, as we shall see below, it may in many cases be a mistake to consider a ring species as a reproductive whole.

divided into seven subspecies and is thought to have begun its ring by migrating from the northern end, since at the southern end there is local sympatry between the two respective subspecies with either very little or no hybridization, depending on the location. In the particular hybrid zone focused on by Wake *et al.*, a zone that is very narrow (approximately 1.4 kilometers), there is low gene flow between the two morphologically distinct subspecies, augmented by predatorial selection against the hybrids (since their coloration, unlike their parental subspecies, neither blends in with their surroundings nor mimics a poisonous related species) and species-level ecological differences (hence competition avoidance) between the two subspecies. Nevertheless, Wake *et al.* refrain from calling the entire complex more than one species (even though at the southern zone and the zone they focused on they admit species-level differentiation), their main reasons being "the complexities associated with the possibility of indirect connection between the [seven] units" (154–155) and their observation that "Speciation seems to have progressed in part" but that "it is still too early to determine" (155). As we shall see again under the next question, the biosimilarity species concept seems to capture quite well the fact that there are incomplete speciation situations where biologists are hesitant to make a taxonomic decision and that rather than being a liability this is an asset of a viable species concept from the evolutionary point of view.

What about messy situations? The term "messy situation" is commonly used to refer to the phenomenon of blurred species boundaries in the horizontal dimension. These can be categorized in a number of ways, but the two most common are in terms of morphology and hybridization. Beginning with morphology, one might take the case of asexual dandelions, which provide a taxonomic mess (cf. Maynard Smith 1975: 225; Ridley 1989: 8; but cf. Richards 1986: 448–453 for an opposing view from a specialist). The question is whether the case of dandelions, assuming a mess, is typical or atypical. Biologists seem to be agreed that it is the latter. Again, as Templeton (1989) pointed out, "the asexual world is for the most part just as well (or even better) subdivided into easily defined biological taxa as is the sexual world" (8; cf. Simpson 1961: 161; Ehrlich and Raven 1969: 61; Mayr 1970: 18–19; Eldredge and Gould 1972: 223; Wiley 1981: 36; Futuyma 1986: 247; Endler 1989: 630; Budd and Mishler 1990: 166; Wilson 1992: 48).[20] As for morphological messy situations in the sexual world, I have treated the issue in chapter 2.4 and I see no reason to repeat the arguments found there. I will add here, however, McDade's (1995) finding in her survey of recent botanical monographs, which delimit and describe a total of 1,790 species, mostly sexual (616). She concluded from her data that "something on the order of 15% of species are, at this horizontal time slice, involved in one or more biological processes

---

[20]   As for a "morphological yardstick" based on interbreeding being objectively applicable to asexual organisms, I should think (contra, e.g., Hull 1965: 215–218 and references therein) that such a concept is entirely invalidated by the existence of sibling species alone, since surely it makes no sense whatsoever to say that there are sibling species of asexual organisms.

that blur species boundaries or lead to infraspecific (and subsequently specific?) differentiation" (614). Given this consensus, then, for both the asexual and sexual worlds, and assuming that it genuinely reflects the biological world, this seems to me most fortunate for the biosimilarity species concept. As I pointed out in chapter 2.4, if messy situations were the norm, then I would seriously doubt that species are real. Conversely, if there were no messy situations, then I would seriously doubt that evolution is a fact. Messy situations, however, do exist although they are not the norm, and evolution is indeed a fact. The biosimilarity species concept, I suggest, accommodates both of these facts as well as can be expected of any species concept.

There still remains the question of messy situations in terms of hybridization. Once again this has been dealt with in chapter 2.4 and I see no reason to repeat the arguments found there. I will once again add here, however, McDade's (1995) finding in her survey of recent botanical monographs. She found that "About 12% of all species treated were identified as participating in hybridization. However, only a dozen or so cases of hybridization were identified as complex or frequent enough that species delimitation was difficult" (614). McDade goes on to argue that for a number of reasons these statistics might possibly underrepresent the actual situation in nature. Nevertheless, none of the reasons she gives provide any reason to think that if there is indeed underrepresentation it is such that hybridization is so wide and intense as to pose a threat to species realism in general. This can only once again bode well for the biosimilarity species concept.

I might also add a few comments here on the issue of semispecies versus multispecies. Recall to mind Templeton's (1989: 10) example of balsam poplars and cottonwoods examined in chapter 3.3 in my discussion on Beckner. On my view it is really a senseless debate whether we should call such forms "multispecies" (following Van Valen 1976: 72 and Templeton above) or "semispecies" (which is the usual view; cf. Mayr 1970: 287–288; Futuyma 1986: 115). Either way these messy situations are examples of cladogenetic speciation which became arrested somewhere along the line in the creation of the two forms and their evolutionary divergence from one another, and it seems to me that a truly viable species concept ought to be capable of capturing the fact of these messy situations rather than having to arbitrate on whether what we have is one species or two. Accordingly, I suggest the neutral term "messyspecies" for these and similar messy situations.

What about the problem of congruence? This is a much more interesting problem for the biosimilarity species concept. As we have seen in the previous section, the biosimilarity species concept requires for its validity a good deal of congruence in the biological world between the causal relations that are taken to delimit species. If that congruence does not obtain but rather the converse is more often the case than not, then the biosimilarity species concept would be invalid. Arguably the most important issue here is between interbreeding relations and ecological relations, in particular selection in the case of the latter. Although it is well known that interbreeding relations might often be able to maintain the integrity of a species in spite of diversifying selection regimes, it is also well known that diversifying selection regimes might often be strong enough to disrupt gene flow. So is congruence the norm in nature, or incongruence? The question, of course, is a matter of evidence, and there is the problem. As Ridley (1993) points out, "it is difficult to test between the ideas, to find out whether gene flow or selection provides a better explanation of species integrity" (395). Nevertheless, Ridley (395–397) provides some interesting examples from recent field studies that support each of the above scenarios. His own conclusion, however, is that "Selection and gene flow are probably not usually opposed forces in nature. Gene flow, tending to break down differences between two forms, and natural selection, working on reproductive behavior, should normally unify the ecological and reproductive species concepts" (397). It might be argued in reply that Ridley's assessment here is heavily biased, since his own species concept (cf. chapter 4.5) requires the said unity. And if I share Ridley's assessment, the charge of bias might be brought against me and my own species concept. Nevertheless, I think that the evidence strongly favors Ridley's assessment, so that the charge of bias is misplaced. Certainly the difficulty of testing particular cases, and then of counting up each of the cases pro and contra, is both serious and cannot ever be expected to be forthcoming. In a sense it is much like the problem of testing the theory of punctuated equilibria (cf. chapter 4.3). Nevertheless, there is a difference that makes all the difference, and this is the evidence from messy situations discussed above. Surely if incongruence were the norm, then messy situations would be the norm. But as we've seen, messy situations are *not* the norm, which is strong evidence that congruence, just like Ridley concluded, is also the norm. The problem of congruence for the biosimilarity species concept, then, finds its solution, interestingly, in the problem of messy situations.

What about endosymbiosis? "Endosymbiosis" may be defined as a mutually beneficial association in which organisms (unicellular or multicellular) of one species live inside (*endo* = inside) organisms of another species at a level of evolved integration (intracellular or intercellular) resulting in a new species of organism (cf. Margulis and Schwartz 1988: 314; Margulis 1993: 171n). The now most famous case of endosymbiosis, championed most vigorously by Lynn Margulis and now part of biological orthodoxy (so that a modern species concept, to be a true contender, can scarcely afford to ignore it), is the case of *serial endosymbiosis* in the origin of eukaryotic (plant, animal, fungal, and protist) cells, in which most notably mitochondria and chloroplasts (and probably also ciliates) are the modern descendants of once free-living bacteria. Theories of endosymbiosis, therefore, are theories of polyphyly. As we have seen in chapter 4.4n46, Margulis (1991) claimed that "rampant polyphyly . . . wreaks havoc with 'cladistics,' . . . [and] invalidates entire 'fields' of study" (10). Since, as we have seen in the previous section, the biosimilarity species concept is not a monophyletic species concept but instead embraces the possibility of polyphyly, there is no problem here.

The problem, however, as we have seen above in §1n9, is with Margulis' (1993) claim that an organism made up of integrated symbionts is a "consortium," that "We people tend to see and name only the most superficial or largest members of the consortium; we then act upon the self-deceptive construction that the consortium is an independent 'individual'" (216). Since the biosimilarity species concept conceives of species as composed of individual organisms and numerous similarity relations between their phenotypes, we might seem to have a serious problem here. At this point it might be useful to consider more closely the quotation from Margulis (1991):

> The "individuals" handled as unities in the population equations are themselves symbiotic complexes involving uncounted members of live entities integrated in diverse ways in an unstudied fashion. In representations of standard evolutionary theory, branches on "family trees" (phylogenies) are allowed only to bifurcate. Yet symbiosis analysis reveals that branches on evolutionary trees are bushy and must anastomose [i.e., reticulate, as with branches of blood vessels]; indeed, every eukaryote, like every lichen, has more than a single type of ancestor. . . . The fact that "individuals"—as the countable unities of population genetics—do not exist wreaks havoc with "cladistics," a science in which common ancestors of composite beings are supposedly rigorously determined. Failure to acknowledge the composite nature of the organisms studied invalidates entire "fields" of study. [10]

Since Margulis mentions lichens (Mycophycophyta), I will focus on them here. Lichens are composite beings with various degrees of endosymbiosis, the symbionts being fungi and algae usually from no more than one species of each (the fungi typically envelope the algae), most unions of which are symbiotic to such an evolved degree that the partners cannot be separated to grow by themselves and accordingly reproduce as single hereditary units. Lichens, moreover, are classified according to their external appearance, which is "morphologically very unlike either the phycobionts (algae) or the mycobionts (fungi) that compose them" (Margulis 1993: 178). According to Margulis and Schwartz (1988: 164), there are approximately 25,000 species of lichens. Lichen species are clearly polyphyletic, with each lichen having at least two genotypes and each lichen species having at least two genomes. As Margulis (1991) put it, "By the opening years of the 20th century, . . . the polygenomic ('dual') nature of lichens was firmly established as biological fact" (3–4). For the biosimilarity species concept, the fact that each lichen has at least two genotypes is not a problem, since as we have seen in the previous section the similarity that constitutes biosimilarity species is not genotypic but phenotypic, and there is no reason why a single lichen cannot be said to have a single phenotype, given its endosymbiosis (cf. Law 1991). So far so good. Indeed when it comes to lichens the biosimilarity species concept fairs better than most. The problem is when Margulis claims that a single lichen is not an individual organism. The difficulty here, I think, is mainly terminological. If the definition of "individual organism" includes the restriction that it must have a single genotype, then a lichen is not an individual organism. But then neither are most reputed organisms, such as any given animal or plant, since the mitochondria and chloroplasts have their own DNA and the former even have their own slightly different genetic code. Interestingly, the term that Margulis

(1993) uses for an endosymbiotic complex is "holobiont" (*holo* = whole), a term she unreservedly uses for lichens (cf. 170, 183, 193) and which she defines when she says "The integrated symbionts (holobionts) become new organisms with a greater level of complexity" (7). Since in many cases a lichen is an integrated, functional, autonomous, and reproductive whole, I suggest that the term "individual organism" be retained without fear. Granting that, the biosimilarity species concept accords amazingly well with the "form taxa" (i.e., not clades)—not just form phyla such as Mycophycophyta but also form species such as lichens—emphasized by Margulis as being the revolutionary taxonomic consequence of modern symbiosis analysis (cf. Margulis and Schwartz 1988: 162, 164; Margulis 1991: 10, 1993: 215–216).

What about the reality of relations? The problem of the reality of relations, of course, is not a problem only for the biosimilarity species concept but for all species concepts that take relations (of whatever kind) as either constitutive or delimitative of species. The problem, it seems to me, is not so much for those who affirm the reality of relations as it is for those who deny it. Since many statements in both science and everyday life involve relation words and more than one subject, those who want to deny that relations are real are placed in the position of having to claim that all sentences expressing a relation between two or more subjects can be reduced to sentences of the subject-predicate form. As we saw in chapter 2.2, Occam not only thought that the statement "Socrates is similar to Plato with regard to whiteness" can be reduced to the conjunction of "Plato is white" and "Socrates is white" but that this is true for all relation statements whatsoever.

An examination of Russell's response to this line of approach is useful here. Russell (1914), interestingly, thought that the reductive approach is initially plausible only for statements of symmetrical relations:

> In the case of symmetrical relations—i.e. relations which, if they hold between A and B, also hold between B and A—some kind of plausibility can be given to this doctrine. A symmetrical relation which is transitive, such as equality, can be regarded as expressing possession of some property, while one which is not transitive, such as inequality, can be regarded as expressing possession of different properties. But when we come to asymmetrical relations, such as before and after, greater and less, etc., the attempt to reduce them to properties becomes obviously impossible. [58]

Although in many of his works Russell argued for the reality of relations using statements of asymmetrical relations, his argument below (Russell 1940) is best suited for my present purpose:

> Such words as "before" and "above," just as truly as proper names, "mean" something which occurs in objects of perception. It follows that there is a valid form of analysis which is not that of whole and part. We can perceive A-before-B as a whole, but if we perceived it *only* as a whole we should not know whether we had seen it or B-before-A. The whole-and-part analysis of the datum A-

before-B yields only A and B, and leaves out "before." In a logical language, therefore, there will be *some* distinctions of parts of speech which correspond to objective distinctions. [344–345]

The point is that if we were to treat A-precedes-B reductively as a whole composed only of the individuals A and B and if we were to try to communicate this whole to someone else, we could not then distinguish A-precedes-B from B-precedes-A. Since the fact which we are trying to express requires this distinction (assuming that the former is true and the latter is false), it would seem to follow that there is something else in the fact which we are trying to communicate besides just the existence of a whole constituted by A and B. That something else is real and cannot be ignored.

This argument, of course, is not ironclad. It presupposes a correspondence theory of truth as well as a belief that syntax may entail ontological commitments. Most biologists, I believe, are knowingly committed to the former, without having given much thought to the latter. The beauty of Russell's argument above is that if we refuse to accept it we shall then be forced to admit that whenever we use relation words (and use we must) we do not really mean what we say! One is certainly free, of course, to take that approach, but at the risk of losing one's credibility.

The different existence conditions for relations on the one hand and properties and physical objects on the other only serves to further distinguish the reality of relations from physical objects and their properties. As William Winslade (1970) put it in his excellent study on Russell's theory of relations, "Properties, but not relations, can exist if only one particular exists. Relations can exist only if two or more relata exist. In this respect all relations differ from all properties" (90). Indeed it seems almost analytic that a property must be a property *of* something. But to be a property it need only have one subject, whereas a relation needs at least two. Or could a relation be real without any relata whatsoever? Although early in his career Russell toyed with the idea that there may be uninstantiated, bare or pure relations, relations without relata (cf. Russell 1913: 88–89), he seems later to have rejected that view as ontologically superfluous and contrary to the desideratum of a minimum vocabulary. As Russell (1959) put it, "We could, at a pinch, imagine a universe consisting only of Alexander or only of Caesar or only of the pair of them. But we cannot imagine a universe consisting only of 'preceded.' . . . Relation-words, it is clear, serve a purpose in enabling us to assert facts which would otherwise be unstatable. So far, I think, we are on firm ground. But I do not think it follows that there is, in any sense whatever, a 'thing' called 'preceding.' A relation-word is only used correctly when relata are supplied" (174–175). Again, "verbs are necessary, but verbal nouns are not" (128).

This argument, of course, supplemented by Winslade's, does not *prove* that relations cannot in some sense exist without relata, only that it is *unnecessary* to think so. Indeed it may even be self-contradictory. Either way, it is important to notice that for the biosimilarity species concept it is an absolute necessity that the relations which constitute a species must in some sense have relata. Otherwise biosimilarity species would exist in some Platonic world and would be incapable of evolving (cf. chapter 3.5). Assuming that relations require relata, however, brings species into *this* world, into the here and now, making them amenable to the empirics of biology, and thus avoiding all the problems of construing them as subsisting in some abstract nether world, or alternatively as being fully physical.

So far so good. Serious problems arise, however, once we begin to question the nature of relational facts such as A-precedes-B. As I pointed out in chapter 1.4, my own, somewhat idiosyncratic view is that the relata are part of what a relation is, that the *something more* in a relation, the something more besides the relata, is both abstract and yet a particular and is just as much a part of the relation as the relata. Russell, however, would not have been in agreement with this view, and it is important to examine why. But before I do this, I should point out that Russell would have agreed that the something more besides the relata is a constituent of a relational fact. As Russell (1914) put it, "a fact . . . is never simple, but always has two or more constituents. When it simply assigns a quality to a thing, it has only two constituents, the thing and the quality. When it consists of a relation between two things, it has three constituents, the things and the relation. When it consists of a relation between three things, it has four constituents, and so on" (60–61).[21]

The difference between Russell's view and the view which I hold is that Russell thought of relations as universals, and in a most idiosyncratic way. Although Russell repeatedly defined universals as "anything which may be shared by many particulars" (Russell 1912: 53), he persistently refused to hold that relations have instances. In his reply to Morris Weitz in the Schilpp volume, in which Weitz (68–69) supposed that Russell later changed his view on this, Russell (1944) stated that "as to relations having no instances. It is a mistake to think that I abandoned this view . . . I have held it continuously since 1902. Nor is there any difference in this respect between relations and qualities. When I say 'A is human' and 'B is human,' there is absolute identity as regards 'human.' One may say that A and B are instances of humanity, and, in like manner, if A differs from B and C differs from D, one may say that the two pairs (A, B) and (C, D) are instances of difference. But there are not two humanities or two differences" (684). According to Winslade (1970), "He thought that if relations had instances, then only similar relations, not the same relation, could hold among different sets of relata. To avoid the nominalism he thought was entailed by denying that the *same* relation can occur in different relational facts, Russell invoked the doctrine that relations are unparticularized. He thought that only if relations are not particularized by their relata, can the same relation occur in different relational facts" (96). Indeed one might sense in this a fear on Russell's part that if one holds that relations have instances, then the similarities between them as instances of one or the

---

[21] I purposely ignore here Russell's later attempt to give reality an extemely close shave by using Occam's Razor to reduce "things" to "bundles of qualities" (cf. Russell 1940: 97–98, 1944: 685–686), since I do not think that such a metaphysics makes for a respectable (at least to biologists) philosophy of biology.

other relation might entail a smeary continuum between relation instances, thus defeating altogether the reality of relation universals (for example that *different* grades into *similar*, that *in* grades into *out*, etc.). At any rate, Winslade (97–99) argues that Russell's view that relation universals do not have instances but are absolutely one and the same wherever they hold is inconsistent with Russell's claim that relations are constituents of facts, that they are shared by many particulars, and that they rely upon relata for their existence. And indeed Winslade does seem to have brought out a fundamental flaw in Russell's metaphysics. In consequence thereof, it seems to me that if relations are to be constituents of facts they must be particularized. But if they must be particularized, and if furthermore they must rely upon relata for their existence, then I don't see how one can avoid the consequence that the relata must be parts of what a relation is. Otherwise what is the nature of the reliance? Any answer that goes beyond the relata themselves (and Russell's theory does this) cannot help but be unsatisfactory from a metaphysical point of view, since it is not parsimonious. Accordingly, *it is the relata that make the relation.* To understand this is to unravel the mystery of relations. It is to understand that a relation is not something *in addition* to the relata, let alone something that makes the relata what they are *qua* relata, but rather simply the relata themselves plus something more about them *in virtue of which* the whole is called a relation. (Whether that something more is itself repeatable is a question I will reserve for when I discuss similarity.)

In arguing that relations are not eliminable in favor of their relata or (as we shall see more fully later) the properties of those relata, one further reductionistic approach must be dealt with before we are finished with our general question on the reality of relations, namely the reputed reduction of relations to classes. The main view here is that all relations are fundamentally dyadic (two-termed) and that all relations can be expressed as classes of ordered pairs. This tends to be the view of those who have a high degree of fondness for set theory. Quine (1987), for example, says "A relation . . . is a class of ordered pairs; thus the uncle relation is the class of all pairs $\langle x, y \rangle$ such that x is uncle of y" (30; cf. 90). Quine (1960) defines the ordered pair as "a device for treating objects two at a time as if we were treating objects of some sort one at a time." "A typical use of the device," he continues, "is in assimilating relations to classes, by taking them as classes of ordered pairs" (257). Further, "it is central to the purposes of the notion of ordered pair to admit ordered pairs as objects. If relations are to be assimilated to classes as classes of ordered pairs, ordered pairs must be available on a par with other objects as members of classes" (258).[22]

---

[22] Reading Quine, I find, is often a slippery, and hence tiring, affair (cf., e.g., chapter 3.5n54). In the above quotations, one might suggest that his use of variations of the word "assimilate" should not be taken to mean reduction-elimination. But this is precisely, it seems to me, what Quine has in mind, although one has to tease this meaning out of his writings. Looking at Quine (1960), we have seen from the quotation at the head of chapter 3 that the universe, on Quine's view, is best thought of as consisting of physical objects and classes of physical objects (as well as, of course, classes of classes). In chapter 3.5 we

Clearly if Quine is right, then the biosimilarity species concept reduces to a class theory and is not really a new paradigm, suffering all the drawbacks of class theories in general.

To remove this difficulty, perhaps one only needs a few examples of relations that are irreducibly more than two-termed. Russell (1940: 45) refers to the relations *jealousy* and *giving* as examples of three-termed relations. Indeed examples abound. Russell's conclusion is that relations "may have three terms, four, or any number" (59). Clearly these cannot be treated simply as classes of ordered pairs. Nevertheless, they can be treated either as classes of ordered triples, ordered quadruples, and so on to n-tuples, or as classes of ordered pairs within ordered pairs (cf. Stoll 1963: 25). Accordingly the relation *giving*, for example, where x gives y to z, could be treated either as the ordered triple $<x, y, z>$ or as the ordered pair $<<x, y>, z>$.

It seems to me, however, that both approaches still suffer from the fact that neither can deal with all relations, in particular relations that are indefinite as to number of terms. Perhaps the clearest example of this is supervenient relations, each of which, as in psychophysical supervenience, is properly an *asymmetrical* relation (cf. Kim 1993: 67). Moving from philosophy of mind to molecular biology, we can clearly see this asymmetry in the supervenience of amino acids on codons (cf. chapters 2.2 and 3.2).[23] Given the codon CGU, for example, one can infer that the amino acid it will code for is arginine, but given the amino acid arginine one cannot infer that it was coded for by CGU, since CGC, CGA, CGG, AGA, and AGG also code for arginine. Since the genetic code for living things on earth is not a law of chemistry but rather an accident of history, a "frozen accident" as Francis Crick (1968) put it, it follows that there may be other genetic codes, possibly not even carried in nucleic acids, so that the subvenient base of codons upon which arginine and all the amino acids supervene is indefinite, perhaps even infinite. This is especially clear for another popular example, namely

---

have seen Quine's approach on reducing attributes in favor of classes. While Quine (1960) does say that "It will suffice for now to cite classes, attributes, propositions, numbers, relations, and functions as typical abstract objects" (233), he also says earlier in the same work that "Once we start admitting abstract objects, there is no end. Not all of them are attributes, at least not *prima facie*; there are or purport to be classes, numbers, functions, geometrical figures, units of measure, ideas, possibilities. Some of these categories are satisfactorily reducible to others, and some are best repudiated" (123). That Quine neglects to include relations in the above list should not dissuade one from concluding that he thought of his assimilation of relations to classes of ordered pairs as anything but reduction-elimination (repudiation, of course, is an unwarranted alternative interpretation, given his emphasis on ordered pairs). As if to confirm this interpretation, he says in the same work that "What classes are to attributes (thus dogkind to caninity), classes of ordered pairs are to relations" (210). And to confirm this even further, he calls his view on attributes and relations a "renunciation" (210). (Cf. Quine 1976: 28, 1987: 9.)

[23] It is important to notice that here the relata are not physical things but rather kinds of physical things, each of the supervenient relata being a kind of amino acid and each of the subvenient relata being a kind of codon.

fitness. Indeed a number of philosophers of biology have focused on fitness as a nonphysical property that supervenes on an indefinite number of physical properties. As Sober (1993) put it, "A cockroach and a zebra differ in numerous ways, but both may happen to have a 0.83 probability of surviving to adulthood" (73; cf. Sober 1984a: 49–51). As Kincaid (1987) put it, "there may well be a possible infinity of physical states realizing a given level of fitness" (346). Indeed Sober (1993) goes so far as to suggest that "*All biological properties supervene on physical properties*" (73). At any rate, the point is that such indefiniteness as to number of terms positively resists the reduction of these relations to classes, whether ordered pairs or n-tuples.

Perhaps, however, the most serious problem for the reduction of relations to classes follows from the fact that class membership is itself a relation (cf. Stoll 1963: 4; Armstrong 1989: 55). Consequently it follows that if relations are to be reduced to classes, then relations of class membership must also be reduced to classes, classes of ordered pairs *sensu* Quine. But now one cannot ever possibly get rid of relations without involving an infinite regress, since every set of relations of class membership entails a new class, a class of ordered pairs, to hold that set of relations, which entails a new class of ordered pairs to hold that set, and so on *ad infinitum*.[24]

In the end, it would seem that relations are more basic than classes. Classes, as Russell has shown, are more easily argued away than relations. Relations clearly seem part of the physical world in a way that classes are not, and great difficulties are incurred in the attempt to reduce them to something else. Moreover, even if in the end it should prove possible to define all relations in terms of classes, the relations themselves still remain. As Armstrong (1989) put it, "although these classes may be fitted to *represent* a relation, . . . yet they do not seem to *be* that relation" (32).

What about the reality of similarity relations? In the above discussion I have confined myself, for the most part, to asymmetrical relations. Accepting that these are ineliminably real, however, does not entail accepting that all relations are real. More specifically, there is no reason why one might not argue that asym-

---

[24] To see this using set theory notation, suppose we have a class with three members, such that $A = \{a, b, c\}$. This means that $a \in A$, $b \in A$, and $c \in A$. In order to deal with each of these relations of class membership, we have to construct the class of ordered pairs $B = \{<a, A>, <b, A>, <c, A>\}$. This means that $<a, A> \in B$, $<b, A> \in B$, and $<c, A> \in B$. In order to deal in turn with each of these relations of class membership, we have to construct the class $C = \{<<a, A>, B>, <<b, A>, B>, <<c, A>, B>\}$. This means that $<<a, A>, B> \in C$, $<<b, A>, B> \in C$, and $<<c, A>, B> \in C$. In order to deal in turn with each of these relations of class membership, we have to . . . . Clearly there is no end to this process. The same holds true, by the way, if one employs the Wiener-Kuratowski device for eliminating ordered pairs (cf. Stoll 1963: 24; Quine 1969c: 58, 259; Armstrong 1989: 31–32), where the ordered pair $<a, A>$, for example, becomes the unordered pair $\{\{a\}, \{a, A\}\}$. This is a nifty device for eliminating ordered pairs in favor of a purer form of set theory (cf. Quine 1960: 260), but it does absolutely nothing to solve the problem above since it retains the relation of class membership.

metrical relations are real but symmetrical relations are not. (We've already seen above a hint of this in Russell's main argument for the reality of relations.) This possibility is especially relevant for the biosimilarity species concept, since the relations which delimit biosimilarity species are all (since they are causal) asymmetrical relations,[25] while the relations which constitute biosimilarity species are symmetrical relations. This latter part follows because similarity is clearly a symmetrical relation (if *a* is similar to *b*, in a certain respect and degree, then so exactly is *b* to *a*). Given this difference, then, a separate discussion is required.

On the topic of similarity I will examine three issues. The first is whether judgments of overall similarity can be objective, or whether they are incurably subjective. The second is whether judgments of similarity are really only judgments of partial identity, in other words whether similarity is nothing more than a matter of properties in common. The third issue is whether similarity relations are repeatable, granting in the first place that they are real.

Beginning with the first issue, one often finds in arguments against the reality of similarity the rejection of similarity based on the rejection of the objectivity of overall similarity. One can see this in Nelson Goodman's (1970) "Seven Strictures on Similarity." Since this is arguably the most widely quoted work on the rejection of similarity, it will be good to look at some of Goodman's arguments. In this paper he tells us right at the beginning that similarity "is insidious," that it is "a pretender, an imposter, a quack," that it is a "false friend" (19). In support of this conclusion, Goodman makes statements such as "Anything is in some way like anything else" (22–23), "no two things have all their properties in common" (26), and "Each two among three or more particulars may be alike (that is, have a quality in common) without all of them having any quality in common" (25).[26] None of these statements, I submit, or any like them, are sufficient to conclude that similarity is not real, or even that all judgments of overall similarity are deluded.

As for the first statement above, I will grant that given any two things there is some kind of similarity between them, if only in one respect and to a certain degree. But this fact is hardly sufficient to entail that similarity is a useless notion, or that no two things can have a greater degree of overall similarity than two other things. I will return to this point in a moment. Goodman's second statement, that no two things have all of their properties in common, is a dogmatic statement that no one can justify, for the simple reason that no one can ever be in a position

---

[25] Certainly ontogenetic and ecological relations, including selection, are asymmetrical. One might think that interbreeding is a symmetrical relation, but this is true only in a superficial sense, a sense that is quickly dispelled once one thinks of all the different kinds of relations that are involved in successful interbreeding in any given sexual species. Colony-formation relations, as with slime molds, would also appear at first to be symmetrical, but they are no more symmetrical than Russell's example of *above* or *precedes*.

[26] In pulling out these statements, I know that I am taking them from different strictures of Goodman's. But Goodman himself admittedly involves in his arguments a good deal of overlap and crisscrossing, so that I don't think my manner of treating his arguments can be used against me.

to know that it is true. It also, of course, depends on how one defines properties. I will return to that later. Granting for the moment, however, what I think most of us will grant, namely that physical objects do have properties, there is logically no reason why two physical objects may not have all of their properties in common. (We need not grant here, of course, that "properties in common" means properties that are qualitatively identical.) And if we grant this, there is then every good reason to say that two physical objects that have all of their properties in common have a greater degree of overall similarity than two physical objects that do not have all of their properties in common. In biology every biologist, I dare say, would agree that two given zebras have a greater degree of overall similarity than the overall similarity between a given zebra and a given cockroach, or that (*ceteris paribus*) two given zebras that are identical twins (clones) have a greater degree of overall similarity than two given zebras that are not identical twins. Why are philosophers less apt to admit the obvious? I can venture no reason except to say that philosophers in their activity as philosophers confine themselves to the realm of speculative thought, whereas the activity of biologists *qua* biologists is much more hands-on.

This is not to say that examples are not ready at hand where the notion of overall similarity clearly appears to be hopelessly subjective. Goodman, as we have seen above, says that any two of three objects may be alike without all of them having any quality in common. He constructs the following example (25), where we have three discs, each with two color qualities (r = red, b = blue, y = yellow):

$$rb \quad by \quad yr$$
$$1 \quad \ 2 \quad \ 3$$

In this example we could easily expand the number of qualities so that it becomes an example with numerous properties but where no two of the objects have any greater overall similarity than any other two. Arguing from examples like this, however, is no more legitimate than arguing for the unreality of all species based on messy situations. To counter generalizations from examples like that above we need only provide examples like the following:

$$rb \quad rb \quad by$$
$$1 \quad \ 2 \quad \ 3$$

In this example, supposing for simplicity that each of the three objects has only two properties, where each kind of letter may represent any kind of property we want, objects 1 and 2 obviously share a greater degree of overall similarity than objects 1 and 3 or objects 2 and 3. In examples such as this the concept of overall similarity, then, I strongly suggest, has objective application. This is true even if we grant that no two of the properties are qualitatively identical. Indeed if we confine ourselves to just the letters themselves, we can see, if only with the aid of a magnifying glass, that no two of the r's are exactly identical, and so on for the rest of the letters. As Denkel (1996) put it, "magnifying every occurrence of the letter 'a' on this page will reveal that no two of them are wholly identical in shape. Aristotelian realism is a simplification, a summary of the actually more complex relation that holds between particulars" (159).

Goodman, of course, is not an Aristotelian realist but rather a nominalist. Interestingly, however, he claims that "any two things have exactly as many properties in common as any other two" (26). I confess I find it difficult to see where he gets this from. Immediately following he claims that "If there are just three things in the universe, then any two of them belong together in exactly two classes and have exactly two properties in common: the property of belonging to the class consisting of the two things, and the property of belonging to the class consisting of all three things" (26). This argument would only seem to work if properties are reducible, *sensu* Quine, to classes. This is problematic enough, as we have seen in chapter 3.5, but even if we grant the reduction, surely there is tomfoolery afoot here. Given a universe consisting only of two identical zebras (clones) and a cockroach, who but a philosopher would ever want to claim that any two of these objects have exactly as many properties in common as any other two of them?

In arguing against the notion of overall similarity, one might simply return with more complicated examples. Goodman (20) gives the following example in his argument that similarity does not pick out tokens of a common type, but it will also serve our present purpose since it serves as an argument against overall similarity as well:

$$a \quad d \quad A$$
$$m \quad w \quad M$$

In reference to this example Goodman tells us that "The idea that inscriptions of the same letter are more alike than inscriptions of different letters evaporates in the glare of such counterexamples as those in Figure 1" (20). In the above example we can ask whether, say, a and d are more or less similar than a and A, whether m and w are more or less similar than m and M, whether a and m are more or less similar than A and M, and so on. In a related stricture Goodman points out that "similarity is relative, variable, culture-dependent" (20). In another stricture he suggests that perhaps some properties are more important than others. This of course is certainly true for many biologists, who typically choose some characters as more important than others for their classifications. For example, we have seen in the previous section that many botanists use functional or adaptive characters as diagnostic of species. As Goodman put it, "More to the point would be counting not all shared properties but rather only *important* properties" (26), which he then quickly dismisses by saying "But importance is a highly volatile matter, varying with every shift of context and interest, and quite incapable of supporting the fixed distinctions that philosophers so often seek to rest upon it" (27). Certainly importance would seem to be relative to matters such as interest, context, and culture, but does it follow that given this relativity all claims about overall similarity are hopelessly subjective and that ontologically there really is no such thing as objective overall similarity?[27] Is it not possible for two or more objects to have more overall similarity between them than two or more other objects no matter what of their properties we take to be more important? To be sure, there are messy situations, but once again messy situations do not negate the possibility of legitimate cases. To return to the example above, in terms of the function of spelling, *a* and A are more similar than *a* and *d*, whereas in terms of the function of emphasis, *a* and *d* are more similar than *a* and A. While we can dismiss the objectivity of each of these greater functional similarities in terms of cultural relatively, no one with eyes to see can dismiss the greater structural similarity of *a* and *d* compared with *a* and A. Moreover, with these examples we can already see that greater functional similarity can sometimes coincide with greater structural similarity. In terms of structure and the function of emphasis, *a* and *d* are more similar than *a* and A. And if emphasis and spelling are the only two functions, then *a* and *d* are more similar overall than *a* and A. And if we change our set of objects above to include more than one of each, who will deny that *a* and *a*, for example, are more similar overall than *d* and M?[28]

In the case of Goodman's arguments we can see that once again the reality of messy situations has been employed to inveigh wholesale against what in many other cases is a perfectly reasonable notion. Given that the same approach does not work against the reality of species, as we have seen in chapter 2.4, philoso-

---

[27] It is interesting to note that not all psychologists seem to think so. In a comprehensive review of the literature, although Medin *et al.* (1993) argue from experimental findings that our similarity judgments (subjective) are to some extent influenced by experience (children tend to judge more holistically, adults more analytically) and context (varying according to different contrast sets and respects), they nevertheless argue that similarity judgments play an important role (among others) in inductive inferences about the world. As they put it, "One function of similarity is to allow people to make educated guesses in the face of limited knowledge. . . . To give a simple example, suppose one sees an unfamiliar type of snake and wonders if it might possibly be dangerous. To the extent that the snake in question resembles a rattlesnake rather than a common garter snake, one might be cautious. Presumably, people's perceptual and conceptual processes have evolved such that information that matters to human needs and goals can be roughly approximated by a similarity heuristic. That is, similar things may behave in similar ways, and the things people tend to be reminded of are useful to an extent that far exceeds what would be expected on the basis of random remindings" (258). Indeed evolution by natural selection well accounts for the ubiquity of (though not always the justification for) stereotyping. If you are in the jungle and you see a tiger but you decide not to stereotype (perhaps because you believe that similarity is a false friend), then you will probably be eaten. In other words, in the biological world stereotyping based on veridical judgments of overall similarity statistically results in greater survival and reproductive success.

[28] There is, of course, a great disanalogy between letters and organisms. Whereas the role or function of a letter in a written language is culturally relative, the role or function of an organism in a ecosystem is not. Gorillas, for example, play in the wild the role of peaceful vegetarian, irrespective of what any human culture prefers to think. In other words, ecological roles are objective in a way that alphabetical roles are not.

phers could well learn from biologists about the folly of using messy situations to reject otherwise valid concepts.[29]

I have dwelt at length on the issue of overall similarity above not because the biosimilarity species concept conceives of species in terms of overall similarity—as we have seen in the previous section, it most certainly does not—but rather (i) because problem examples with that concept are often used to reject it altogether, and (ii) because the biosimilarity species concept is not averse to two or more organisms having a greater degree of overall similarity than two or more other organisms. Indeed nature routinely produces such groups horizontally conceived, and many of these groups are united with other groups by relevant evolutionary causal relations. This is something which I strongly suspect most biologists take to be a matter of objective fact. The biologists who do not take it to be a matter of objective fact, most notably cladists, reflect the skepticism of philosophers. But in this skepticism neither these philosophers nor their biologist counterparts do any justice to the topic, and it is important that other biologists know that not all philosophers are skeptical about the reality of similarity relations, even overall similarity, and also why they are not skeptical.

The rejection of overall similarity by philosophers such as Goodman leads to the next issue, which is that perhaps the only respectable notion of similarity is similarity in terms of specified respects. Goodman says "One might argue that what counts is not degree of similarity but rather similarity in a certain respect" (20–21). Philosophers who are critical of objective similarity, however, are usually not content to stop there; instead they tend to think that similarity in terms of respects makes similarity superfluous. It is clear that Goodman takes this view when he says

> when to the statement that two things are similar we add a specification of the property they have in common, we again remove an ambiguity; but rather than

---

[29]   Popper is another example. In his 1959 New Appendix X to Popper (1934), Popper argues, with the aid of combinations of geometrical shapes and shadings much to the same effect as Goodman's Figure 1 above, that repetitions are "only *more or less similar*" (420), that "Two things which are similar are always similar in *certain respects*" (420–421), that similarity is characterized by "relativity" (420), that "any two things which are from one point of view similar may be dissimilar from another point of view" (421), that "similarity, and with it repetition, always presuppose the adoption of *a point of view*" (421), so that "it is logically necessary that points of view, or interests, or expectations, are logically prior, as well as temporally (or causally or psychologically) prior, to repetition" (422). Once again, one could reply with counterexamples such as I have provided above against Goodman's related argument. One could also add here the evidence from biology and cross-cultural tests discussed in chapter 2.4. What these examples show is what scientists and laymen alike both know, namely that we may arrive at the same results in spite of different interests, in other words that different points of view may sometimes result in the same conclusions. The ultimate lesson to be learned is that it need not be the case that prior interests always result in subjective, conventional results. Rather, it need only be the case that objective results cannot be obtained without prior interests (cf. Stamos 1996b: 189n6).

supplementing our initial statement, we render it superfluous. For, as we have already seen [cf. 21], to say that two things are similar in having a specified property in common is to say nothing more than that they have that property in common. . . . "is similar to" functions as little more than a blank to be filled. [27]

This is a view, as we have seen in chapter 2.2, that was shared by Occam. And indeed still today not only many philosophers but many biologists (e.g., Simpson 1961: 77; Sokal and Sneath 1963: 18, 63–64; Ghiselin 1987a: 141) treat similarity reductively as a shorthand for properties in common. But as Denkel (1996) put it, "Surely, 'similar in a respect' does not mean 'identical in a respect' or 'possessing a common respect'" (161).[30] Indeed to Goodman and others we may make the reply that from an empirical point of view two respects (properties) that are being compared will probably not be qualitatively identical (just compare any two noses), in which case similarity returns.

Goodman has no answer to this, but it can be found in philosophers who would reduce similarity in respects to partial possession of common microproperties. This is the view that similarity is really only "partial identity." As Denkel (1996) put it, "By this philosophers understand something like identity infused with impurities: theoretically, by purifying such a relation from the residues it contains we would obtain sheer identity" (163). One can see this view in, for example, Quine's famous paper "Natural Kinds" (Quine 1969a), discussed in chapter 3.2. In this paper Quine argues that as humans progress (both as a species and in individual development) their standards of similarity become more refined, moving from "primitive inductions and expectations" to a system of kinds that "changes and even turns multiple as one matures," then to "the remoter objectiv-

---

[30] I should perhaps at this point distance myself from Denkel's concept of properties. In attempting to take similarity seriously as extra-mentally (ontologically) real, Denkel (or at least Denkel 1989) seems to me to go too far by reducing properties to similarity complexes. Denkel (1989) defends the view, as pointed out earlier, "that the 'common aspects' observed among things in nature, such as properties or kinds, are a matter of resemblance rather than identity. It is owing to such resemblances in different degrees that we are able to speak about 'common' properties and natures and thus classify the world accordingly. From the point of view of the RT [Resemblance Thesis], Aristotelian Realism is a simplification, a summary of the actually more complex relation which holds between particulars" (37). What this means for Denkel, in other words, is that "properties are objective resemblances" (40n11; cf. 43), such that "there are indeed relations of resemblance between every blue object and every other, which are absent between blue objects and objects of other colours" (44). Given, however, the different existence condition for properties, as Winslade, as we saw earlier, rightly pointed out, it follows that properties cannot be reduced to or eliminated in favor of similarity relations. In other words it is logically possible to have a universe with only one individual such that that individual is blue (assuming for the sake of argument that blue is a primary quality; a much better example, of course, would be something such as a nose). In such a case it becomes clear that a property cannot be a similarity complex between particular respects. My ontology therefore includes properties in addition to relations and individuals, although it is not committed to the view that common properties need be qualitatively identical (exactly similar).

ity of a similarity determined by scientific hypotheses and posits and constructs" (133). Not content to stop there, however, Quine argues that the mark of a mature science is when it gets rid of its former standards of similarity altogether in favor of common microproperties, "and so by-pass the similarity" (135). One science that has already reached this stage, according to Quine, is chemistry:

> Comparative similarity of the sort that matters for chemistry can be stated outright in chemical terms, that is, in terms of chemical composition. Molecules will be said to *match* if they contain atoms of the same elements in the same topological combinations. Then, in principle, we might get at the comparative similarity of objects *a* and *b* by considering how many pairs of matching molecules there are, one molecule from *a* and one from *b* each time, and how many unmatching pairs. The ratio gives even a theoretical measure of relative similarity, and thus abundantly explains what it is for *a* to be more similar to *b* than to *c*. Or we might prefer to complicate our definition by allowing also for degrees in the matching of molecules; molecules having almost equally many atoms, or having atoms whose atomic numbers or atomic weights are almost equal, could be reckoned as matching better than others. [134–135]

Quine advocates the same kind of reductionism for similarity in biology. Although he at first suggests reducing organism similarity in terms of "family trees," more specifically "proximity and frequency of their common ancestors," he apparently decides in favor of genes in common, when he states that "a more significant concept of degree of similarity might be devised in terms of genes" (136). That Quine favors the genetic approach becomes clear when he discusses the desideratum of unified science, which can be considered accomplished "only insofar as their [the different branches of science] several similarity concepts were *compatible*; capable of meshing, that is, and differing only in the fineness of their discriminations" (137). "In this career of the similarity notion," continues Quine, "starting in its innate phase, developing over the years in the light of accumulated experience, passing then from the intuitive phase into theoretical similarity, and finally disappearing altogether, we have a paradigm of the evolution of unreason into science" (138).

But science itself rebels, at least in biology, and what appears as unreason to Quine may be reason after all. To speak of reducing organism similarity to genes, of course, is at least proximately to speak of DNA, since properties such as noses, for example, are coded for by DNA. This causal-reductive argument, however, would only work if there is a one-one or one-many relation between genotypes and phenotypes. But of course there isn't. Given the synonymy of codons and the supervenience of amino acids (the very building blocks of phenotypes), as we have seen in chapters 2.2 and 3.2, phenotypic structures cannot be reduced to genotypic ones, anymore than fitness can be reduced to physical properties or mind to matter. We would seem in biology, then, to be stuck with phenotypic similarity as an irreducible *real* relation.

Indeed from a biological point of view, similarity is hardly a "false friend." Not only as we have seen are there good reasons to believe that phenotypic similarity is objectively real, but without it and the recognition of it (*sensu* Paterson and what every ecologist knows) we and other species could not possibly survive, let alone even be here in the first place. Even Quine (1987) felt forced to recognize that "Our [everyday] standard of similarity, for all its subjectivity, is remarkably attuned to the course of nature. For all its subjectivity, in short, it is remarkably objective. In the light of Darwin's theory of natural selection we can see why this might be. Veridical expectation has survival value in the wild" (160; cf. Quine 1969a: 126–127, 133). Had Quine known a little more biology perhaps even he would have come to admit that natural selection has equipped us not only to perceive individuals and their properties but also various relations and relation complexes, including real similarity, of which biological individuals are a part. And indeed so important are these relations from a scientific point of view that I can't see them ever being denied by biology. What could be more obvious in this regard, to give a simple example, than the role of these relations in *discovery* when scientists or laymen discover a new species? It is that role that makes me believe not only that species are real but that the different kinds of relevant relations must somehow be either constitutive or delimitative of species. When, for instance, what is thought to be a new species of bird is discovered, say on a newly discovered island, and then what is thought to be another new species of bird is discovered on that same island, what unites what was thought of as two species into one is if it is further discovered that all the birds from the one group are male, all the others from the second group are female, and that the two groups interbreed. In other words it is particular relations of a certain kind that unite what was thought to be two species into one. (Lepidoptera, of course, would provide another good example; cf. Hull 1978: 301.) But similarity is just as important, for otherwise no two organisms would have correctly been thought to be conspecific in the first place.

But perhaps this homage to biological similarity is misplaced. Perhaps instead we should look at all similarity, phenotypic included, from the point of view of physics (including chemistry), as reducing to the sharing or not sharing of the elementary constituents of matter (from structural molecules to atoms, to electrons, protons, and neutrons, to quarks, and all the way down to superstrings), so that the subjective similarity that Quine admits is nothing but a loose though extremely useful approximation to the real thing, which is not similarity at all. Armstrong (1989) provides a good example of this line of thinking. Taking the view that "physics is *the* fundamental science" (87), he holds the view that "As resemblance of properties gets closer and closer, we arrive in the limit at identity. Two become one. This suggests that as resemblance gets closer, more and more constituents of the resembling properties are identical, until all the constituents are identical and we have identity rather than resemblance" (106). This is a reply to my argument from the supervenience of phenotypes that many will share. But now we have descended into a deep metaphysical basement, where the lights are not on but many are home. The difficulty is that it might just be the case that the microelements of physics (whatever they are) share more similarity than identity. A postulate to the contrary is neither proof nor an argument, and until either is supplied the possibility remains open that, just as macrostructures may supervene on microstructures,[31] similarity at higher levels might supervene on similarity at lower levels, perhaps as an emergent from similarity at the lowest level.

We turn now to the final issue, which is whether similarity relations are repeatable, granting they are real. The issue would seem to be important for my position (cf. the end of chapter 4.4) that the extinction of a species is not necessarily forever. This position is a desideratum, it seems to me, not so much because of movies such as *Jurassic Park* and the fact that many biologists such as Richard Dawkins take the possibility seriously, but because it goes hand in hand with the issue of multiple origins (particularly as it is found in botany) and recent developments in symbiosis theory, as we have seen in this and the previous section.

Let's return to the issue of the nature of similarity itself. One point in favor of the logical possibility of the reappearance of an extinct species, where species are conceived in the sense of the biosimilarity species concept, is that similarity relations are spatiotemporally unrestricted. Not all relations, of course, are spatiotemporally unrestricted. A particular interbreeding relation, for example, is spatiotemporally restricted, a matter that hardly needs further elaboration. But similarity relations are not the same. To see this one need only conduct the following thought experiment: Take, for example, any two pens from your desk. They have a certain degree of overall similarity. Assuming throughout the experiment that neither of the pens change intrinsically, take one of the pens and relocate it to a new location in space, let's say 100 million light years away. The overall similarity relation remains exactly the same. Now bring back that pen and place it in a new location in time, let's say 100 million years into the past. Once again, the overall similarity relation remains exactly the same. It is this feature of similarity relations that allows for species extinction, from the point of view of the biosimilarity species concept, to be not necessarily forever.

But of course if a species that went extinct is to exist once again, it will not be by bringing organisms from the past into the present. It will be, if it is to be at all, by other means, possibly genetic engineering, possibly repeated polyploidy, etc.

---

[31] This would make sense of Denkel's (1996) claim that "two resembling complexes don't have to have a common part" (165). Denkel, however, defends his claim with a Wittgensteinian "family resemblance" type of approach, when he says "the noses of the sultans of the Ottoman Dynasty are strikingly alike in shape, though not, for instance, in size. If this is partial identity, then which is the feature that persists identically in each of the noses? There is no guarantee that there exists such an identity, for if Wittgenstein is right, there may not even be one aspect common to all the noses" (169). There may be common genes, however, not in the sense of "one gene, one nose," but in the sense of multifactorial inheritance, i.e., "many pairs of alleles at many different loci, each allele difference by itself producing only a small effect" (Maynard Smith 1975: 66). Denkel would therefore do better in defending his claim not by appealing to Wittgenstein's theory of family resemblances (which is after all a cluster class theory) but by appealing to supervenience. To use Goodman's (1970: 28) example of two glasses filled with a colorless liquid, one might be filled with water and the other with hydrochloric acid. In this case the two exactly similar respects supervene on very different microstructures.

From the point of view of the biosimilarity species concept, then, it might appear that a further feature of similarity is required, namely repeatability.

To say that a relation, whether similarity or any other, is repeatable is to make a claim about identity or sameness, which interestingly brings us back to the problem of universals discussed at the beginning of this book. Russell (1912), as we have seen above, defined universals as "anything which may be shared by many particulars" (53). As regards similarity, he provided two arguments to prove that similarity is a universal (Russell 1912: 55 and 1940: 346–347), the earlier of which is the most famous, but I see no reason to go through them here.[32] As we have seen, Russell held the idiosyncratic view that a universal does not have instances but is numerically identical (one and the same) wherever it is shared. We have also seen that this view is apparently incoherent with his view that things, qualities, and relations are constituents of facts. To have similarity relations as constituents of facts, we need similarity relations that are at least particularized, if not also tokens or instances.

I have taken the idiosyncratic view that the relata are part of what a relation is. This certainly allows for particularized similarity relations, and accordingly for those relations to be constituents of facts. It is also essential that I take this view for the biosimilarity species concept, since it makes organisms, in virtue of their phenotypic properties, constitutive of species. This is essential because it allows us to make sense of what biologists say of species, including that they evolve.

But now it would seem that similarity relations on my view cannot possibly be repeatable. This is because organisms, as the relata in biosimilarity complexes, are themselves not repeatable. The individuality theorists examined in chapter 4 are absolutely correct about this. A copy of me, no matter how exact, is not *me*. Interestingly, however, asymmetrical relations, on my view of relations, might well be considered repeatable. For example, if *a* precedes *b* at some point in time, and then—let's say *a* and *b* are billiard balls— we rack them up and it so happens that *a* precedes *b* yet once again, in both relations the *a*'s and *b*'s are numerically identical, and for the sake of argument let's say the *something more* in virtue of which they are relata in the relations exactly resemble each other. Has then not the relation, even on my view of relations, been repeated?

The issue now is the all-important issue in the problem of universals, namely whether qualitative identity can ever be something more, namely numerical iden-

---

[32]   I should point out, however, that a major difference between the 1912 argument and the 1940 argument is that the latter explicitly allows for the fact that similarity is not a matter of all or nothing but is a matter of different degrees, a point that Russell also acknowledged earlier in the same work (24–25). This acknowledgment, however, which is likewise acknowledged by virtually everyone who is a similarity realist, creates further problems for Russell's view that relations do not have instances, for how can all the similarity relations which are constitutive of facts and which are of different degrees nevertheless be equally one with the one and only similarity universal? Based on Russell's definition of universals, it would seem to follow that each different degree of similarity requires a separate universal, which means an infinity of similarity universals.

tity.[33] For two entities to be instances of a universal, qualitative identity between them is not enough. What is required is numerical identity. As Armstrong (1989) put it, "The whole point of a universal, . . . is that it should be identical in its different instances" (82), in other words something "is not a universal, . . . [if] it is not something repeated or repeatable" (37). Or as Denkel (1996) put it, "If such ubiquitous repetition is an objective fact, the logical implication is that properties, relations and kinds apply multiply; they each exist simultaneously, . . . an identity in a plurality" (153).[34] To return to similarity as I conceive it, although an individual organism, for example, cannot recur while retaining its numerical identity, perhaps each of its properties can. And if each of its properties can, then perhaps the *something more* in each of the similarity relations of which they are a part can also recur while retaining their numerical identity. But is the recurrence, given that it is exact, really numerical identity? Denkel (1989) allows that "two things may be said to possess aspects which resemble each other exactly, without having to possess or share something identical" (51). This is because of "the empirically observed fact of resemblance," while identities are "abstract and often empirically unobservable" (46). What this view must allow, however, is that if two things may be said to have aspects which exactly resemble, then it must also be possible for three or more things to have aspects which exactly resemble. And if this is possible with exact resemblance, then it must also be possible for every degree of resemblance below exact resemblance. But then we seem in each of these cases to have a recurring resemblance relation. Or do we? While Denkel's view that exact resemblance does not entail that the relata have any properties in common is defensible, the epistemological basis for the defense cannot rule out actual (ontological) identities. But now we seem to have arrived at an impasse. Both nominalists and realists allow for two or more similarity relations, constitutive of facts, to exactly resemble each other. But the nominalist refuses to go further and call the two resemblances (also the property relata of the resemblances) numerically identical, while the realist insists on taking that further step. The one refuses to call it repetition, the other insists that it is repetition. The one calls it numerically different relations in different places, the other calls it numerically the same relation in different places. Both agree on the facts, but both disagree on the interpretation. (The analogy to moral debates should not be overlooked here.) At this point I think we may safely conclude that the debate is entirely verbal.

On my view, then, although I hold relata to be part of what a relation is, I do allow for the possibility of qualitatively identical properties between numerically different individuals (even for all their properties) and so also for qualitative identity between particular similarity relations. Whether one wishes to call this a uni-

---

[33] The issue, of course, is not always only about qualitative identity. But by focusing on qualitative identity we bring into the brightest possible light the essence of the problem of universals, which then serves to illuminate the cases less than qualitative identity.

[34] An identity in plurality, a one in many, is, of course, an Aristotelian conception of universals, whereas the Platonic conception is one over many (cf. chapter 3.1), but we can overlook this qualification for present purposes.

versals realism with regard to similarity, or rather more specifically with regard to the *something more* in similarity relations, is to me a debate of no account.

It is also of no account for the question of whether from the perspective of the biosimilarity species concept the extinction of a species is necessarily forever. As pointed out in §2n17 above, in the case of the reappearance of an extinct language no one would require that the words either written or spoken must be qualitatively identical with any of those from the days of the extinct language. Indeed even within a contemporaneous language community we do not require that the same words must be qualitatively identical. What matters instead is a certain degree of functional identity.[35] For the same reason, then, in the case of species I do not see that either qualitative identity or repeatability in the *something more* in the similarity relations that constitute biosimilarity species must be necessary for the logical possibility of the reappearance of an extinct species. All that matters, instead, is that there be a sufficient degree of overall similarity compared with the extinct organisms. This answer, of course, is somewhat vague, but after all it reflects, I submit, the nature of the beast.

5.4   Concluding Remarks

The revolution inaugurated by Darwin is arguably the most momentous of all the great conceptual revolutions in science. Like no other the *Origin* is truly the book that shook the world. So great and so many are its consequences that it has become almost a truism today in educated circles that very little makes sense except in the light of evolution. The species category, of course, was and continues to be at the very core of the Darwinian revolution. Surprisingly, although Darwin amassed an enormous amount of evidence to raise evolution from the status of mere speculation to that of fact, and although he argued strongly for natural selection as the main mechanism of evolutionary change, he virtually left it to others to determine the ontological status of species in accordance with his views. As for his reason, perhaps there was more than a joke implied in one of his letters to Lyell (February 23, 1860) when he wrote "I have not metaphysical Head" (Burkhardt *et al.* 1993: 103). Indeed in his *Autobiography* (1876) he confided that "My power to follow a long and purely abstract train of thought is very limited; I should, therefore, never have succeeded with metaphysics or mathematics" (84).

---

[35]   Indeed a possibility that Denkel overlooks is functional identity. Functional identity may be said to supervene on structures which may range in their degree of physical similarity. This is indeed, as we saw in chapter 3.1n8, largely Aristotle's sense of universals, or identity in different things. And indeed when it comes to functional identity we tend to overlook the inexact physical resemblances that Denkel focuses on with the letter "a" and the microscope example. What matters more to us is functional identity, and when we think we have it any physical dissimilarities become irrelevant. A hammer, for example, is just as good as another, providing that they are of the same size, shape, weight, and hardness, even though the handle of each hammer might be scratched in different places. Nevertheless with functional identity the problem still arises: Are the particular functions merely qualitatively identical or are they also numerically identical? Yet once again it does not seem to matter.

Perhaps there is a false modesty here, but the fact remains that it took over a hundred years following the *Origin* for the species problem to finally come of age. Since then, in the past 40 years or so a rich diversity of solutions to the species problem has emerged. Why has it taken so long? And what is the reason for the diversity?

These are no doubt questions that require long and involved answers. Perhaps the best that can be said by way of a short answer to the first question is that the maturation of the species problem required the Modern Synthesis to first take hold. I do not think it is a mere coincidence that the enormity of the debate on the nature of species emerged not shortly after the *Origin* but only shortly after the establishment of the Modern Synthesis.

The second question still remains, however. Why the diversity? A skeptical answer might be that the biologists and philosophers who write on the species problem are not so much interested in pursuing truth as they are in occupying every possible theoretical niche so as to make a name for themselves. But this fails to account for the timing. I think a more truthful answer, and certainly a more charitable one, is that the science of biology itself underwent a profound change. In ushering in an age of synthesis between the previously disparate biological sciences, the Modern Synthesis raised biology from the level of an adolescent to that of a mature science. And in becoming a mature science, it became a science of many specializations. Gone were the days when there could be a Darwin, someone who could undertake research in a variety of fields and keep abreast of the latest developments in all of the different subdisciplines. Instead now it was more than enough to just confine oneself to one specialization. Polymaths became dinosaurs. With a mature science it cannot be any other way. As a consequence of this specialization, different species concepts arose depending on the kind of research one was doing and the kinds of organisms one focused on. As Mishler and Donoghue (1982) put it, "It seems clear that the group of organisms on which one specializes strongly influences the view of 'species' that one develops" (123). (I would add that the kind of metaphysics one studies—and doesn't study—also has a lot to do with it.)

At any rate, granting the diversity in species concepts, and its possible causes, what are we to make of it? One rather obvious way to look at it is from the viewpoint of theory competition. In the biological world, competition is the main driving force of evolution (this is not to overlook symbiosis, which I will return to). In the biological world, then, competition may be said to be a good thing. And if competition is a good thing, then so is diversity, for it both follows from and in turn feeds competition. So maybe we should look at the various solutions to the species problem, particularly from the past forty years, as a good thing. In his most famous and influential work, *The Logic of Scientific Discovery* (1934), Karl Popper thought that he found in Darwinian natural selection a theoretical and by implication an empirical analogue for theory competition and progress in science. According to Popper "what characterizes the empirical method is its manner of exposing to falsification, in every conceivable way, the system to be tested. Its aim is not to save the lives of untenable systems but, on the contrary, to select the one which is by comparison the fittest, by exposing them all to the fiercest struggle for survival" (42).

No doubt this model of theory competition may well account for what Stephen Jay Gould (1983) has called "names and nastiness" (357) with regard to the tenor of theory competition in taxonomy. Similarly, with regard to competing species concepts David Johnson (1990) commented on "the passion with which some people argue in favour of their preferred version" (67). And maybe this is good. Maybe we should welcome "philosophy of biology red in tooth and claw" the same way we accept this kind of competition in nature.

But this kind of competition can easily go too far. As Gould (1980) so poignantly put it, "We who revel in nature's diversity and feel instructed by every animal tend to brand *Homo sapiens* as the greatest catastrophe since the Cretaceous extinction" (289). The rise to full consciousness of the goal of the preservation of species other than our own and of the preservation of nature's diversity is itself only a recent phenomenon in our history. The tendency to exterminate is still the more basic. Those who hunt with guns and those who poach continue to far outnumber those who hunt with cameras and those who try to preserve or restore. It is a tendency, equally unfortunate, that finds expression at higher levels, at the level of theories and theory competition. For example, Philip Kitcher (1987) moved that the concept of abstract class "be banned from any future discussion on the topic of species" (186). More recently Ernst Mayr (1996), in an attack on what he calls "armchair taxonomists"—authors who "have never personally analyzed any species populations or studied species in nature," so that "they lack any feeling for what species actually are"—quoted in obvious approbation a passage in one of Darwin's earlier letters in which he says "no one has hardly a right to examine the question of species who has not minutely described many" (262).[36] Analyze, criticize, in painstaking minutiae even yes, but ban or censor certain kinds of theorists or theories? Even if they take both biologists and evolution seriously? One might look at proposals such as Kitcher's and Mayr's as admirable attempts to put into effect the philosophical equivalent of biological control. But surely, given both the empirical and conceptual nature of the species problem, such proposals seem highly inappropriate to say the least.

Perhaps, then, in concert with recent developments in evolutionary biology,

---

[36]   The quotation is from Darwin's letter to Hooker of September 10, 1845 (Burkhardt and Smith 1987: 253). From the context of the letter, however, it is clear that "the question of species" is not the question of the ontology of species taxa, but instead the much narrower question of whether species taxa (whatever they may be) are indefinitely immutable. Indeed that was the nature of the debate in Darwin's time. Granted that the correct answer to that narrower question has been firmly settled by modern biologists, does it follow that only modern biologists intimately acquainted with species taxa are qualified to answer the further question of the ontology of those taxa? I say no. As I have shown in this book, that further question requires *philosophical* erudition, though of course any answer to it is bound to be utterly worthless without a sound grounding in the relevant facts and current theories of biology.

we should take a more symbiotic approach to species concept diversity, where the diversity is to be positively cherished. Interestingly, Aristotle taught that "we should venture on the study of every kind of animal without distaste; for each and all will reveal to us something natural and something beautiful" (*Pa. An.* 645^{a}21–23). And indeed there is "something natural and something beautiful" in each and every species concept, which taken together in their diversity reveal a conceptual world as rich and as breathtaking, in its own way, as anything to be found in the biological world. And if symbiosis in the biological world is truly a source of evolutionary innovation, there is no reason why it cannot also be so in the conceptual world of theories. All the more reason, then, to value rather than slash and burn the diversity of solutions to the species problem, for out of that diversity endosymbiotic innovations may be born. And indeed we have already seen this in the case of some species concepts, such as Ridley's version of the cladistic species concept and my own biosimilarity species concept.

But this attitude can also go too far. To cherish the great diversity of species concepts the way an Aristotle or a modern naturalist might cherish the great diversity of species is to support an extreme form of species pluralism. And nothing serves more to highlight the great disanalogy between species and species concepts than the fact that cherishing species diversity presupposes species realism while cherishing species concept diversity presupposes species nominalism (cf. chapter 2.4). Moreover, while endosymbiosis among theories might well result in conceptual innovations that truly represent progress, it must never be forgotten that symbiosis in nature, of whatever degree, does not really eliminate or avoid competition but rather only defers it. As Darwin said in the case of William Harvey's *Begonia frigida* (cf. chapter 4.1)—though not a case of endosymbiosis, but supposedly of saltational innovation, the concept still applies—"if only a few (as he supposes) of the seedlings inherited his monstrosity natural selection would be necessary to select & preserve them" (Burkhardt *et al.* 1993: 93). As in nature, so perhaps also in theory competition.

Yet another way to look at the species problem (and one might feel this especially after reading this book) is to say that it has gone too far and we should simply try, as Robert O'Hara (1993) put it, "not to solve the species problem, but rather to get over it" (232). This echoes the pragmatist John Dewey's (1910) famous claim about philosophical and scientific change, given in the context of his discussion on the history of the concept of species or *eidos* (which he did not explicitly apply to its solution):

> But in fact intellectual progress usually occurs through sheer abandonment of questions together with both of the alternatives they assume—an abandonment that results from their decreasing vitality and a change of urgent interest. We do not solve them: we get over them. Old questions are solved by disappearing, evaporating, while new questions corresponding to the changed attitude of endeavor and preference take their place. Doubtless the greatest dissolvent in contemporary thought of old questions, . . . is the one effected by the scientific revolution that found its climax in the "Origin of Species." [19]

Should biologists, then, with their Darwinian and post-Darwinian interests, try their best to forget about the species problem and just get on with their lives *qua* biologists? Some might feel this way, but I do not think this approach will ever become popular, for the simple reason that most subdisciplines in biology require a species concept and, as Joel Cracraft (1989a) put it, "Different species concepts organize the world differently and conflictingly" (40). But more importantly, as Cracraft elsewhere points out (Cracraft 1997: 332–337), the increasing environmental crisis has been putting an increasing pressure on the related fields of taxonomy and conservation biology to solve the species problem. This pressure is unlike anything the science of biology has seen before. Unlike a failed romance, then, I do not think the species problem will go away simply by trying to ignore it and by moving on to something else. It cannot be ignored! Instead it will only go away when it is solved. Should the day ever come when it is generally agreed that the species problem has indeed been solved—and there really are no theoretical reasons preventing this—the Modern Synthesis will then be complete. As for how it will be solved, I share the view of Melissa Luckow (1995), namely that "the species problem will be solved by the continued collection and analysis of data, the clarification of issues and terms, and the application of new ideas" (600). Toward this end this book, this "one long argument" (to borrow Darwin's description of the *Origin*), was written and designed. Whether it will have succeeded in its purpose is something, of course, that only time will tell.

References

—all references are to reprints where indicated—

Abercrombie, M., *et al.*, eds. (1990). *The New Penguin Dictionary of Biology*. London: Penguin Books.

Agassi, Joseph (1968). *The Continuing Revolution*. New York: McGraw-Hill.

———. (1975). *Science in Flux*. Dordrecht: D. Reidel Publishing Company.

Agassiz, Louis (1860a). "Minutes of Meeting of March 13." *Proceedings of the American Academy of Arts and Sciences* (4: 410).

———. (1860b). "On the Origin of Species." *American Journal of Science and Arts* (2d ser., 30: 142–154).

Andersson, Lennart (1990). "The Driving Force: Species Concepts and Ecology." *Taxon* (39: 375–382).

Andersson, Malte (1994). *Sexual Selection*. Princeton: Princeton University Press.

Armstrong, D.M. (1989). *Universals: An Opinionated Introduction*. Boulder: Westview Press.

Ashton, Paul A., and Abbott, Richard J. (1992). "Multiple Origins and Genetic Diversity in the Newly Arisen Allopolyploid Species, *Senecio Cambrensis* Rosser (Compositae)." *Heredity* (68: 25–32).

Atran, Scott (1999). "The Universal Primacy of Generic Species in Folkbiological Taxonomy: Implications for Human Biological, Cultural, and Scientific Evolution." In Wilson, ed. (1999: 231–261).

Austin, J.L. (1964). *Sense and Sensibilia*. New York: Oxford University Press.

Avise, John C., and Ball, R. Martin, Jr. (1990). "Principles of Genealogical Concordance in Species Concepts and Biological Taxonomy." *Oxford Surveys in Evolutionary Biology* (7: 45–67).

Ayala, Francisco J. (1983). "Beyond Darwinism? The Challenge of Macroevolution to the Synthetic Theory of Evolution." In P. Asquith and T. Nickles, eds. (1983: 275–291). *PSA 1982*. Volume II. East Lansing, Mich.: Philosophy of Science Association. Reprinted in Ruse, ed. (1989a: 118–133).

Ayers, Michael R. (1981). "Locke versus Aristotle on Natural Kinds." *The Journal of Philosophy* (78: 247–272).

———. (1991). *Locke: Epistemology and Ontology*. Two Volumes. London: Routledge.

Balme, D.M. (1962). "*ΓΕΝΟΣ* and *ΕΙΔΟΣ* in Aristotle's Biology." *The Classical Quarterly* (XII: 81–98).

———. (1980). "Aristotle's Biology Was Not Essentialist." *Archiv für Geschichte der Philosophie* (62: 1–12). Reprinted in Allan Gotthelf and James G. Lennox, eds. (1987: 291–302). *Philosophical Issues in Aristotle's Biology*. Cambridge: Cambridge University Press.

Bambrough, Renford (1960–61). "Universals and Family Resemblances." *Proceedings of the Aristotelian Society* (61: 207–222). Reprinted in Andrew B. Schoedinger, ed. (1992: 266–279). *The Problem of Universals*. New Jersey: Humanities Press.

Barnes, Jonathan, ed. (1984). *The Complete Works of Aristotle*. Two Volumes. Princeton: Princeton University Press.

———., ed. (1995a). *The Cambridge Companion to Aristotle*. Cambridge: Cambridge

University Press.

———. (1995b). "Metaphysics." In Barnes, ed. (1995a: 66–108).

Barrett, Paul H., ed. (1977). *The Collected Papers of Charles Darwin*. Two Volumes. Chicago: The University of Chicago Press.

Barrett, Paul H., *et al.*, eds. (1987). *Charles Darwin's Notebooks, 1836–1844*. Ithaca, N.Y.: Cornell University Press.

Barrett, Spencer C.H. (1989). "Mating System Evolution and Speciation in Heterostylous Plants." In Otte and Endler, eds. (1989: 257–283).

Bates, Henry Walter (1862). "Contributions to an Insect Fauna of the Amazon Valley. Lepidoptera: Heliconidæ." *Transactions of the Linnean Society* (23: 495–566).

Baum, David A. (1992). "Phylogenetic Species Concepts." *Trends in Ecology and Evolution* (7: 1–2).

———. (1998). "Individuality and the Existence of Species Through Time." *Systematic Biology* (47: 641–653).

Baum, David A., and Donoghue, Michael J. (1995). "Choosing among Alternative 'Phylogenetic' Species Concepts." *Systematic Botany* (20: 560–573).

Baum, David A., and Shaw, Kerry L. (1995). "Genealogical Perspectives on the Species Problem." In Peter C. Hoch and A.G. Stephenson, eds. (1990: 289–303). *Experimental and Molecular Approaches to Plant Biosystematics*. St. Louis: Missouri Botanical Garden.

Beatty, John (1985). "Speaking of Species: Darwin's Strategy." In Kohn, ed. (1985: 265–281).

Beckner, Morton (1959). *The Biological Way of Thought*. New York: Columbia University Press.

Bendall, D.S., ed. (1983). *Evolution from Molecules to Men*. Cambridge: Cambridge University Press.

Bickerton, Derek (1990). *Language & Species*. Chicago: The University of Chicago Press.

Black, Max (1971). "The Elusiveness of Sets." *The Review of Metaphysics* (24: 614–636).

Bock, Walter J. (1994). "Ernst Mayr, Naturalist: His Contributions to Systematics and Evolution." *Biology & Philosophy* (9: 267–327).

Bowler, Peter J. (1985). "Scientific Attitudes to Darwinism in Britain and America." In Kohn, ed. (1985: 641–681).

———. (1989). *Evolution: The History of an Idea*. Berkeley: University of California Press.

Brandon, Robert N., and Burian, Richard M., eds. (1984). *Genes, Organisms, Populations*. Cambridge: The MIT Press.

Budd, Ann F., and Mishler, Brent D. (1990). "Species and Evolution in Clonal Organisms—Summary and Discussion." *Systematic Botany* (1990: 166–171).

Bunge, Mario (1981). "Biopopulations, Not Biospecies, Are Individuals and Evolve." *The Behavioral and Brain Sciences* (4: 284–285).

Buffon, Georges Louis-Leclerc Comte de (1749–1804). *Histoire Naturelle*. Forty-four Volumes. Paris: l'Imprimerie Royale.

Burkhardt, Richard W. (1987). "Lamarck and Species." In Roger and Fischer, eds. (1987: 161–180).

Burkhardt, Frederick, and Smith, Sydney, eds. (1987). *The Correspondence of Charles Darwin: Volume 3, 1844–1846*. Cambridge: Cambridge University Press.

———. (1990). *The Correspondence of Charles Darwin: Volume 6, 1856–1857*. Cambridge: Cambridge University Press.

———. (1991). *The Correspondence of Charles Darwin: Volume 7, 1858–1859*. Cambridge: Cambridge University Press.

Burkhardt, Frederick, *et al.*, eds. (1993). *The Correspondence of Charles Darwin: Volume 8, 1860*. Cambridge: Cambridge University Press.

Burma, Benjamin H. (1949). "The Species Concept: A Semantic View." *Evolution* (3: 369–370).

———. (1954). "Reality, Existence, and Classification: A Discussion of the Species Problem." *Madroño* (12: 193–209).

Cain, A.J. (1954). *Animal Species and Their Evolution*. London: Hutchinson's University Library. Reprinted with an Afterword (1993). Princeton: Princeton University Press.

———. (1993). "Afterword (1993)." In Cain (1954: 187–201).

Caplan, Arthur L. (1980). "Have Species Become Déclassé?" In P. Asquith and R. Giere, eds. (1980: 71–82). *PSA 1980*. Volume 1. East Lansing, Mich.: Philosophy of Science Association. Reprinted in Ruse, ed. (1989a: 156–166).

———. (1981). "Back to Class: A Note on the Ontology of Species." *Philosophy of Science* (48: 130–140).

Carr, Thomas D. (1999). "Craniofacial Ontogeny in Tyrannosauridae (Dinosauria, Coelurosauria)." *Journal of Vertebrate Paleontology* (19: 497–520).

Cheetham, Alan H. (1986). "Tempo of Evolution in a Neogene Bryozoan: Rates of Morphologic Change Within and Across Species Boundaries." *Paleobiology* (12: 1190–1202).

Cherrett, J.M., ed. (1989). *Ecological Concepts*. Oxford: Blackwell Scientific Publications.

Chin, Karen (1995). "Lessons from Leavings." *Natural History* (104: 67).

Churchland, Paul M. (1985). "Conceptual Progress and Word/World Relations: In Search of the Essence of Natural Kinds." *Canadian Journal of Philosophy* (15: 1–17).

Claridge, M.F., Dawah, H.A., and Wilson, M.R., eds. (1997). *Species: The Units of Biodiversity*. London: Chapman & Hall.

Colwell, Robert K. (1992). "Niche: A Bifurcation in the Conceptual Lineage of the Term." In Keller and Lloyd, eds. (1992: 241–248).

Copi, Irving M. (1954). "Essence and Accident." *The Journal of Philosophy* (51: 706–719). Reprinted in Stephen P. Schwartz, ed. (1977: 176–191). *Naming, Necessity, and Natural Kinds*. Ithaca: Cornell University Press.

Cowley, Fraser (1991). *Metaphysical Delusion*. Buffalo: Prometheus Books.

Coyne, Jerry A., and Charlesworth, Brian (1997). "Response." *Science* (276: 339–341).

Coyne, Jerry A., and Orr, H. Allen (1989). "Two Rules of Speciation." In Otte and Endler, eds. (1989: 180–207).

Coyne, Jerry A., Orr, H. Allen, and Futuyma, Douglas J. (1988). "Do We Need a New Species Concept?" *Systematic Zoology* (37: 190–200).

Cracraft, Joel (1983). "Species Concepts and Speciation Analysis." *Current Ornithology* (1: 159–187). Reprinted in Ereshefsky, ed. (1992: 93–120).

———. (1987). "Species Concepts and the Ontology of Evolution." *Biology & Philosophy* (2: 329–346).

———. (1989a). "Species as Entities of Biological Theory." In Ruse, ed. (1989b: 31–52).

———. (1989b). "Speciation and Its Ontology: The Empirical Consequences of Alternative Species Concepts for Understanding Patterns and Processes of Differentiation." In Otte and Endler, eds. (1989: 28–59).

———. (1997). "Species Concepts in Systematics and Conservation Biology—An Ornithological Viewpoint." In Claridge *et al.*, eds. (1997: 325–339).

Crick, Francis (1968). "The Origin of the Genetic Code." *Journal of Molecular Biology* (38: 367–379).

Cronquist, Arthur (1978). "Once Again, What Is a Species?" *Biosystematics in Agriculture: Beltsville Symposia in Agricultural Research* (2: 3–20).

Damuth, John (1992). "Extinction." In Keller and Lloyd, eds. (1992: 106–111).

Darwin, Charles (1851). *A Monograph on the Sub-Class Cirripedia: The Lepadidae*. London: The Ray Society.

———. (1859). *On the Origin of Species*. London: John Murray. Facsimile edition (1964). Cambridge: Harvard University Press.

———. (1871). *The Descent of Man*. Part I. London: John Murray. Facsimile edition (1981). Princeton: Princeton University Press.

———. (1876). *Autobiography*. Reprinted in Gavin de Beer, ed. (1983). *Charles Darwin and Thomas Henry Huxley: Autobiographies*. Oxford: Oxford University Press.

Davis, Jerrold I (1995). "Species Concepts and Phylogenetic Analysis—Introduction." *Systematic Botany* (20: 555–559).

Dawkins, Richard (1976). *The Selfish Gene*. Oxford: Oxford University Press. Reprinted (1989).

———. (1982). *The Extended Phenotype*. Oxford: Oxford University Press.

———. (1983). "Universal Darwinism." In Bendall, ed. (1983: 403–425).

———. (1986). *The Blind Watchmaker*. Harlow, England: Longman Scientific & Technical.

Denkel, Arda (1989). "Real Resemblances." *The Philosophical Quarterly* (39: 36–56).

———. (1996). *Object and Property*. Cambridge: Cambridge University Press.

de Queiroz, Kevin (1999). "The General Lineage Concept of Species and the Defining Properties of the Species Category." In Wilson, ed. (1999: 49–89).

de Queiroz, Kevin, and Donoghue, Michael J. (1988). "Phylogenetic Systematics and the Species Problem." *Cladistics* (4: 317–338).

Desmond, Adrian, and Moore, James (1991). *Darwin: The Life of a Tormented Evolutionist*. London: Penguin Books. Reprinted (1994). New York: W.W. Norton & Company.

De Vries, Hugo (1912). *Species and Varieties: Their Origin by Mutation*. Third edition. Chicago: Open Court. Reprinted (1988). New York: Garland Publishing, Inc.

Dewey, John (1910). *The Influence of Darwin on Philosophy and Other Essays*. New York: H. Holt and Co.

Dobzhansky, Theodosius (1937). *Genetics and the Origin of Species*. New York: Columbia University Press.

———. (1973). "Nothing in Biology Makes Sense Except in the Light of Evolution." *American Biology Teacher* (35: 125–129).

Donoghue, Michael J. (1985). "A Critique of the Biological Species Concept and Recommendations for a Phylogenetic Alternative." *The Bryologist* (88: 172–181).

Doyle, Jeff J. (1995). "The Irrelevance of Allele Tree Topologies for Species Delimitation, and a Non-Topological Alternative." *Systematic Botany* (20: 574–588).

Dressler, Robert L. (1990). "Species—What Causes the Patterns?" *Taxon* (39: 448).

Dupré, John (1981). "Natural Kinds and Biological Taxa." *The Philosophical Review* (90: 66–90).

———. (1993). *The Disorder of Things*. Cambridge: Harvard University Press.

———. (1999). "On the Impossibility of a Monistic Account of Species." In Wilson, ed. (1999: 3–22).

Eaton, Ralph M. (1931). *General Logic*. New York: Charles Scribner's Sons.

Eccles, John C. (1989). *Evolution of the Brain*. London: Routledge.

Echelle, Anthony A. (1990). "In Defense of the Phylogenetic Species Concept and the Ontological Status of Hybridogenetic Taxa." *Herpetologica* (46: 109–113).

Eddington, Arthur S. (1928). *The Nature of the Physical World*. Cambridge: Cambridge University Press.

Ehrlich, Paul R., and Holm, Richard W. (1962). "Patterns and Populations." *Science* (137:

652–657).

Ehrlich, Paul R., and Raven, Peter H. (1969). "Differentiation of Populations." *Science* (165: 1228–1232). Reprinted in Ereshefsky, ed. (1992b: 57–67).

Eigen, Manfred (1983). "Self-Replication and Molecular Evolution." In Bendall, ed. (1983: 105–130).

———. (1992). *Steps Towards Life*. Oxford: Oxford University Press.

———. (1993). "Viral Quasispecies." *Scientific American* (269: 42–49).

———. (1995). "What Will Endure of 20th Century Biology?" In Michael P. Murphy and Luke A.J. O'Neill, eds. (1995: 5–23). *What is Life? The Next Fifty Years*. Cambridge: Cambridge University Press.

Eigen, Manfred, McCaskill, John, and Schuster, Peter (1988). "Molecular Quasi-Species." *Journal of Physical Chemistry* (92: 6881–6891).

Eisner, Thomas, Lubchenco, Jane, Wilson, Edward O., Wilcove, David S., and Bean, Michael J. (1995). "Building a Scientifically Sound Policy for Protecting Endangered Species." *Science* (269: 1231–1232).

Eldredge, Niles (1985a). *Time Frames*. New York: Simon & Schuster, Inc. Reprinted (1989). Princeton: Princeton University Press.

———. (1985b). *Unfinished Synthesis*. New York: Oxford University Press.

———. (1989). *Macroevolutionary Dynamics*. New York: McGraw-Hill.

———. (1995). *Reinventing Darwin*. New York: John Wiley & Sons, Inc.

Eldredge, Niles, and Cracraft, Joel (1980). *Phylogenetic Patterns and the Evolutionary Process*. New York: Columbia University Press.

Eldredge, Niles, and Gould, Stephen Jay (1972). "Punctuated Equilibria: An Alternative to Phyletic Gradualism." In T.J.M. Schopf, ed. (1972: 82–115). *Models in Paleobiology*. San Francisco: Cooper and Co. Reprinted in Eldredge (1985a: 193–223).

Ellegård, Alvar (1958). *Darwin and the General Reader*. Stockholm: Almqvist and Wiksell. Reprinted (1990). Chicago: The University of Chicago Press.

Endler, John A. (1989). "Conceptual and Other Problems in Speciation." In Otte and Endler, eds. (1989: 625–648).

Ereshefsky, Marc (1988a). "Axiomatics and Individuality: A Reply to Williams' 'Species Are Individuals.'" *Philosophy of Science* (55: 427–434).

———. (1988b). "Individuality and Macroevolutionary Theory." In A. Fine and J. Leplin, eds. (1988: 216–222). *PSA 1988*. Volume I. East Lansing, Mich.: Philosophy of Science Association.

———. (1992a). "Eliminative Pluralism." *Philosophy of Science* (59: 671–690).

———., ed. (1992b). *The Units of Evolution: Essays on the Nature of Species*. Cambridge: The MIT Press.

———. (1998). "Species Pluralism and Anti-Realism." *Philosophy of Science* (65: 103–120).

Erwin, Douglas H., and Anstey, Robert L., eds. (1995). *New Approaches to Speciation in the Fossil Record*. New York: Columbia University Press.

Fales, Evan (1982). "Natural Kinds and Freaks of Nature." *Philosophy of Science* (49: 67–90).

Farber, Paul L. (1972). "Buffon and the Concept of Species." *Journal of the History of Biology* (5: 259–284).

Feder, Kenneth L., and Park, Michael Alan (1993). *Human Antiquity*. Mountain View, Calif.: Mayfield Publishing Company.

Fenster, Eugene J., and Sorhannus, Ulf (1991). "On the Measurement of Morphological Rates of Evolution: A Review." In Max Hecht, *et al.*, eds. (25: 375–410). *Evolutionary Biology*. New York: Plenum Press.

Flew, Antony, ed. (1956). *Essays in Conceptual Analysis*. London: Macmillan. Reprinted (1981). Westport: Greenwood Press.

———. (1984a). *God, Freedom, and Immortality*. Buffalo: Prometheus Books.

———. (1984b). *Darwinian Evolution*. London: Paladin Books.

Fodor, Jerry A. (1975). *The Language of Thought*. New York: Thomas Y. Crowell Co., Inc. Reprinted (1980). Cambridge: Harvard University Press.

Foottit, R.G. (1997). "Recognition of Parthenogenetic Insect Species." In Claridge *et al.*, eds. (1997: 291–307).

Fristrup, Kurt (1992). "Character: Current Usages." In Keller and Lloyd, eds. (1992: 45–51).

Frost, Darrel R., and Wright, John W. (1988). "The Taxonomy of Uniparental Species, with Special Reference to Parthenogenetic *Cnemidophorus* (Squamata: Teiidae)." *Systematic Zoology* (37: 200–209).

Futuyma, Douglas J. (1986). *Evolutionary Biology*. Sunderland, Mass.: Sinauer Associates, Inc.

Gaffney, Eugene S., Dingus, Lowell, and Smith, Miranda K. (1995). "Why Cladistics?" *Natural History* (104: 33–35).

Gasking, Douglas (1960). "Clusters." *The Australasian Journal of Philosophy* (38: 1–36).

Gayon, Jean (1996). "The Individuality of Species: A Darwinian Theory?—from Buffon to Ghiselin, and Back to Darwin." *Biology & Philosophy* (11: 215–244).

Ghiselin, Michael T. (1966). "On Psychologism in the Logic of Taxonomic Controversies." *Systematic Zoology* (15: 207–215).

———. (1969). *The Triumph of the Darwinian Method*. Berkeley: University of California Press.

———. (1974). "A Radical Solution to the Species Problem." *Systematic Zoology* (23: 536–544). Reprinted in Ereshefsky, ed. (1992b: 279–292).

———. (1981). "Taxa, Life, and Thinking." *The Behavioral and Brain Sciences* (4: 303–310).

———. (1987a). "Species Concepts, Individuality, and Objectivity." *Biology & Philosophy* (2: 127–143).

———. (1987b). "Response to Commentary on the Individuality of Species." *Biology & Philosophy* (2: 207–212).

———. (1988). "Species Individuality Has No Necessary Connection with Evolutionary Gradualism." *Systematic Zoology* (37: 66–67).

———. (1989). "Individuality, History and Laws of Nature in Biology." In Ruse, ed. (1989b: 53–66).

———. (1995). "Ostensive Definitions of the Names of Species and Clades." *Biology & Philosophy* (10: 219–222).

———. (1997). *Metaphysics and the Origin of Species*. Albany: State University of New York Press.

Giddings, Luther Val, Kaneshiro, Kenneth Y., and Anderson, Wyatt W. (1989). *Genetics, Speciation, and the Founder Principle*. New York: Oxford University Press.

Gill, Douglas E. (1989). "Fruiting Failure, Pollinator Inefficiency, and Speciation in Orchids." In Otte and Endler, eds. (1989: 458–481).

Gilmour, J.S.L. (1940). "Taxonomy and Philosophy." In Julian Huxley, ed. (1940: 461–474). *The New Systematics*. London: Clarendon Press.

Goodman, Nelson (1970). "Seven Strictures on Similarity." In Lawrence Foster and J.W. Swanson, eds. (1970: 19–29). *Experience & Theory*. Amherst: The University of Massachusetts Press.

Gotthelf, Allan, ed. (1985). *Aristotle on Nature and Living Things*. Pittsburgh: Mathesis

Publications, Inc.

Gould, Stephen Jay (1977). *Ever Since Darwin.* New York: W.W. Norton & Company.

———. (1980). *The Panda's Thumb.* New York: W.W. Norton & Company.

———. (1981). *The Mismeasure of Man.* New York: W.W. Norton & Company.

———. (1982). "Darwinism and the Expansion of Evolutionary Theory." *Science* (216: 380–387). Reprinted in Ruse, ed. (1989a: 100–117).

———. (1983). *Hen's Teeth and Horse's Toes.* New York: W.W. Norton & Company.

———. (1992). "Punctuated Equilibrium in Fact and Theory." In Somit and Peterson, eds. (1992: 54–84).

Gould, Stephen Jay, and Eldredge, Niles (1977). "Punctuated Equilibria: The Tempo and Mode of Evolution Reconsidered." *Paleobiology* (3: 115–151).

Grene, Marjorie (1989a). "A Defense of David Kitts." *Biology & Philosophy* (4: 69–72).

———. (1989b). "Interaction and Evolution." In Ruse, ed. (1989b: 67–73).

Griesemer, James R. (1992). "Niche: Historical Perspectives." In Keller and Lloyd, eds. (1992: 231–240).

Griffiths, Paul E. (1994). "Cladistic Classification and Functional Explanation." *Philosophy of Science* (61: 206–227).

——— . (1999). "Squaring the Circle: Natural Kinds with Historical Essences." In Wilson, ed. (1999: 209–228).

Guthrie, W.K.C. (1978). *A History of Greek Philosophy V, The Later Plato and the Academy.* Cambridge: Cambridge University Press.

———. (1981). *A History of Greek Philosophy VI, Aristotle.* Cambridge: Cambridge University Press.

Haack, Susan (1978). *Philosophy of Logics.* Cambridge: Cambridge University Press.

Haldane, J.B.S. (1956). "Can a Species Concept be Justified?" In P.C. Sylvester-Bradley, ed. (1956: 95–96). *The Species Concept in Palaeontology.* London: The Systematics Association.

Hamilton, Edith, and Cairns, Huntington, eds. (1961). *The Collected Dialogues of Plato.* Princeton: Princeton University Press.

Hankinson, R.J. (1995). "Philosophy of Science." In Jonathan Barnes, ed. (1995a: 109–139).

Harrison, Richard G. (1998). "Linking Evolutionary Pattern and Process: The Relevance of Species Concepts for the Study of Speciation." In Howard and Berlocher, eds. (1998: 19–31).

Hattiangadi, J.N. (1987). *How is Language Possible?* La Salle, Ill.: Open Court.

Haynes, Robert H. (1987). "The 'Purpose' of Chance in Light of the Physical Basis of Evolution." In John M. Robson, ed. (1987: 1–31). *Origin and Evolution of the Universe.* Kingston and Montreal: McGill-Queen's University Press.

———. (1990). "Ethics and Planetary Engineering." In Don MacNiven, ed. (1990: 161–183). *Moral Expertise.* London: Routledge.

Hempel, Carl G. (1966). *Philosophy of Natural Science.* Englewood Cliffs, N.J.: Prentice-Hall, Inc.

Hengeveld, R. (1987). "Mayr's Ecological Species Criterion." *Systematic Zoology* (37: 47–55).

Hennig, Willi (1966). *Phylogenetic Systematics.* D. Dwight Davis and Rainer Zangerl, trans. Urbana: University of Illinois Press.

Herbert, Sandra, and Barrett, Paul H. (1987). "Introduction to Notebook M." In Barrett *et al.*, eds. (1987: 517–519).

Hoffman, Antoni (1989). *Arguments On Evolution.* New York: Oxford University Press.

————. (1992). "Twenty Years Later: Punctuated Equilibrium in Retrospect." In Somit and Peterson, eds. (1992: 121–138).

Hogan, Melinda (1992). "Natural Kinds and Ecological Niches—Response to Johnson's Paper." *Biology & Philosophy* (7: 203–208).

Holsinger, Kent E. (1984). "The Nature of Biological Species." *Philosophy of Science* (51: 293–307).

Hospers, John (1956). "What is Explanation?" In Flew, ed. (1956: 94–119).

Howard, Daniel J. (1998). "Unanswered Questions and Future Directions in the Study of Speciation." In Howard and Berlocher, eds. (1998: 439–448).

Howard, Daniel J., and Berlocher, Stewart H., eds. (1998). *Endless Forms: Species and Speciation.* New York: Oxford University Press.

Howard, Jonathan (1982). *Darwin.* Oxford: Oxford University Press.

Hull, David L. (1965). "The Effect of Essentialism on Taxonomy: Two Thousand Years of Stasis." *British Journal for the Philosophy of Science* (15: 2–32, 16: 1–18). Reprinted in Ereshefsky, ed. (1992b: 199–225).

————. (1970). "Contemporary Systematic Philosophies." *Annual Review of Ecology and Systematics* (1: 19–54). Reprinted in Sober, ed. (1994: 295–330).

————. (1976). "Are Species Really Individuals?" *Systematic Zoology* (25: 174–191).

————. (1977). "The Ontological Status of Species as Evolutionary Units." In R. Butts and J. Hintikka, eds. (1977: 91–102). *Foundational Problems in Special Sciences.* Dordrecht: Reidel Publishing Company. Reprinted in Hull (1989: 79–88).

————. (1978). "A Matter of Individuality." *Philosophy of Science* (45: 335–360). Reprinted in Ereshefsky, ed. (1992b: 293–316).

————. (1980). "Individuality and Selection." *Annual Review of Ecology and Systematics* (11: 311–332).

————. (1981a). "Kitts and Kitts and Caplan on Species." *Philosophy of Science* (48: 141–152).

————. (1981b). "Units of Evolution: A Metaphysical Essay." In U.L. Jensen and R. Harré, eds. (1981: 23–44). *The Philosophy of Evolution.* Brighton: Harvester Press. Reprinted in Brandon and Burian, eds. (1984: 142–160).

————. (1983). "Karl Popper and Plato's Metaphor." In N. Platnick and V. Funk, eds. (1983: 177–189). *Advances in Cladistics.* Volume 2. New York: Columbia University Press. Reprinted in Hull (1989: 144–161).

————. (1987). "Genealogical Actors in Ecological Roles." *Biology & Philosophy* (2: 168–184).

————. (1988). *Science as a Process.* Chicago: The University of Chicago Press.

————. (1989). *The Metaphysics of Evolution.* Albany: State University of New York Press.

————. (1992). "Individual." In Keller and Lloyd, eds. (1992: 180–187).

————. (1997). "The Ideal Species Concept—And Why We Can't Get It." In Claridge *et al.*, eds. (1997: 357–380).

Hume, David (1748/51). *Enquiries Concerning Human Understanding and Concerning the Principles of Morals.* P.H. Nidditch, ed. (1975). Oxford: Oxford University Press.

Hunter, R.S.T., Arnold, A.J., and Parker, W.C. (1988). "Evolution and Homeomorphy in the Development of the Paleocene *Planorotalites pseudomenardii* and the Miocene *Globorotalia (Glororotalia) maragritae* lineages." *Micropaleontology* (34: 181–192).

Huxley, Julian, (1912). *The Individual in the Animal Kingdom.* Cambridge: Cambridge University Press. Reprinted (1995). Woodbridge, Conn.: Ox Bow Press.

Hyman, Arthur, and Walsh, James J., eds. (1973). *Philosophy in the Middle Ages.* Indianapolis: Hackett Publishing Company.

Johnson, David M. (1990). "Can Abstractions Be Causes?" *Biology & Philosophy* (5: 63–77).

Keller, Evelyn Fox, and Lloyd, Elisabeth A., eds. (1992). *Keywords in Evolutionary Biology*. Cambridge: Harvard University Press.

Kim, Jaegwon (1993). *Supervenience and Mind*. Cambridge: Cambridge University Press.

Kincaid, Harold (1987). "Supervenience Doesn't Entail Reducibility." *The Southern Journal of Philosophy* (25: 343–356).

Kitcher, Philip (1984a). "Species." *Philosophy of Science* (51: 308–333).

———. (1984b). "Against the Monism of the Moment: A Reply to Elliott Sober." *Philosophy of Science* (51: 616–630).

———. (1987). "Ghostly Whispers: Mayr, Ghiselin and the 'Philosophers' on the Ontology of Species." *Biology & Philosophy* (2: 184–192).

———. (1989). "Some Puzzles About Species." In Ruse, ed. (1989b: 183–208).

Kitts, David B. (1987). "Plato on Kinds of Animals." *Biology & Philosophy* (2: 315–328).

Kitts, David B., and Kitts, David J. (1979). "Biological Species as Natural Kinds." *Philosophy of Science* (46: 613–622).

Kluge, Arnold G. (1990). "Species as Historical Individuals." *Biology & Philosophy* (5: 417–431).

Kohn, David, ed. (1985). *The Darwinian Heritage*. Princeton: Princeton University Press.

Kornet, D.J. (1993). "Permanent Splits as Speciation Events: A Formal Reconstruction of the Internodal Species Concept." *Journal of Theoretical Biology* (164: 407–435).

Kripke, Saul A. (1972). "Naming and Necessity." In Donald Davidson and Gilbert Harman, eds. (1972: 254–355). *Semantics of Natural Language*. Dordrecht: D. Reidel. Reprinted (1980) as *Naming and Necessity*. Cambridge: Harvard University Press.

Lamarck, Jean Baptiste de (1809). *Philosophie Zoologique*. Paris: Dentu. Hugh Elliot, trans. (1963). *Zoological Philosophy*. New York: Hafner Publishing Company.

Lambert, D.M., Michaux, B., and White, C.S. (1987). "Are Species Self-Defining?" *Systematic Zoology* (36: 196–205).

Lambert, David M., and Spencer, Hamish G., eds. (1995). *Speciation and the Recognition Concept*. Baltimore: The Johns Hopkins University Press.

Larson, James L. (1968). "The Species Concept of Linnaeus." *Isis* (59: 291–299).

Law, Richard (1991). "The Symbiotic Phenotype: Origins and Evolution." In Margulis and Fester, eds. (1991: 57–71).

Law, R., and Watkinson, A.R. (1989). "Competition." In Cherrett, ed. (1989: 243–284).

Lawton, J.H. (1989). "Food Webs." In Cherrett, ed. (1989: 43–78).

Leikola, Anto (1987). "The Development of the Species Concept in the Thinking of Linnaeus." In Roger and Fischer, eds. (1987: 45–59).

Lennox, James G. (1985). "Are Aristotelian Species Eternal?" In Gotthelf, ed. (1985: 67–94).

Levin, Donald A. (1979). "The Nature of Plant Species." *Science* (204: 381–384).

———. (1983). "Polyploidy and Novelty in Flowering Plants." *The American Naturalist* (122: 1–25).

Lide, David R., *et al.*, eds. (1995). *Handbook of Chemistry and Physics*. New York: CRC Press.

Lidén, Magnus, and Oxelman, Bengt (1990). "Species—Evolutionary Actors or Evolutionary Products?" *Taxon* (39: 449).

Lipton, Peter (1991). *Inference to the Best Explanation*. London: Routledge.

Locke, John (1700). *An Essay Concerning Human Understanding*. Fourth edition. Peter H. Nidditch, ed. (1975). Oxford: Clarendon Press.

Losos, Jonathan B. (1996). "Phylogenies and Comparative Biology, Stage II: Testing Causal Hypotheses Derived from Phylogenies with Data from Extant Taxa." *Systematic Biology* (45: 259–260).

Lovejoy, Arthur O. (1936). *The Great Chain of Being*. Cambridge: Harvard University Press.

———. (1959). "Buffon and the Problem of Species." In Bentley Glass, *et al.*, eds. (1959: 84–113). *Forerunners of Darwin*. Baltimore: The Johns Hopkins Press.

Luckow, Melissa (1995). "Species Concepts: Assumptions, Methods, and Applications." *Systematic Botany* (20: 589–605).

Lyell, Charles (1830). *Principles of Geology*. Volume I. London: John Murray.

———. (1832). *Principles of Geology*. Volume II. London: John Murray.

———. (1838). *Elements of Geology*. London: John Murray.

———. (1863). *The Geological Evidences of the Antiquity of Man, with Remarks on Theories of the Origin of Species by Variation*. London: John Murray.

Lyon, John, and Sloan, Phillip R., eds. (1981). *From Natural History to the History of Nature: Readings from Buffon and His Critics*. Notre Dame: University of Notre Dame Press.

Mallet, James (1995). "A Species Definition for the Modern Synthesis." *Trends in Ecology and Evolution* (10: 294–299).

Mahner, Martin (1993). "What is a Species?" *Journal for General Philosophy of Science* (24: 103–126).

Mahner, Martin, and Bunge, Mario (1997). *Foundations of Biophilosophy*. Berlin: Springer-Verlag.

Marcus, Ruth Barcan (1974). "Classes, Collections, and Individuals." *American Philosophical Quarterly* (11: 227–232). Reprinted as "Classes, Collections, Assortments, and Individuals" in Ruth Barcan Marcus (1993: 89–100). *Modalities*. New York: Oxford University Press.

Margulis, Lynn (1991). "Symbiogenesis and Symbionticism." In Margulis and Fester, eds. (1991: 1–14).

———. (1993). *Symbiosis in Cell Evolution*. New York: W.H. Freeman and Company.

Margulis, Lynn, and Fester, René, eds. (1991). *Symbiosis as a Source of Evolutionary Innovation*. Cambridge: The MIT Press.

Margulis, Lynn, and Schwartz, Karlene V. (1988). *Five Kingdoms*. New York: W.H. Freeman and Company.

Marsh, Robert Charles, ed. (1956). *Logic and Knowledge*. London: Allen & Unwin. Reprinted (1992). London: Routledge.

Masters, J.C., and Spencer, H.G. (1989). "Why We Need a New Genetic Species Concept." *Systematic Zoology* (38: 270–279).

Mayden, R.L. (1997). "A Hierarchy of Species Concepts: the Denouement in the Saga of the Species Problem." In Claridge *et al.*, eds. (1997: 381–424).

Maynard Smith, John (1975). *The Theory of Evolution*. Harmondsworth: Penguin Books. Reprinted (1993). Cambridge: Cambridge Univeristy Press.

Mayr, Ernst (1942). *Systematics and the Origin of Species*. New York: Columbia University Press.

———. (1949). "The Species Concept: Semantics Versus Semantics." *Evolution* (3: 371–372).

———. (1954). "Change of Genetic Environment and Evolution." In Julian Huxley, *et al.*, eds. (1954: 157–180). *Evolution as a Process*. London: Allen & Unwin.

———. (1964). Introduction to Darwin (1859).

———. (1970). *Populations, Species, and Evolution*. Cambridge: Harvard University Press.

————. (1972). "Lamarck Revisited." *Journal of the History of Biology* (5: 55–94). Reprinted (revised) in Mayr (1976: 222–250). *Evolution and the Diversity of Life*. Cambridge: Harvard University Press.

————. (1975). "The Unity of the Genotype." *Biologisches Zentralblatt* (94: 377–388). Reprinted in Brandon and Burian, eds. (1984: 70–84).

————. (1982). *The Growth of Biological Thought*. Cambridge: Harvard University Press.

————. (1987a). "The Ontological Status of Species: Scientific Progress and Philosophical Terminology." *Biology & Philosophy* (2:145–166).

————. (1987b). "Answers to These Comments." *Biology & Philosophy* (2: 212–220).

————. (1988a). *Toward a New Philosophy of Biology*. Cambridge: Harvard University Press.

————. (1988b). "A Response to David Kitts." *Biology & Philosophy* (3: 97–98).

————. (1988c). "The Why and How of Species." *Biology & Philosophy* (3: 431–441).

————. (1991). *One Long Argument*. Cambridge: Harvard University Press.

————. (1992). "A Local Flora and the Biological Species Concept." *American Journal of Botany* (79: 222–238).

————. (1996). "What is a Species, and What is Not?" *Philosophy of Science* (63: 262–277).

Mayr, Ernst, and Ashlock, Peter D. (1991). *Principles of Systematic Zoology*. Second edition. New York: McGraw-Hill, Inc.

Mayr, Ernst, and Short, Lester L. (1970). *Species Taxa of North American Birds*. Cambridge: Nuttall Ornithological Club.

McDade, Lucinda A. (1995). "Species Concepts and Problems in Practice: Insight from Botanical Monographs." *Systematic Botany* (20: 606–622).

McEvey, Shane F., ed. (1993). *Evolution and the Recognition Concept of Species*. Baltimore: The Johns Hopkins University Press.

McIntosh, Robert (1992). "Competition: Historical Perspectives." In Keller and Lloyd, eds. (1992: 61–73).

McKitrick, Mary C., and Zink, Robert M. (1988). "Species Concepts in Ornithology." *The Condor* (90: 1–14).

Medawar, P.B., and Medawar, J.S. (1983). *Aristotle to Zoos*. Cambridge: Harvard University Press.

Medin, Douglas L., Goldstone, Robert L., and Gentner, Dedre (1993). "Respects for Similarity." *Psychological Review* (100: 254–278).

Mill, John Stuart (1881). *A System of Logic*. Eighth edition. London: Longmans, Green. Reprinted (abridged). Ernest Nagel, ed. (1950). *John Stuart Mill's Philosophy of Scientific Method*. New York: Hafner Press.

Miller, Kathleen (1994). "Abstractions Can Be Causes: A Response to Professor Hogan." *Biology & Philosophy* (9: 99–103).

Mishler, Brent D. (1999). "Getting Rid of Species?" In Wilson, ed. (1999: 307–315).

Mishler, Brent D., and Brandon, Robert N. (1987). "Individuality, Pluralism, and the Phylogenetic Species Concept." *Biology & Philosophy* (2: 397–414).

Mishler, Brent D., and Donoghue, Michael J. (1982). "Species Concepts: A Case for Pluralism." *Systematic Zoology* (31: 491–503). Reprinted in Ereshefsky, ed. (1992b: 121–137).

Moore, G.E. (1922). "External and Internal Relations." In *Philosophical Studies*. London: Routledge & Kegan Paul Ltd. Reprinted in Thomas Baldwin, ed. (1993: 79–105). *G.E. Moore: Selected Writings*. London: Routledge.

Nagel, Ernest. (1961). *The Structure of Science*. New York: Harcourt, Brace & World, Inc.

Nixon, Kevin C., and Wheeler, Quentin D. (1990). "An Amplification of the Phylogenetic

Species Concept." *Cladistics* (6: 211–223).

O'Hara, Robert J. (1993). "Systematic Generalization, Historical Fate, and the Species Problem." *Systematic Biology* (42: 231–246).

Olmstead, Richard G. (1995). "Species Concepts and Plesiomorphic Species." *Systematic Botany* (20: 623–630).

Otte, Daniel, and Endler, John A., eds. (1989). *Speciation and Its Consequences*. Sunderland, Mass.: Sinauer Associates, Inc.

Panchen, Alec L. (1992). *Classification, Evolution, and the Nature of Biology*. Cambridge: Cambridge University Press.

Parsons, Charles (1975). "What Is the Iterative Conception of Set?" In Robert E. Butts and Jaakko Hintikka, eds. (1977: 335–367). *Logic, Foundations of Mathematics, and Computability Theory*. Dordrecht: D. Reidel. Reprinted in Paul Benacerraf and Hilary Putnam, eds. (1983: 503–529). *Philosophy of Mathematics*. Cambridge: Cambridge University Press.

Paterson, H.E.H. (1978). "More Evidence against Speciation by Reinforcement." *South African Journal of Science* (74: 369–371). Reprinted in McEvey, ed. (1993: 20–31).

———. (1980). "A Comment on 'Mate Recognition Systems.'" *Evolution* (34: 330–331). Reprinted in McEvey, ed. (1993: 32–34).

———. (1981). "The Continuing Search for the Unknown and Unknowable: A Critique of Contemporary Ideas on Speciation." *South African Journal of Science* (77: 113–119). Reprinted in McEvey, ed. (1993: 35–57).

———. (1982a). "Perspective on Speciation by Reinforcement." *South African Journal of Science* (78: 53–57). Reprinted in McEvey, ed. (1993: 76–91).

———. (1982b). "Darwin and the Origin of Species." *South African Journal of Science* (78: 272–275). Reprinted in McEvey, ed. (1993: 92–103).

———. (1985). "The Recognition Concept of Species." In E.S. Vrba, ed. (1985: 21–29). *Species and Speciation*. Pretoria: Transvaal Museum Monograph No. 4. Reprinted in McEvey, ed. (1993: 136–157).

———. (1986). "Environment and Species." *South African Journal of Science* (82: 62–65). Reprinted in McEvey, ed. (1993: 158–167).

———. (1988). "On Defining Species in Terms of Sterility: Problems and Alternatives." *Pacific Science* (42: 65–71). Reprinted in McEvey, ed. (1993: 168–179).

———. (1989). "A View of Species." In B. Goodwin, *et al.*, eds. (1989: 77–88). *Dynamic Structures in Biology*. Edinburgh: Edinburgh University Press. Reprinted in McEvey, ed. (1993: 180–193).

———. (1991). "The Recognition of Cryptic Species among Economically Important Insects." In M.P. Zalucki, ed. (1991: 1–10). *Research Methods and Prospects*. New York: Springer-Verlag. Reprinted in McEvey, ed. (1993: 199–211).

Paterson, H.E.H., and Macnamara, M. (1984). "The Recognition Concept of Species: Macnamara Interviews Paterson." *South African Journal of Science* (80: 312–318). Reprinted in McEvey, ed. (1993: 104–123).

Patterson, Richard (1985). *Image and Reality in Plato's Metaphysics*. Indianapolis: Hackett Publishing Co., Inc.

Pellegrin, Pierre (1982). *La Classification des Animaux chez Aristotle*. Paris: Les Belles Lettres. Revised edition (1986). Anthony Preus, trans. *Aristotle's Classification of Animals*. Berkeley: University of California Press.

———. (1985). "Aristotle: A Zoology without Species." In Gotthelf, ed. (1985: 95–115).

Popper, Karl R. (1934). *The Logic of Scientific Discovery*. English edition (1959). London: Hutchinson. Reprinted (1992). London: Routledge.

———. (1936). "The Poverty of Historicism." *Economica* (1944–1945). Reprinted and revised in Karl Popper (1960). *The Poverty of Historicism*. London: Routledge & Kegan Paul, Ltd.

————. (1963). *Conjectures and Refutations*. London: Routledge & Kegan Paul, Ltd. Reprinted (1992). London: Routledge.

————. (1972). *Objective Knowledge: An Evolutionary Approach*. Oxford: Oxford University Press. Reprinted and revised (1979).

Popper, Karl R., and Eccles, John C. (1977). *The Self and Its Brain*. Berlin: Springer-Verlag. Reprinted (1983). London: Routledge & Kegan Paul, Ltd.

Portin, Petter (1993). "The Concept of the Gene: Short History and Present Status." *The Quarterly Review of Biology* (68: 173–223).

Preus, Anthony (1979). "*Eidos* as Norm in Aristotle's Biology." *Nature and System* (1: 79–101). Reprinted in John P. Anton and Anthony Preus, eds. (1983: 340–363). *Essays in Ancient Greek Philosophy II*. Albany: State University of New York Press.

Putnam, Hilary (1970). "Is Semantics Possible?" In H.E. Kiefer and M.K. Munitz, eds. (1970: 50–63). *Language, Belief and Metaphysics*. Albany: State University of New York Press. Reprinted in Putnam (1975b: 139–152).

————. (1973). "Explanation and Reference." In Glen Pearce and Patrick Maynard, eds. (1973: 199–221). *Conceptual Change*. Dordrecht: D. Reidel. Reprinted in Putnam (1975b: 196–214).

————. (1975a). "The Meaning of 'Meaning.'" In Keith Gunderson, ed. (1975: 131–193). *Language, Mind and Knowledge*. Minneapolis: University of Minnesota Press. Reprinted in Putnam (1975b: 215–271).

————. (1975b). *Mind, Language and Reality*. Cambridge: Cambridge University Press.

Quine, W.V.O. (1960). *Word and Object*. Cambridge: The MIT Press.

————. (1961). *From a Logical Point of View*. Cambridge: Harvard University Press.

————. (1965). *Elementary Logic*. Cambridge: Harvard University Press. Revised (1980).

————. (1969a). "Natural Kinds." In Quine (1969b: 114–138).

————. (1969b). *Ontological Relativity and Other Essays*. New York: Columbia University Press.

————. (1969c). *Set Theory And Its Logic*. Cambridge: Harvard University Press.

————. (1976). *The Ways of Paradox*. Cambridge: Harvard University Press.

————. (1981). *Theories and Things*. Cambridge: Harvard University Press.

————. (1982). *Methods of Logic*. Cambridge: Harvard University Press.

————. (1986). *Philosophy of Logic*. Cambridge: Harvard University Press.

————. (1987). *Quiddities*. Cambridge: Harvard University Press.

Quinton, Anthony (1989). "Universals." In J.O. Urmson and Jonathan Rée, eds. (1989: 317–318). *The Concise Encyclopedia of Western Philosophy and Philosophers*. London: Unwin Hyman.

Ramsey, F.P. (1925). "Universals." *Mind* (34: 401–417). Reprinted in D.H. Mellor, ed. (1990: 8–30). *F.P. Ramsey: Philosophical Papers*. Cambridge: Cambridge University Press.

Raup, David M. (1991). *Extinction: Bad Genes or Bad Luck?* New York: W.W. Norton & Company.

Reichenbach, Hans (1951). *The Rise of Scientific Philosophy*. Berkeley: University of California Press.

Richards, A.J. (1986). *Plant Breeding Systems*. London: Chapman & Hall.

Richards, Robert J. (1992). *The Meaning of Evolution*. Chicago: The University of Chicago Press.

Ridley, Mark (1989). "The Cladistic Solution to the Species Problem." *Biology & Philosophy* (4: 1–16).

————. (1990). "Comments on Wilkinson's Commentary." *Biology & Philosophy* (5: 447–450).

————. (1993). *Evolution*. Boston: Blackwell Scientific Publications.

Rieppel, Olivier (1986). "Species Are Individuals: A Review and Critique of the Argument." In Max K. Hecht, *et al.*, eds. (20: 283–317). *Evolutionary Biology*. New York: Plenum Press.

Roger, Jacques, and Fischer, Jean-Louis, eds. (1987). *Histoire du Concept d'Espèce dans les Sciences de la Vie*. Paris: Fondation Singer-Polignac.

Rosen, David (1978). "The Importance of Cryptic Species and Specific Identifications as Related to Biological Control." *Biosystematics in Agriculture: Beltsville Symposia in Agricultural Research* (2: 23–35).

Rosen, Donn E. (1978). "Vicariant Patterns and Historical Explanation in Biogeography." *Systematic Zoology* (27: 159–188).

————. (1979). "Fishes from the Uplands and Intermontane Basins of Guatemala: Revisionary Studies and Comparative Geography." *Bulletin of the American Museum of Natural History* (162: 267–376).

Rosenberg, Alexander (1985). *The Structure of Biological Science*. Cambridge: Cambridge University Press.

————. (1987). "Why Does the Nature of Species Matter? Comments on Ghiselin and Mayr." *Biology & Philosophy* (2: 192–197).

————. (1994). *Instrumental Biology, or The Disunity of Science*. Chicago: The University of Chicago Press.

Ruse, Michael (1979). *The Darwinian Revolution*. Chicago: The University of Chicago Press.

————. (1987). "Biological Species: Natural Kinds, Individuals, or What?" *British Journal for the Philosophy of Science* (38: 225–242). Reprinted in Ereshefsky, ed. (1992b: 343–363).

————. (1988). *Philosophy of Biology Today*. Albany: State University of New York Press.

————., ed. (1989a). *Philosophy of Biology*. New York: Macmillan Publishing Company.

————., ed. (1989b). *What the Philosophy of Biology Is: Essays Dedicated to David Hull*. Dordrecht: Kluwer Academic Publishers.

————. (1992). "Is the Theory of Punctuated Equilibria a New Paradigm?" In Somit and Peterson, eds. (1992: 139–167).

————. (1993). "Were Owen and Darwin *Naturphilosophen*?" *Annals of Science* (50: 383–388).

Russell, Bertrand (1900). *A Critical Exposition of the Philosophy of Leibniz*. Cambridge: Cambridge University Press. Reprinted (1992). London: Routledge.

————. (1908). "Mathematical Logic as Based on the Theory of Types." *American Journal of Mathematics* (30: 222–262). Reprinted in Marsh, ed. (1956: 59–102).

————. (1910). *Philosophical Essays*. London: Allen & Unwin. Reprinted (1994). London: Routledge.

————. (1912). *The Problems of Philosophy*. London: William & Norgate. Reprinted (1967). Oxford: Oxford University Press.

————. (1913). *Theory of Knowledge*. First published (1984). *The Collected Papers of Bertrand Russell, Volume 7*. London: Allen & Unwin. Reprinted (1992). London: Routledge.

————. (1914). *Our Knowledge of the External World*. London: Allen & Unwin. Reprinted (1993). London: Routledge.

————. (1918). "The Philosophy of Logical Atomism." *Monist* (28–29). Reprinted in

Marsh, ed. (1956: 177–281).

———. (1919). *Introduction to Mathematical Philosophy.* London: Allen & Unwin. Reprinted (1993). London: Routledge.

———. (1940). *An Inquiry Into Meaning and Truth.* London: Allen & Unwin. Reprinted (1992). London: Routledge.

———. (1944). "Reply to Criticisms." In Paul Arthur Schilpp, ed. (1944: 679–741). *The Philosophy of Bertrand Russell.* La Salle: Open Court.

———. (1946). *A History of Western Philosophy.* London: Allen & Unwin. Reprinted (1991). London. Routledge.

———. (1959). *My Philosophical Development.* London: Allen & Unwin. Reprinted (1993). London: Routledge.

Ryder, Oliver A. (1986). "Species Conservation and Systematics: the Dilemma of Subspecies." *Trends in Ecology and Evolution* (1: 9–10).

Ryle, Gilbert (1949). *The Concept of Mind.* London: Hutchinson.

Salthe, Stanley N. (1985). *Evolving Hierarchical Systems.* New York: Columbia University Press.

Sampson, Scott D. (1995). "Horns, Herds, and Hierarchies." *Natural History* (104: 36–40).

Schoener, T.W. (1989). "The Ecological Niche." In Cherrett, ed. (1989: 79–113).

Sharples, R.W. (1985). "Species, Form, and Inheritance: Aristotle and After." In Gotthelf, ed. (1985: 116–128).

Shaw, Alan B. (1969). "Adam and Eve, Paleontology, and the Non-Objective Arts." *Journal of Paleontology* (43: 1085–1098).

Simpson, George Gaylord (1961). *Principles of Animal Taxonomy.* New York: Columbia University Press.

Sloan, Phillip R. (1987). "From Logical Universals to Historical Individuals: Buffon's Idea of Biological Species." In Roger and Fischer, eds. (1987: 101–140).

Sneath, Peter H.A., and Sokal, Robert R. (1973). *Numerical Taxonomy.* San Francisco: W.H. Freeman and Company.

Sober, Elliott (1980). "Evolution, Population Thinking, and Essentialism." *Philosophy of Science* (47: 350–383).

———. (1984a). *The Nature of Selection.* Cambridge: The MIT Press. Reprinted (1993). Chicago: The University of Chicago Press.

———. (1984b). "Sets, Species, and Evolution: Comments on Philip Kitcher's 'Species.'" *Philosophy of Science* (51: 334–341).

———. (1993). *Philosophy of Biology.* Boulder: Westview Press.

———., ed. (1994). *Conceptual Issues in Evolutionary Biology.* Cambridge: The MIT Press.

Sober, Elliott, and Lewontin, Richard C. (1982). "Artifact, Cause and Genic Selection." *Philosophy of Science* (49: 157–180).

Sokal, Robert R. (1973). "The Species Problem Reconsidered." *Systematic Zoology* (22: 360–374).

———. (1986). "Phenetic Taxonomy: Theory and Methods." *Annual Review of Ecology and Systematics* (17: 423–442).

Sokal, Robert R., and Crovello, Theodore J. (1970). "The Biological Species Concept: A Critical Evaluation." *American Naturalist* (104: 127–153). Reprinted in Ereshefsky, ed. (1992b: 27–55).

Sokal, Robert R., and Sneath, Peter H.A. (1963). *Principles of Numerical Taxonomy.* San Francisco: W.H. Freeman and Company.

Somit, Albert, and Peterson, Steven A., eds. (1992). *The Dynamics of Evolution.* Ithaca, N.Y.: Cornell University Press.

Sonea, Sorin (1991). "Bacterial Evolution without Speciation." In Margulis and Fester, eds. (1991: 95–105).

Sonea, Soren, and Panisset, Maurice (1989). *A New Bacteriology*. Boston: Jones and Bartlett.

Song, Keming, and Osborn, Thomas C. (1992). "Polyphyletic Origins of *Brassica napus:* New Evidence Based on Organelle and Nuclear RFLP Analyses." *Genome* (35: 992–1001).

Splitter, Laurance J. (1988). "Species and Identity." *Philosophy of Science* (55: 323–348).

Stamos, David N. (1996a). "Was Darwin Really a Species Nominalist?" *Journal of the History of Biology* (29: 127–144).

———. (1996b). "Popper, Falsifiability, and Evolutionary Biology." *Biology & Philosophy* (11: 161–191).

———. (1998). "Buffon, Darwin, and the Non-Individuality of Species—A Reply to Jean Gayon." *Biology & Philosophy* (13: 443–470).

———. (1999). "Darwin's Species Category Realism." *History and Philosophy of the Life Sciences* (21: 137–186).

———. (2002). "Species, Languages, and the Horizontal/Vertical Distinction." *Biology & Philosophy* (17: 171–198).

Stanford, P. Kyle (1995). "For Pluralism and Against Realism About Species." *Philosophy of Science* (62: 70–91).

Stebbins, G. Ledyard (1989). "Plant Speciation and the Founder Principle." In Giddings *et al.*, eds. (1989: 113–125).

Stevens, Peter F. (1980). "Evolutionary Polarity of Character States." *Annual Review of Ecology and Systematics* (11: 333–358).

———. (1992). "Species: Historical Perspectives." In Keller and Lloyd, eds. (1992: 302–311).

Stoll, Robert R. (1963). *Set Theory and Logic*. San Francisco: W.H. Freeman and Company. Reprinted (1979). New York: Dover Publications, Inc.

Strawson, P.F. (1959). *Individuals: An Essay in Descriptive Metaphysics*. London: Methuen. Reprinted (1990). London: Routledge.

Sulloway, Frank J. (1979). "Geographic Isolation in Darwin's Thinking: The Vicissitudes of a Crucial Idea." In William Coleman and Camilla Limoges, eds. (1979: 23–65). *Studies in History of Biology*. Volume 3. Baltimore: The Johns Hopkins University Press.

———. (1982). "Darwin and His Finches: The Evolution of a Legend." *Journal of the History of Biology* (15: 1–53).

Sytsma, Kenneth J. (1990). "DNA and Morphology: Inference of Plant Phylogeny." *Trends in Ecology and Evolution* (5: 104–110).

Templeton, Alan R. (1989). "The Meaning of Species and Speciation: A Genetic Perspective." In Otte and Endler, eds. (1989: 3–27).

———. (1998). "Species and Speciation: Geography, Population Structure, Ecology, and Gene Trees." In Howard and Berlocher, eds. (1998: 32–43).

Thayer, H.S. (1953). *Newton's Philosophy of Nature*. New York: Hafner Press.

Thompson, Paul (1989). *The Structure of Biological Theories*. Albany: State University of New York Press.

Urmson, J.O. (1990). *The Greek Philosophical Vocabulary*. London: Duckworth.

Uvarov, E.B., *et al.*, eds. (1979). *The Penguin Dictionary of Science*. Harmondsworth, England: Penguin Books.

Van Regenmortal, M.H.V. (1997). "Viral Species." In Claridge *et al.*, eds. (1997: 17–24).

Van Valen, Leigh (1973). "A New Evolutionary Law." *Evolutionary Theory* (1: 1–30).

———. (1976). "Ecological Species, Multispecies, and Oaks." *Taxon* (25: 233–239). Reprinted in Ereshefsky, ed. (1992b: 69–78).

Vermeij, Geerat J. (1987). *Evolution and Escalation*. Princeton: Princeton University Press.

Vrba, Elisabeth S. (1995). "Species as Habitat-Specific, Complex Systems." In Lambert *et al.*, eds. (1995: 3–44).

Wake, David B., Yanev, Kay P., and Frelow, Monica M. (1989). "Sympatry and Hybridization in a 'Ring Species': the Plethodontid Salamander *Ensatina eschscholtzii*." In Otte and Endler, eds. (1989: 134–157).

Wallace, Alfred Russell (1858). "On the Tendency of Varieties to Depart Indefinitely from the Original Type." *Journal of the Proceedings of the Linnean Society (Zoology)* (3: 53–62). Reprinted in Barrett, ed. (1977: II: 10–19).

Ward, Peter Douglas (1992). *On Methuselah's Trail*. New York: W.H. Freeman and Company.

Warnock, G.J. (1956). "Metaphysics in Logic." In Flew, ed. (1956: 75–93).

Waters, C. Kenneth (1994). "Genes Made Molecular." *Philosophy of Science* (61: 163–185).

Wayne, Robert K., and Gittleman, John L. (1995). "The Problematic Red Wolf." *Scientific American* (273: 36–39).

Weinberg, Steven (1992). *Dreams of a Final Theory*. New York: Pantheon Books. Reprinted with an Afterword (1994). New York: Vintage Books.

Weishampel, David B. (1995). "Designer Jaws." *Natural History* (104: 64–66).

Weisz, Paul B. (1967). *The Science of Biology*. New York: McGraw-Hill, Inc.

Werth, Charles R., Guttman, Sheldon I., and Eshbaugh, W. Hardy (1985). "Recurring Origins of Allopolyploid Species in *Asplenium*." *Science* (228: 731–733).

Wheeler, Quentin D., and Meier, Rudolf, eds. (2000). *Species Concepts and Phylogenetic Theory: A Debate*. New York: Columbia University Press.

Whewell, William (1837). *History of the Inductive Sciences*. London: Parker.

———. (1840). *Philosophy of the Inductive Sciences*. London: Parker.

Wiley, Edward O. (1978). "The Evolutionary Species Concept Reconsidered." *Systematic Zoology* (27: 17–26). Reprinted in Ereshefsky, ed. (1992: 79–92).

———. (1981). *Phylogenetics*. New York: John Wiley & Sons, Inc.

———. (1989). "Kinds, Individuals and Theories." In Ruse, ed. (1989b: 289–300).

Wilkinson, Mark (1990). "A Commentary on Ridley's Cladistic Solution to the Species Problem." *Biology & Philosophy* (5: 433–446).

Williams, George C. (1966). *Adaptation and Natural Selection*. Princeton: Princeton University Press.

———. (1992). *Natural Selection*. Oxford: Oxford University Press.

Williams, Mary B. (1985). "Species Are Individuals: Theoretical Foundations for the Claim." *Philosophy of Science* (52: 578–590).

———. (1987). "A Criterion Relating Singularity and Individuality." *Biology & Philosophy* (2: 204–206).

Wilson, Bradley E. (1995). "A (Not-so-Radical) Solution to the Species Problem." *Biology & Philosophy* (10: 339–356).

Wilson, Edward O. (1992). *The Diversity of Life*. New York: W.W. Norton & Company.

Wilson, Robert, ed. (1999). *Species: New Interdisciplinary Essays*. Cambridge: The MIT Press.

Winslade, William J. (1970). "Russell's Theory of Relations." In E.D. Klemke, ed. (1970: 81–101). *Essays on Bertrand Russell*. Urbana: University of Illinois Press.

Wittgenstein, Ludwig (1922). *Tractatus Logico-Philosophicus*. D.F. Pears and B.F. Mc-

Guinness, trans. London: Routledge & Kegan Paul. Reprinted (1961).

———. (1958). *Philosophical Investigations*. G.E.M. Anscombe, trans. Third edition. Englewood Cliffs, N.J.: Prentice Hall.

Wyatt, Robert, Odrzykoski, Ireneusz J., and Stoneburner, Ann (1992). "Isozyme Evidence of Reticulate Evolution in Mosses: *Plagiomnium medium* is an Allopolyploid of *P. ellipticum* × *P. insigne*." *Systematic Botany* (17: 532–550).

Index

About the Author

David N. Stamos was born in Toronto, Canada, in 1957. All of his postsecondary education was at York University in Toronto, where he received a Ph.D. in philosophy in 1996, and where he currently teaches in two departments. In addition to having a number of papers published on various aspects of the species problem, including the first genuine reconstruction of Darwin's species concept, he is also published on Descartes, Hume, and Popper, as well as on the relation between quantum chance and point mutations (a cutting-edge semi-scientific paper published in *Philosophy of Science* 2001). He is currently working on a prequel to *The Species Problem*, titled *Darwin and the Modern Species Problem*, as well as an anthology titled *Evolution and Philosophical Issues: A Reader*. His wide interests in philosophy and science reflect an eclectic essence that developed in stages. At the age of fourteen he became interested in and began playing pipe organ music, especially Bach, a love that has continued to the present day. At the age of sixteen he began bodybuilding, at first to combat asthma, but since then he has remained an avid bodybuilder due mainly to numerous other benefits, both physical and mental. At the age of twenty he fell in love with philosophy, when he first read Plato's *Apology* in a university course on ancient Greek and Roman literature, and though that love has never abated his favorite philosopher long remains Bertrand Russell. At the age of twenty-six, although always an animal lover, he became and remains a vegetarian for ethical reasons, mainly as a result of having read a few years prior Albert Schweitzer's *The Teaching of Reverence for Life* and Peter Singer's *Animal Liberation*. At the age of thirty-three he fell in love with Jeeps and currently owns a modified 1992 Jeep Renegade, which he loves taking for drives in the country and to beaches in the summer. Finally, although he has long felt like a species of one, and partly because of it has had more short-term relationships than he cares to count, he continues to search for his other half, the love of his life, a belief (however irrational) that began when he was sixteen, and which (unlike religion, which he quit when he was twenty-six) he has been unable to shake off ever since.